A-Z

HANDBOOK 4TH EDITION

Biology

Bill Indge

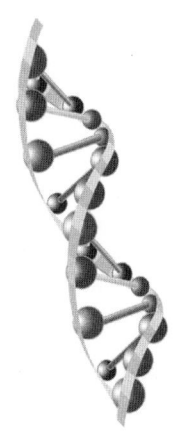

DIGITAL
EDITION

PHILIP ALLAN
UPDATES

Philip Allan Updates, an imprint of Hodder Education, an Hachette UK company, Market Place, Deddington, Oxfordshire OX15 0SE

Orders

Bookpoint Ltd, 130 Milton Park, Abingdon, Oxfordshire OX14 4SB
tel: 01235 827720
fax: 01235 400454
e-mail: uk.orders@bookpoint.co.uk

Lines are open 9.00 a.m.–5.00 p.m., Monday to Saturday, with a 24-hour message answering service. You can also order through the Philip Allan Updates website: www.philipallan.co.uk

ISBN 978-0-340-99099-5

First published 1997
Second edition 2000
Third edition 2003
Fourth edition 2009

Impression number   5   4   3   2
Year   2014   2013   2012   2011   2010

Typeset by Macmillan, India.

Printed by CPI Antony Rowe, Chippenham, Wiltshire.

Environmental information

Hachette UK's policy is to use papers that are natural, renewable and recyclable products and made from wood grown in sustainable forests. The logging and manufacturing processes are expected to conform to the environmental regulations of the country of origin.

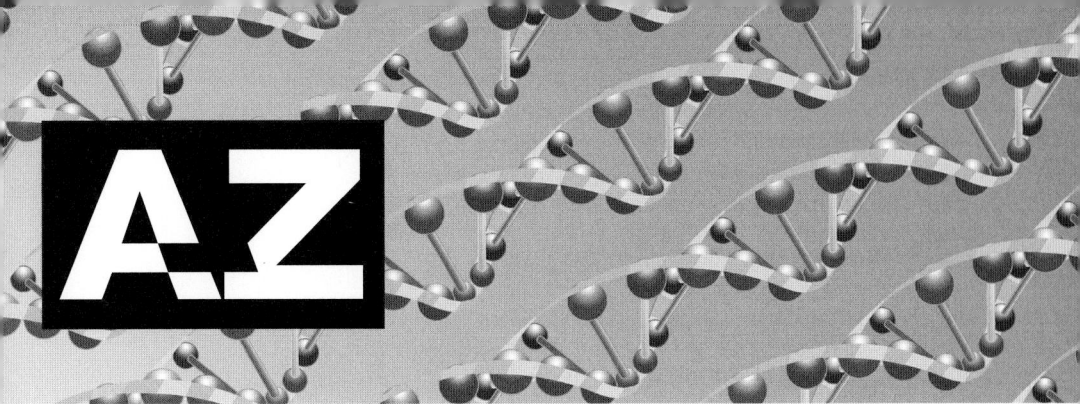

Contents

How to use this book

The *A–Z Biology Handbook* explains the main terms and important ideas in A-level biology specifications. It does not set out to give complete definitions of every technical word you might meet. In writing the handbook, we have selected the terms that you are likely to come across if you are following an A-level course. The entries cover the subject content of both the AS and A-level specifications and include not only biology but also human biology.

Each item starts with a simple, one-line definition and goes on to explain the term in a little more detail, showing how it relates to other areas of biology. We have tried to strike a balance between providing a simple explanation that should be understood by someone encountering the term for the first time, and producing an entry that is sufficiently detailed to answer an examination question that asks 'What is meant by…?' Entries are extensively cross-referenced. Bold italics have been used to identify terms which have separate entries.

During your course, the *A–Z Biology Handbook* is essentially a book to pick up and put down. It is a book to browse through and use to add to your understanding of basic ideas. It is also a book which will enable you to clarify new ideas. In addition, it should prove useful at the end of the course when you come to revise. To help you use it effectively during your revision, appendices have been added:

- **Biology revision lists**. Found on pages 306–09, this section identifies the key terms that you will need to understand for an AS or A-level unit test. For each concept, there is a list of key words which can act as starting points for your revision. Cross-references will then allow you to build on this understanding. You can also use the website that accompanies this handbook to access revision lists specific to your exam board unit. You should find these lists valuable when you come to do your revision.
- **Revising biology**. Different people revise in different ways, but however you revise, there are some strategies that you should think about adopting. This section, on pages 310–11, lists some general points that you should consider in order to make your programme of revision as effective as possible.
- **Examiners' terms**. Many candidates fail to do themselves justice in examinations, because they do not follow the instructions on the paper. This section, on pages 312–14, is a glossary of instructions that examiners use. You should keep this beside you throughout the course. If you become familiar with these terms from the beginning, you should be successful in turning your hard-earned knowledge into examination marks.

A–Z Online

This new digital edition of the *A–Z Biology Handbook* includes free access to a supporting website and a free desktop widget to make searching for terms even quicker. Log on to **www.philipallan.co.uk/a-zonline** and create an account using the unique code provided on the inside front cover of this book.

Once you are logged on, you will be able to:
- search the entire database of terms in this handbook
- print revision lists specific to your exam board
- get expert advice from examiners on how to get an A* grade
- create a personal library of your favourite terms
- expand your vocabulary with our word of the week

You can also add the other *A–Z Handbooks (digital editions)* that you own to your personal library on A–Z Online.

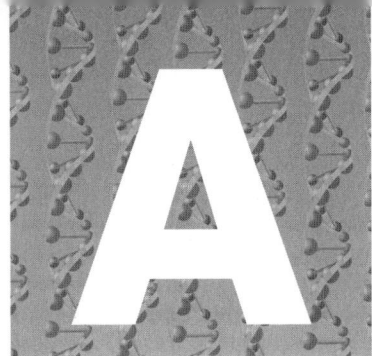

A band: one of the dark bands that runs across a *myofibril* in a *skeletal muscle*. It corresponds to filaments of the protein *myosin*.

abiotic: an ecological factor concerned with the non-living part of the *environment*. Rainfall, soil pH and temperature are examples of abiotic factors that affect the distribution of various organisms. *Competition* and predation, on the other hand, are concerned with the living part of the environment and are *biotic* factors. Abiotic factors are usually *density-independent factors*, that is, it doesn't matter how many organisms you have in a particular area, abiotic factors will affect them all. If conditions become hot and dry, for example, it is likely that all the plants of a particular species growing in the area concerned will be affected. It will not matter if the species is common or rare.

ABO blood groups: see *blood group*.

abscisic acid (ABA): a *plant growth substance* that acts mainly as a growth inhibitor. It is present in large quantities in dormant buds. It is also involved in controlling the opening of *stomata*. A rise in the concentration of abscisic acid is associated with the closing of stomata.

absolute growth rate: see *growth rate*.

absorption: the process by which small, soluble molecules produced by *digestion* are taken up from the gut. There are three basic stages involved. These are:

1 movement of molecules to the wall of the intestine. This is helped by the mixing of the gut contents. Muscles in the gut wall bring about movements of the gut that achieve this
2 transport across the *cell-surface membrane* into the *cytoplasm* of the cells which line the gut. A number of mechanisms are involved
3 transport away from the gut by blood and *lymph*. Molecules such as simple sugars and *amino acids* are absorbed into the blood. They are then transported in the hepatic portal vein to the *liver* where they are processed. The products of *fat* digestion enter the *lacteals*, branches of the *lymphatic system* in the *small intestine*.

Most absorption in the gut of a mammal takes place in the small intestine. This organ shows a number of adaptations. The surface area is increased enormously by the possession of small finger-like processes known as *villi*. In addition, the epithelial cells that line these are covered in *microvilli*. The cytoplasm of the epithelial cells contains large numbers of *mitochondria*, which provide the *ATP* necessary for *active transport* of substances from the *epithelium* of the gut.

absorption spectrum: a graph which shows the relative amounts of light of different wavelengths that are absorbed by a particular pigment. The diagram shows the absorption spectrum for one particular type of **chlorophyll** called chlorophyll a.

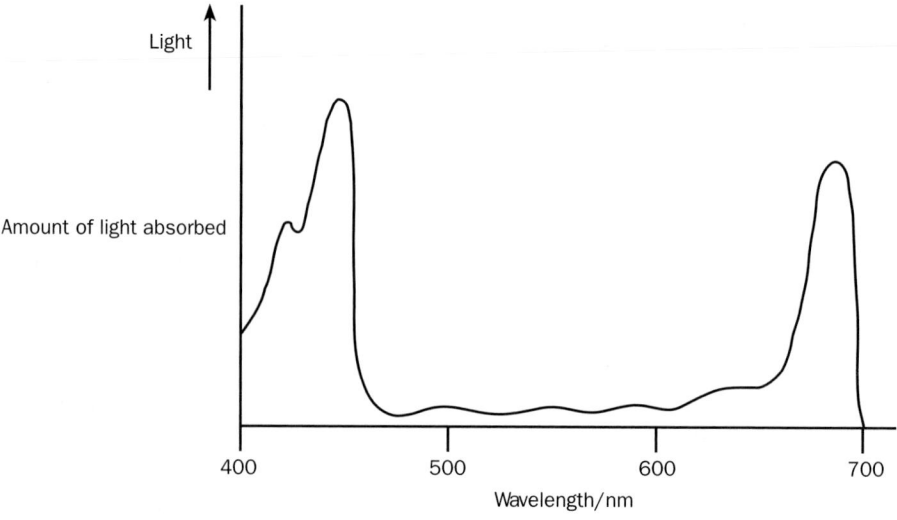

Absorption spectrum for chlorophyll a

The curve shows two peaks, one between 400 and 500 nm, and the other between 600 and 700 nm. This is because chlorophyll a strongly absorbs light at the blue and red ends of the spectrum. Most plant leaves appear green in colour because chlorophyll does not absorb green light; it reflects it.

accommodation: the ability of the eye to change focus so that near or distant objects can be seen clearly. When a person looks at a distant object, the rays of light are bent or refracted by the **cornea** and the **lens** and focused on the **retina**. If the object is brought closer, the light rays have to be refracted more if they are still to be focused. This is done by changing the shape of the lens. The **ciliary muscles** contract and the lens becomes more convex as a result.

acetylation: a chemical reaction in which a hydrogen atom is replaced by an acetyl group (CH_3CO). **Histones** are **proteins** associated with **DNA**. Acetylation of histones plays a part in gene regulation. Decreased acetylation represses the process of **transcription**.

acetylcholine (ACh): a substance responsible for the transmission of a **nerve impulse** across a **synapse**. Acetylcholine is an important **neurotransmitter** in the **central nervous system** and in synapses in the parasympathetic branch of the autonomic nervous system. Synapses in which acetylcholine is the neurotransmitter are called **cholinergic** synapses. When a nerve impulse arrives at the presynaptic membrane of a cholinergic synapse, acetylcholine is released into the synaptic cleft. It diffuses across the cleft where it binds to receptors on the postsynaptic membrane. These trigger another nerve impulse. (See also **parasympathetic nervous system**.)

acetylcholinesterase: an enzyme found in **synapses**, which causes the rapid breakdown of **acetylcholine (ACh)**. Acetylcholine must be removed rapidly from a synapse once it has triggered a *nerve impulse* or there would be continuous transmission along the postsynaptic neurone.

acetyl coA: see *acetylcoenzyme A*.

acetylcoenzyme A: a substance produced in the *link reaction* of *respiration* when coenzyme A accepts the two-carbon fragment formed from pyruvate. As well as being the molecule that feeds this two-carbon fragment into the *Krebs cycle*, acetylcoenzyme A is important in other biochemical pathways involving *fatty acids* and *amino acids*.

acid rain: rain that is more acidic than normal because various gases from the atmosphere have dissolved in it. Because of dissolved carbon dioxide, normal unpolluted rain water is slightly acid with a pH of about 5.6. In industrial areas, however, the combustion of fossil fuels releases oxides of nitrogen and sulphur into the air. These produce nitric and sulphuric acids, which may lower the pH of the rain to around 4. Although we do not understand all the ways in which acid rain affects *ecosystems*, we know that it is associated with the following:

- a fall in pH of freshwater lakes. This leads to the death of fish and many invertebrates
- death of trees, particularly of conifers, where the first signs involve die-back of the crowns
- release of more aluminium into solution. Aluminium is toxic to many organisms.

acne vulgaris: a skin condition in which the *sebaceous glands* on the face, shoulders and back become inflamed. This causes spots and pustules to develop. The ducts of the glands are often blocked. Bacteria may colonise the ducts. They produce substances that lead to further *inflammation*. Many factors cause acne but it is often linked with the increased secretion of *androgen hormones* around *puberty*.

acoelomate: an animal without a *coelom* (body cavity). *Flatworms* and tapeworms belong to the phylum *Platyhelminthes*. They are examples of acoelomates.

acrosome: a cap-like *vesicle* at the front of a *sperm*, which contains *enzymes*. Just before *fertilisation* the membrane that surrounds this vesicle bursts and the enzymes are released. Their digestive action helps to separate the cells which still surround the egg. They also make it possible for the sperm to penetrate the membranes surrounding the egg and fertilisation to take place.

acrosome reaction: the process that takes place in the *acrosome* of a sperm cell as it comes into contact with an egg. The membrane surrounding the acrosome fuses with the *cell-surface membrane* of the sperm. This releases *enzymes* that digest the membranes surrounding the egg cell. The sperm then penetrates the egg cell and *fertilisation* takes place.

actin: a protein found in many cells. It is very important in muscle. In the *myofibrils* of skeletal muscles, actin molecules form the thin filaments in the *I band*. These thin actin filaments slide between thicker *myosin* filaments when a muscle contracts.

action potential: the change which occurs in the electrical charge across the *cell-surface membrane* of a nerve cell during the passage of a *nerve impulse*. When a nerve cell is stimulated electrically, sodium *ion-channel proteins* in the membrane open and sodium ions enter. As the inside of the cell is normally negatively charged, the entry of these positive sodium ions makes it less negative. In other words, there is a decrease in the potential difference across the membrane. This change in potential difference allows even more sodium ions to enter, changing the potential difference still more. The end result is that the potential difference changes from its **resting potential** of around −60 mV to a value of approximately +40 mV. This is summarised in the flow chart.

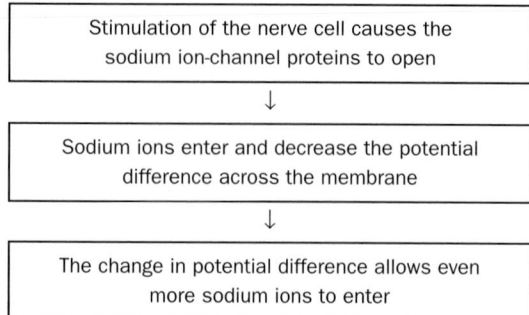

Once it has reached +40 mV, the sodium ion-channel proteins close leaving potassium ion-channel proteins fully open. Sodium ions can no longer enter but potassium ions can leave. The result of positive potassium ions moving out is that the potential difference falls back to its resting level. The overall change in the electrical charge across the membrane, lasting only a few milliseconds, is called an action potential. The spread of an action potential along the nerve cell is the basis of a nerve impulse. After the impulse passes, the sodium ions which entered and the potassium ions which left the nerve cell during the action potential are pumped back to where they began by **active transport**.

action spectrum: a graph that shows the proportion of light of different wavelengths that is used in a particular process. The action spectrum for **photosynthesis** is similar to the **absorption spectrum** for **chlorophyll**. This shows the importance of chlorophyll in absorbing the light used in photosynthesis.

activation energy: the amount of energy necessary to bring molecules together so that they will react with each other. The graph shows the changes in energy level during a chemical reaction.

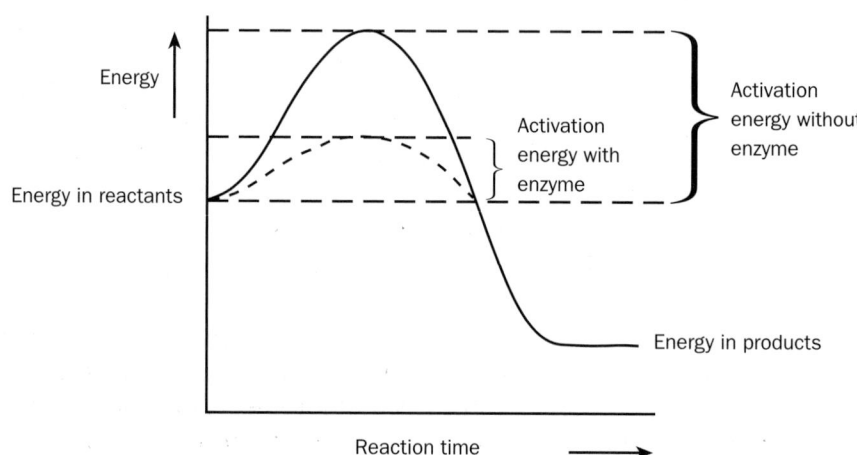

Think about burning some ethanol. This is a chemical reaction in which ethanol and oxygen combine to produce carbon dioxide and water. There is a lot of energy in ethanol and that is why it is a good fuel. There is a lot less energy in the products, carbon dioxide and water. However, ethanol does not combine with oxygen to produce carbon dioxide and water on its own. To make it do this we need to supply some extra energy, the activation energy, in the form of heat. **Enzymes** work by lowering the activation energy by splitting the reaction into stages. In this way the reaction can take place at the temperatures and pressures which exist inside living cells.

active immunity: a form of immunity that results when a person encounters **antigens** from a **pathogen** responsible for a particular disease. **Lymphocytes** in the blood multiply rapidly and produce **plasma cells** and **memory cells**. The plasma cells produce **antibodies**. The memory cells remain in the blood after the original infection has gone but they respond rapidly if they encounter the same antigen again. Response to a second infection is therefore much more rapid and results in immunity. Active immunity occurs naturally but it also occurs artificially when **vaccines** are used to protect against disease. Compare active immunity with **passive immunity**.

active site: part of an **enzyme** molecule into which a **substrate** molecule fits during a biochemical reaction. Enzymes are **proteins**. Each enzyme has its own sequence of **amino acids**, resulting in the molecule folding into a particular shape. In other words, each enzyme has a specific **tertiary structure**. A small group of the amino acids which make up the molecule is important in forming the active site of the enzyme. This is the place where the substrate molecule will fit to produce an **enzyme–substrate complex**. Only a substrate with the complementary shape will fit the active site and this results in enzymes being specific. In addition, anything which alters the shape of the active site will mean that the enzyme will not function. High temperatures, changes in pH and non-competitive inhibitors affect enzymes in this way (see **non-competitive inhibition**).

active transport: the movement of a substance from where it is less concentrated to where it is more concentrated; in other words, it is the movement of a substance against a concentration gradient. Active transport uses specific **carrier proteins** in **plasma membranes** to transport the molecules or ions concerned. Moving a substance against a concentration gradient requires an input of energy. This is supplied by ATP produced by **respiration**. Because of this, cells in which a lot of active transport occurs contain many **mitochondria**. Active transport is extremely important in living organisms. Most cells take up substances that are only present at low concentrations in their **environment**. For example, the uptake of ions by the cells of a plant root, loading of sugars into **phloem** sieve tubes (see **sieve tube element**) and reabsorption in the **kidney** all involve movement against a concentration gradient by active transport.

actomyosin: a protein formed when **actin** binds to **myosin** during muscle contraction. Muscle contraction involves the formation of cross-bridges between the filaments of actin and myosin. The head of the myosin molecule attaches to a particular place on an actin molecule forming an actinomyosin bridge. The formation, changing position and detaching of these bridges causes the muscle to contract. (See also **sliding-filament model**.)

acuity: the ability of the eye to distinguish between objects close together. It depends in part on the connections which the *cone cells* in the *fovea* of the eye have with *sensory neurones*. Each cone cell is associated with a single sensory neurone. Because of this, stimulation of cone cells, even if they are close together, results in separate *nerve impulses* in separate neurones in the *optic nerve*. Objects that are close together can therefore be seen as separate. The idea of acuity also applies to other senses such as touch.

acute disease: a disease that develops rapidly but lasts a relatively short time. Acute disease may be compared with *chronic disease*.

adaptive radiation: the way in which a single ancestral type of organism can give rise to different forms, each occupying a different *niche*. Adaptive radiation is often seen among species that live on islands. A familiar example is provided by Darwin's finches, a group of birds found on the Galapagos islands. These are volcanic islands about 960 km off the coast of Ecuador. It is thought that, at some stage in the past, these islands were colonised by finch-like birds from the South American mainland. There were no other small land birds on these islands so there was no *competition* from other species. Over a period of time, different groups of finches adapted to different ecological niches. As a result, there are now 13 different species of finch found on the Galapagos islands.

addition: a type of *gene mutation* in which one or more bases are added to the DNA that forms an organism's genetic material.

adenine: one of the *nucleotide bases* found in *nucleic acid* molecules. Adenine is a *purine* which means that it has two rings of *atoms* in each of its molecules. When two polynucleotide chains come together in a molecule of *DNA*, adenine always bonds with the base *thymine*. The atoms of these two bases are arranged in such a way that two *hydrogen bonds* are able to form between them.

adenosine diphosphate: see *ADP*.

adenosine triphosphate: see *ATP*.

ADH: see *antidiuretic hormone (ADH)*.

ADP: adenosine diphosphate, a substance found in all living cells, that is involved in the transfer of energy. A molecule of ADP is made up from a molecule of *adenine* joined to the five-carbon sugar *ribose* and two phosphate groups. ADP can join with a phosphate group to produce *ATP*.

adrenal gland: one of a pair of *hormone*-secreting *glands* which are near the *kidneys*. Each adrenal gland consists of two parts, an outer cortex and an inner *medulla*. The outer cortex produces *steroid* hormones, which are mainly involved in the control of *carbohydrates*, *proteins* and mineral ions. The medulla secretes *adrenaline*, a hormone that is released at times of stress. Adrenaline affects many organs and helps the body respond to emergencies.

adrenaline: a *hormone* produced by the *adrenal glands* and released at times of stress. It affects many organs and helps the body to respond to an emergency. Some of these effects are shown opposite.

Effect on the body	Advantages in preparing the body for action
Increases **stroke volume** and heart rate	Increases the supply of substances such as **glucose** and oxygen to muscles
Constricts **blood vessels** to many of the internal organs	Increases blood pressure
Dilates blood vessels supplying the muscles	Increases the supply of glucose and oxygen to the muscles and removes waste products from them
Stimulates the conversion of **glycogen** to glucose	Makes more glucose available for increased **respiration**
Causes contraction of the muscles at the base of hairs	Makes the hair stand on end which, in many animals, may make them look bigger and fiercer than they actually are

adrenergic: a *synapse* in which the *neurotransmitter* is *noradrenaline*. In mammals and other vertebrate animals, adrenergic synapses are found in the *sympathetic nervous system*. Here, the neurotransmitter in the synapse between the *motor neurone* and the *effector* to which it goes is noradrenaline. The sympathetic nerves that supply the heart, for example, release noradrenaline which increases heart rate.

aerenchyma: a tissue found in the stems of plants, such as rice and water lilies, that grow in water. It contains a mixture of cells and large, air-filled spaces. These spaces have two important functions:
- The air they contain provides buoyancy. This helps to ensure that the leaves and stems reach the surface where they will receive sufficient light for *photosynthesis*.
- They form a pathway allowing the *diffusion* of oxygen and carbon dioxide to parts of the plant that are under water.

aerobic exercise: a form of exercise that can be carried out at a moderate level over a considerable period of time. The energy supply to the muscles during aerobic exercise comes from *aerobic respiration*. Aerobic exercise therefore requires an increased supply of oxygen to the muscles. This is produced by an increase in breathing and heart rate. Regular aerobic exercise results in increased cardiovascular fitness and has been associated with a lower risk of developing cardiovascular disease, *diabetes* and some forms of cancer. (See also *tumour*.)

aerobic respiration: *respiration* that requires the presence of oxygen. The equation that represents this process is:

$$C_6H_{12}O_6 + 6O_2 \rightarrow 6H_2O + 6CO_2$$

This equation can be rather misleading as it only summarises the process. Respiration is much more complex than this and consists of the series of biochemical reactions shown in the diagram overleaf.

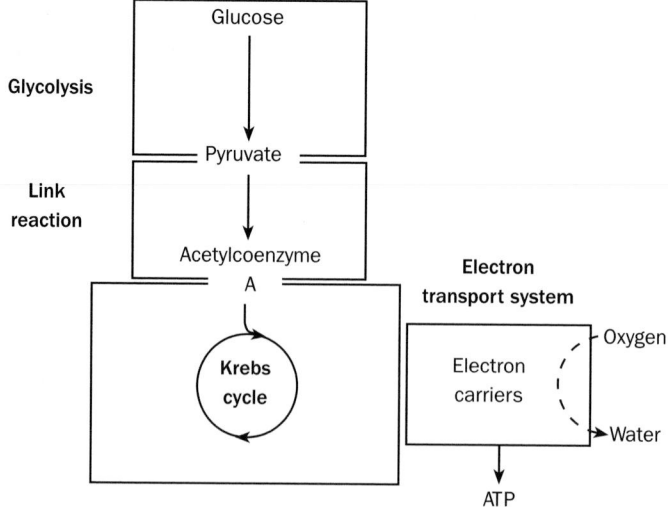

The biochemical pathway of aerobic respiration

The six-carbon **glucose** molecule is first split into two molecules of pyruvate. Each of these contains three carbon **atoms**. This process, known as **glycolysis**, generates a small amount of **ATP**. The pyruvate produced in glycolysis is converted into acetylcoenzyme A in the **link reaction**. Acetylcoenzyme A then enters a cycle of biochemical reactions known as the **Krebs cycle**. In the Krebs cycle, electrons and protons are lost and passed along the **electron transport chain** or respiratory chain. At the end they combine with oxygen to produce water. During the passage along this chain a considerable amount of energy is released and this is used to produce ATP.

afferent: means 'going to'. In a **reflex arc**, for example, the afferent neurone transmits a **nerve impulse** to the **spinal cord**. In the **kidney**, the afferent **arteriole** takes blood to the **glomerulus**.

agar: a substance obtained from seaweed. It is important in making solid media for the culture of **microorganisms**. After heating and mixing with various nutrients, it cools to form a stiff jelly on which microorganisms such as bacteria and fungi may be grown. Agar is an inert substance and therefore provides no nutrients. Different nutrients may be added to make media suitable for different microorganisms. Blood agar, for example, is made by mixing horse blood with agar and is used for culturing the bacteria that destroy blood cells. As few microorganisms have the ability to break down agar, the jelly stays firm but, at the same time, it allows substances to diffuse freely.

age population pyramid: see **population pyramid**.

agglutination: the sticking together or clumping of red blood cells. Red blood cells have **proteins** on their **cell-surface membranes**. These proteins act as **antigens**. Group B people, for example, have red blood cells with antigen B. Suppose that, during a **blood transfusion**, group B blood was wrongly given to someone with group A blood. The person with group A blood would have **antibody** B in his blood **plasma**. Antigen B would combine with antibody B, causing the red blood cells to stick to each other. This is agglutination and it can lead to blocked **blood vessels** and death. Agglutination must be distinguished from the completely different process of **blood clotting**.

agricultural ecosystem: an *ecosystem* based on domestic animals or crop plants. The principles that apply to natural ecosystems also apply to those based on agriculture. Energy is transferred from producers to *consumers* and nutrients are cycled in the same way. Agricultural ecosystems, however, have a number of characteristics:

- Agricultural ecosystems have a low species diversity. Modern agriculture is often based on *monoculture*, where large areas of land are devoted to single crops. *Pesticides* also reduce the numbers of other species of organisms present.
- There is a lack of genetic diversity. Selection for greater productivity in domesticated animals and crop plants has resulted in less genetic variation.
- Productivity is increased by managing *abiotic* conditions. This may be done, for example, by applying *fertiliser* and irrigating field crops, by enhancing light, temperature and carbon dioxide concentration in glasshouse crops, or by keeping domestic animals in heated sheds or barns and feeding them on controlled amounts of concentrate. However, this increase in productivity usually involves single species. The productivity of the ecosystem as a whole may be lower in agricultural ecosystems because fewer ecological *niches* are filled.
- Agriculture demands a high energy input. The manufacture of agricultural machinery, its use in cultivating crops, the production of fertilisers and pesticides and transport all add to the real energy cost of growing a crop.

Agrobacterium: a *genus* of bacteria found naturally in the soil. These bacteria enter plant cells and cause cancer-like growths known as crown galls to form (see *tumour*). *Agrobacterium* possesses tumour-forming genes that are contained in small rings of DNA known as *plasmids*. The plasmids are incorporated into the DNA of the plant and the tumour-forming genes bring about abnormal growth of the infected cells.

Genetic engineers can use *enzymes* to cut open these plasmids and insert particular genes such as that, for example, which gives *resistance* to the *herbicide* glyphosate. When the bacterium invades the plant cells, it introduces this gene. The plant will then be resistant to the herbicide. This means that *weeds* can be controlled by spraying fields with glyphosate. The weeds will be killed but the crop will be resistant and, therefore, it will not be affected.

AIDS/HIV: see *HIV*.

albinism: a condition in which an organism does not produce a particular pigment. In mammals, albinos do not produce a dark brown pigment called *melanin*. Because of this, they are light skinned and have white hair and pink eyes. The condition is inherited and controlled by a gene that codes for an enzyme involved in the synthesis of melanin. The *allele* for albinism is *recessive* to that for normal pigmentation.

algal bloom: the rapid growth of algae resulting from an increased nutrient supply. Pollution of freshwater lakes and rivers may result in an increase in the concentration of phosphate and nitrate ions. A low concentration of these ions normally limits the growth of the microscopic algae which float at the surface and form the plankton. The addition of more phosphates and nitrates results in very rapid growth of the algal population. Some species of algae produce toxic substances so algal blooms may prove dangerous to the health of both humans and domestic animals. More frequently, however, they lead to an increase in the activity of *decomposers* and the depletion of dissolved oxygen. (See also *eutrophication*.)

alimentary canal: an alternative name for the gut of an animal. It runs from the mouth at one end to the *anus* at the other. The alimentary canal has a number of different functions. These are:

- *ingestion* – taking food into the mouth
- *digestion* – breaking down of large insoluble molecules like *starch* and protein by *enzymes* into smaller soluble ones such as *glucose* and *amino acids*
- *absorption* – the process by which the soluble molecules which result from digestion pass through the wall of the alimentary canal into the body
- *egestion* – the removal of waste material from the gut in the form of *faeces*.

In humans, the main regions of the alimentary canal are the *oesophagus, stomach, small intestine* and *large intestine*. Although all these regions have a similar basic structure, there are obvious variations. Each has a structure which is related to its function. In addition, the detailed structure of the gut is different in different species. Even among mammals, these differences can be marked. The gut of a *ruminant* like a cow, for example, is quite unlike that of a carnivore such as a dog.

all or nothing: a term used to describe a *nerve impulse*. When an *action potential* is produced in a nerve cell, it is always the same size. It does not matter how big the initial stimulus, the action potential will always involve the same change in potential difference across the *cell-surface membrane*. Because of this, the only way that information about the strength of a stimulus can be carried is by changing the number of nerve impulses in a given time. It is not possible to have a small action potential for a small stimulus and a larger one for a large stimulus.

allele: one of the different forms of a particular *gene*. For example, snails that belong to the species *Cepea nemoralis* have shells which may be either plain or banded. This shell pattern is controlled by a gene. There are two different forms of this gene. One form or allele produces a plain shell whereas the other produces a banded snail. If the letter B is used to represent the gene, then a capital B represents the *dominant* allele which results in a plain shell and a small b represents the recessive allele which produces a banded shell. Although many genes have two different alleles, it is possible for a single gene to have more than two forms. These are known as *multiple alleles*.

allergen: a foreign substance or *antigen* capable of stimulating an allergic reaction. (See also *allergy*.)

allergy: a condition in which the body becomes extremely sensitive to a particular substance that produces characteristic symptoms every time it is encountered. People encounter many foreign substances, bearing *antigens*, during the course of their lives. Under normal conditions, *antibodies* in the blood and tissues get rid of these antigens from the body and there are no side effects. In a person suffering from an allergy, however, the reaction between a particular antigen (known as an *allergen*) and antibodies in the tissues leads to the release of substances such as *histamine*. These substances produce the characteristic symptoms of the particular allergy. Some allergens produce localised effects such as occur with *hay fever* and *asthma*. The effect with others is more generalised and can be extremely serious. Anaphylactic shock, which may occur, for example, in people allergic to nuts, affects the *lungs* and circulatory system. Unless medical help is given immediately death may result.

allopatric speciation: the formation of new *species* which occurs when *populations* of a particular species become separated from each other by some form of geographical barrier. If a species is spread over a wide area, then features such as mountain ranges, seas, rivers or deserts may divide it into a number of populations which will be unable to breed with each other. These populations can be described as being reproductively isolated from each other. *Natural selection* will result in the evolution of slight differences in the characteristics of the organisms in the different populations. Gradually, over a period of time, these differences will become more and more pronounced. Finally, the separated populations will be sufficiently different to be recognised as separate species. Even if they mix with each other, individuals will be unable to interbreed. Allopatric *speciation* is shown in the diagram.

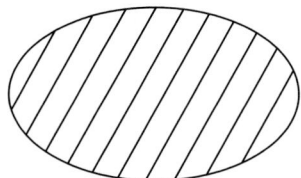

Single species spread over
wide area

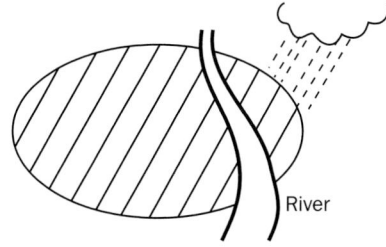

River

Geographical features divide species
into different populations. Each
population adapts to its environment

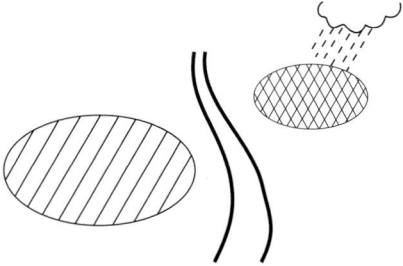

The separate populations are so
different that they may be
considered as different species

Allopatric speciation

allosteric: refers to a molecule whose shape can be altered by something in its *environment*. This usually leads to a change in its function. Allosteric changes are important in enzyme-controlled reactions as they allow the rate of the reaction to be controlled. *Enzymes* have *active sites* which bind to *substrate* molecules. There are also other places on the surface of the enzyme to which different molecules can bind. Binding with one of these other molecules alters the shape of the active site so that the enzyme is stopped from working. This is an example of *non-competitive inhibition*.

Another molecule that undergoes an allosteric change is *haemoglobin*. A change in pH of the blood alters the shape of the molecule and affects the way in which it binds with oxygen. This gives rise to the *Bohr effect*.

α-glucose (alpha glucose): a form or *isomer* of *glucose* in which the –H and –OH groups on carbon 1 are arranged as shown in the diagram.

α-glucose

Molecules of α-glucose link by *glycosidic bonds* formed by *condensation* to produce *maltose* and *starch*.

α-helix (alpha helix): a type of *secondary structure* found in *protein* molecules. In an α-helix, the *polypeptide* chain is coiled in a spiral shape. It makes one complete turn every 3.6 *amino acids*. The coils in the spiral are held in place by *hydrogen bonds* which form between amino acids in the helix. The α-helix is a common type of secondary structure but it is not found in all proteins. Many structural proteins are coiled in this way. An example is keratin, a protein found in hair and skin.

alpha-1-antitrypsin: a substance produced by the body. It inhibits the action of several protein-digesting *enzymes*. The body produces a number of these protein-digesting enzymes and they help to protect it from harmful *proteins*. These enzymes must be carefully regulated, however, or they could damage healthy tissue. This is the role of alpha-1-antitrypsin.

Alpha-1-antitrypsin deficiency is a common inherited condition. People with this condition have low concentrations of alpha-1-antitrypsin in their blood. Smokers with alpha-1-antitrypsin deficiency are very likely to develop *emphysema*.

alternation of generations: the alternation of two distinct stages in the life cycle of a plant. One of these stages has the *haploid* number of *chromosomes*, the other has the *diploid* number. The diagram opposite shows this basic cycle with the haploid *gamete*-producing stage known as the *gametophyte* alternating with the diploid *spore*-producing stage called the *sporophyte*.

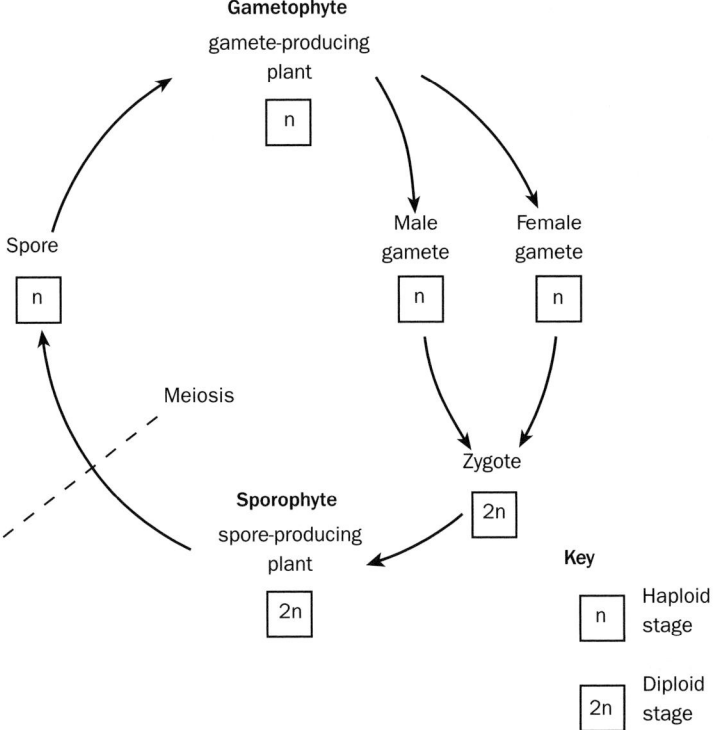

Gametophyte
gamete-producing
plant

n

Spore

n

Male
gamete

n

Female
gamete

n

Meiosis

Zygote

Sporophyte
spore-producing
plant

2n

2n

Key

n | Haploid stage

2n | Diploid stage

In the simplest plants, such as mosses, the gametophyte is the dominant stage. The adult green plant is a gametophyte which will produce male and female gametes. In *flowering* plants, on the other hand, the adult plant, whether it a daisy or an oak tree, is a sporophyte. The gametophyte stages in flowering plants are reduced to tiny structures found within the *flowers*. The change from the gametophyte as the most important stage to the sporophyte being dominant is an important evolutionary trend in plants.

alveoli: the air sacs in the *lungs* where gas exchange takes place. In the alveoli (singular alveolus), oxygen enters the blood, and carbon dioxide is removed from it by *diffusion*. The alveoli show a number of adaptations which make them efficient gas exchange surfaces:

- There is a large surface area over which diffusion can take place. This is provided by the huge number of alveoli. There are an estimated 300 million alveoli in each human lung giving a total surface area of between 50 and 80 m².
- There is a large difference in the concentration of respiratory gases. Oxygen, for example, is continually being removed by combining with *haemoglobin*. *Ventilation* replaces the air so that the concentration of oxygen in the alveoli remains high.
- The diffusion pathway is very short. The alveoli are lined with thin, flat epithelial cells known as *squamous epithelium*. The total distance from the air in the alveoli, through these epithelial cells and those in the *capillary* wall, into the blood is less than one micrometre.

Alzheimer's disease: a disease that occurs mainly in older people and involves the progressive deterioration of mental processes. It is a form of senile *dementia*. In its early stages, it is characterised by loss of short-term memory and mental confusion, but as the disease runs its course, these effects become more and more pronounced. Post-mortem examinations of affected patients show that the condition is associated with structural and chemical changes in the **brain** tissue. These include the presence of plaques of an abnormal protein which form outside the brain cells and of tangled fibres of another sort of protein found inside them. Although the precise cause of Alzheimer's disease is unknown, it is thought that certain people inherit genes that make them more susceptible to the environmental factors which lead to development of the condition.

amino acid: the basic unit or **monomer** from which **proteins** are formed. The diagram shows the general structure of an amino acid.

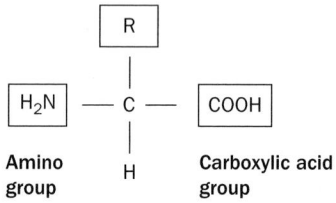

The basic structure of an amino acid

There are 20 different **amino acids** that may be linked together by **condensation** to form proteins. All of these amino acids have the same basic chemical structure. It is only the R-group that differs. In the amino acid glycine, for example, it is H while in alanine it is CH_3. The amino acids cysteine and methionine are a little different as their R-groups contain sulphur. This is important in forming bonds that help to give proteins their **tertiary structure**.

Plants are able to make amino acids by combining some of the molecules produced by **photosynthesis** with nitrogen obtained from nitrates taken up from the soil. Animals are unable to do this and need to obtain many of the amino acids they require from their diet. From these, they can produce others by the process of **transamination**. Animals cannot store excess amino acids in their bodies until they are required. The surplus molecules are broken down by **deamination** and converted to other products that are excreted.

ammonification: the stage in the **nitrogen cycle** in which saprobiotic (see **saprobiont**) **microorganisms** break down organic, nitrogen-containing substances such as **proteins** and produce ammonium compounds. These are then oxidised to nitrites and nitrates in the process of **nitrification**.

amniocentesis: a technique used for the **genetic screening** of a **fetus** while still inside the uterus of its mother. **Ultrasound** is used to determine the precise position of the fetus and the **placenta** within the uterus. A fine needle is then inserted through the abdominal wall of the mother into the amniotic cavity. A sample of amniotic fluid is removed. This will contain some fetal cells. These cells can be examined and their **chromosomes** observed or the DNA that they contain investigated.

amnion: one of the membranes that surrounds a developing mammalian **fetus** in the **uterus** of its mother. The cells that make up the amnion secrete amniotic fluid. This fluid fills the space between the membrane and the fetus where it provides protection and support for the delicate fetal tissues. There are always some fetal cells in the amniotic fluid. *Amniocentesis* allows a sample of these cells to be removed and used for *genetic screening* of the fetus.

amylase: an enzyme that digests *starch*. Like other digestive *enzymes*, amylases are *hydrolases*. They break down starch into soluble sugars by means of *hydrolysis*. Although amylases are important mammalian digestive enzymes found in *saliva* and pancreatic juice, they are also found in many *microorganisms* and in plant tissues.

amylopectin: a *polysaccharide* that is an important constituent of *starch*. It consists of branched chains of *α-glucose* molecules joined together by *condensation*.

amyloplast: an *organelle* found in plant cells, which stores *starch*. Large numbers of amyloplasts can be found in storage organs such as potato tubers.

amylose: a *polysaccharide* that is an important constituent of *starch*. It consists of long straight chains of *α-glucose* molecules joined together by *condensation*.

anabolic reaction: a chemical reaction in which smaller molecules combine to form larger ones. Examples are the reactions involved in *protein synthesis* and *photosynthesis*. Anabolic reactions require energy. In living organisms, this is usually provided by *ATP*.

Hormones such as *testosterone* work by switching on the genes that are responsible for particular anabolic reactions. The increased muscle and *bone* development at puberty, for example, is controlled by testosterone in this way. Testosterone is an anabolic *steroid* because it is a steroid that influences the building up of substances in the body. Anabolic steroids have unfortunately been the subject of abuse by some athletes.

anaemia: a condition in which there is a reduced amount of *haemoglobin* in the blood. People with anaemia get tired easily. They have pale skin and get out of breath if they exert themselves. There are a number of causes of anaemia. These include:
- a shortage of *iron* in the diet; iron is an important part of haemoglobin molecules
- conditions that lead to loss of blood. This might either be the result of an accident or come from something like an ulcer which results in a long-term loss of blood
- conditions that result in the destruction of red blood cells. *Sickle-cell disease* is an inherited condition in which the affected person has a different type of haemoglobin present in his or her body that is less efficient at transporting oxygen. The red blood cells of someone with sickle-cell anaemia have a shorter life-span than normal red blood cells.

anaerobic respiration: *respiration* that takes place in the absence of oxygen. Most organisms rely on oxygen in order to respire. Occasionally, however, there is too little oxygen available to meet their respiratory requirements and the whole organism, or part of it, is able to respire anaerobically for a period of time. Anaerobic respiration relies on *glycolysis* to produce *ATP*. Since the net amount of ATP produced in glycolysis is only 2 molecules for each molecule of *glucose* as opposed to the 38 molecules of ATP that may be produced in *aerobic respiration*, it can be seen that the anaerobic process is not very efficient.

In glycolysis the coenzyme *NAD* is reduced. Reduced NAD is normally reconverted to NAD in the reactions of aerobic respiration. If no oxygen were present, the NAD would run out. The biochemical pathways involved in anaerobic respiration in animals and in plants and *microorganisms* such as yeast are shown overleaf.

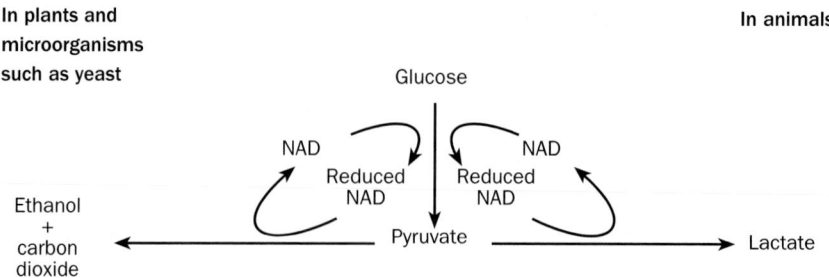

In plants and microorganisms such as yeast

In animals

Glucose

NAD — Reduced NAD — Reduced NAD — NAD

Ethanol + carbon dioxide ← Pyruvate → Lactate

You will see that, in the conversion of pyruvate to *lactate* or to ethanol and carbon dioxide, NAD is produced from reduced NAD. This allows glycolysis to continue in the absence of oxygen.

anaphase: in *mitosis*, the stage when sister *chromatids* separate and move towards the poles. One chromatid from each pair goes to each pole. This is the result of shortening of *spindle* fibres. In *meiosis*, anaphase is similar but, in the first division, it is pairs of homologous *chromosomes* that separate.

anaphylaxis: an abnormal reaction by the body to a particular *antigen*. *Histamine* is released. This may cause local effects such as those associated with *hay fever*. Occasionally, the symptoms are more widespread and a condition known as anaphylactic shock may result. This may produce narrowing of the *bronchioles*, heart failure and death unless treated.

androgen: a *hormone* that stimulates or controls the development of male characteristics, such as *testosterone*.

aneuploidy: an abnormal number of *chromosomes*. The most common form of aneuploidy in humans is *Down syndrome* where there are three copies of chromosome 21 in each cell instead of the normal two copies.

aneurysm: a weakening of the wall of an *artery*, which produces a balloon-like swelling. The consequences for the patient are very serious if it bursts.

angiosperm: see *Angiospermophyta*.

Angiospermophyta: the plant phylum that contains the *flowering* plants. It includes a great variety of plants ranging from daisies to oak trees and includes most plants that are of economic importance. Members of the Angiospermophyta share the following features:
● the reproductive organs are found within *flowers*
● *seeds* are formed as a result of *fertilisation* and these seeds are enclosed in *fruits*.

angina: pain in the centre of the chest, usually triggered by exercise and relieved by resting. It results when insufficient oxygen is brought to the heart muscle by the *coronary arteries*. Angina is often caused by *atheroma* narrowing the coronary arteries.

angioplasty: the repair of damaged *blood vessels*. Coronary angioplasty is often used to repair *coronary arteries* narrowed by *atheroma*. A hollow tube called a catheter is inserted into the *artery*. This catheter has a tiny balloon inside it. When the catheter is in the right place, the balloon is inflated. This pushes the artery walls outwards and makes the artery wider.

animal: a member of the *kingdom Animalia*.

Animalia: the *kingdom* containing animals (see *classification*). Animals share the following features:
- they are multicellular organisms whose cells do not possess *cell walls*
- they obtain their food by taking it into a digestive cavity in a process known as *ingestion*. Once inside the digestive cavity, food is digested and absorbed
- after *fertilisation* a *zygote* is formed. This zygote develops into a hollow ball of cells called a *blastula*.

annelid: see *Annelida*.

Annelida: the animal phylum that contains segmented worms such as earthworms and leeches. Members of the Annelida share the following features:
- they have worm-shaped bodies that are divided into segments
- they possess small bristles or chaetae on their body surfaces, which assist with locomotion
- they have a fluid-filled *coelom*, which helps to form a skeleton against which the locomotory muscles can contract when the animal moves.

anorexia nervosa: a disorder in which patients starve, use laxatives or make themselves vomit in order to lose weight. The underlying causes are psychological and lead to a fear of becoming fat. The symptoms include severe loss of weight and, in female patients, stopping of menstrual periods (see *menstrual cycle*).

antagonistic muscles: a pair of muscles which, when they contract, produce opposite effects to each other. Muscles can only do work by shortening or contracting. Therefore, if a limb has to be moved, two muscles are required to do this. One will bend the limb when it contracts, the other will straighten it. The two muscles are said to be antagonistic. The term does not apply only to muscles which involve the movement of limbs. The earthworm has two sets of muscles in its body wall which are antagonistic. Circular muscles will make the animal longer and thinner when they contract; longitudinal muscles will make it shorter and fatter.

anther: the part of the *stamen* of a *flower* in which the *pollen* develops and from which it is released.

antheridium: the male sex organ in plants such as mosses, liverworts and ferns. It contains male *gametes*. The female equivalent is the *archegonium*.

antibiotic: a substance produced by one type of *microorganism* which kills or stops the growth of another. There are a number of exceptions to this definition, however, so it is difficult to produce a general definition of the term. Substances that have an antibiotic effect have now also been found in a wide range of different animals and plants, ranging from toads to snowdrops. Antibiotics used in medicine are often altered chemically to make them more effective and some are made entirely by chemical means.

There are a number of small but important differences between the biochemical processes that take place in microorganisms and those that take place in mammals. Antibiotics work by affecting those processes that are found only in microorganisms. Penicillin, for example, prevents chemical bonds forming that strengthen bacterial *cell walls*. Since human cells do not have *cell walls*, penicillin has no effect on them. The diagram overleaf shows the effects of some antibiotics on bacteria.

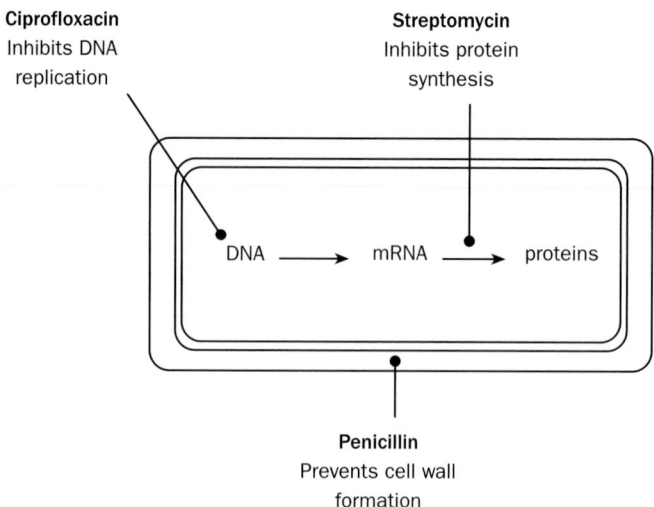

The effects of some antibiotics on bacteria

If an antibiotic kills bacteria, it is a **bactericidal** (biocidal) antibiotic. If the bacteria are simply prevented from reproducing, then it is a **bacteriostatic** (biostatic) antibiotic.

antibody: a molecule secreted by certain **lymphocytes** in response to stimulation by the appropriate **antigen**. The term describes what the molecule does. It acts against foreign bodies. Antibodies are types of **proteins** known as immunoglobulins. The way in which these molecules work is closely related to their structure. This is explained in the diagram.

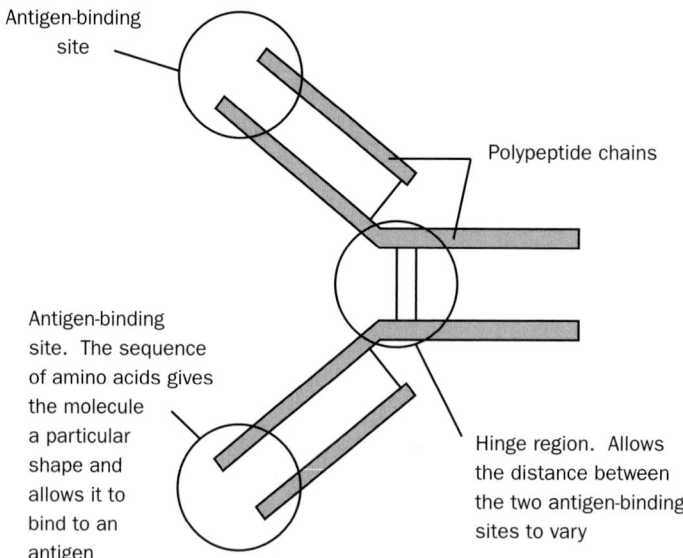

The structure and function of an antibody

anticoagulant: a drug that prevents **blood clotting**. It may be used in the treatment of conditions such as **deep vein thrombosis (DVT)**. Anticoagulants include heparin and warfarin. Heparin inhibits **thrombin**, whereas warfarin stops the **liver** from producing substances necessary for the clotting of blood.

anticodon: the sequence of three bases on a molecule of transfer RNA (*tRNA*) which corresponds to a particular *codon* on an *mRNA* molecule. During *translation* a molecule of mRNA becomes attached to a *ribosome*. The ribosome is just the right size to hold two codons of mRNA. Suppose one of these codons has the sequence of bases GCU which codes for the amino acid alanine. Only a tRNA molecule with the complementary anticodon, CGA, will be able to bind to this codon. A tRNA molecule with this anticodon will always carry alanine.

antidiuretic hormone (ADH): a *hormone* involved in the control of water balance in the body. Antidiuretic hormone is produced by a group of cells in the *hypothalamus* of the *brain*. It is stored in the posterior lobe of the *pituitary gland*. The flow chart summarises the way in which this hormone helps to control water balance when the *water potential* of the blood falls. This is particularly important in hot conditions.

A fall in the water potential of the blood is detected by osmoreceptors in the hypothalamus
↓
Nerve impulses from the hypothalamus bring about the release of ADH from where it is stored in the posterior lobe of the pituitary gland
↓
ADH is released into the blood. It travels to the kidney where it makes the cells of the second convoluted tubule and the collecting duct more permeable to water
↓
More water is reabsorbed and less is lost in the urine

The release of ADH is inhibited when the water potential of the blood increases. The cells of the second convoluted tubule and the *collecting duct* become less permeable to water. Therefore, less water is reabsorbed and large quantities of dilute *urine* are produced. Removal of excess water in dilute urine results in the water potential of the blood returning to its normal level.

antigen: a molecule that triggers an immune response by the *lymphocytes*. Although usually *proteins*, other types of molecule such as *polysaccharides*, *lipids* and even *nucleic acids* can act as antigens. In some circumstances, inorganic substances such as nickel can bind to proteins in the body and can also act as antigens. This is what happens, for example, in a person who develops an *allergy* to metal earrings. The one feature that all antigens have in common is the large size of their molecules.

antigen–antibody complex: formed when an *antibody* binds to an *antigen*. Antibodies are *proteins*. They have specific antigen-binding sites. The sequence of *amino acids* that forms these sites gives them a specific shape. This allows binding to an antigen that has a complementary shape. During an infection, the formation of an antigen–antibody complex is the first stage in the destruction of a cell containing a non-self antigen.

antigen presentation: a process that takes place in the immune system in which foreign *antigens* are presented to *T cells* in a way that they can recognise. T cells continually monitor the body for infection. They must distinguish between self-antigens present on the host's own cells and foreign antigens present on *pathogens*. They only recognise foreign antigens when these antigens are bound to appropriate molecules on the surface of the antigen-presenting cell. Most cells can present antigens but some of the cells in the immune system are specifically adapted for this purpose.

antigenic variability: a change in an *antigen* present on the surface of a *microorganism*. In *influenza* viruses, antigenic variation results in new strains of the virus. People who have previously had influenza may not be immune to a new strain. Some parasites, such as those that cause *malaria* and sleeping sickness, continually change the antigens in their surface coats. In this way they survive the immune response of their hosts.

antihistamine: a drug used in the treatment of allergic reactions such as *urticaria* and *hay fever*. Antihistamines bind to histamine receptors and block the action of *histamine*.

antihypertensive: a drug used in medicine to treat high blood pressure. Antihypertensives lower blood pressure either by reducing the volume of the blood or by dilating *blood vessels*.

antioxidant: a substance that slows down or prevents the oxidation of another substance. In living organisms, oxidation reactions produce free radicals and these may damage cells. This damage may increase the risk of developing *coronary heart disease* and cancer (see *tumour*). Some people think that food supplements containing antioxidants are beneficial to health, but clinical trials have so far failed to prove this.

antiparallel: molecules, or parts of molecules, that are arranged side by side but which run in opposite directions. A *DNA* molecule consists of two antiparallel *polynucleotide* chains. One runs in one direction; the other runs in the opposite direction.

antiseptic: a substance that kills *microorganisms* but can be used safely on the skin or the *mucous membranes* of the body. Antiseptics are like *disinfectants* in that they kill most of the harmful microoganisms present.

anus: the opening of the lower end of the *alimentary canal* through which waste material in the form of *faeces* is passed out of an animal. The anus is normally closed by two rings of muscle called *sphincter muscles*. These sphincter muscles open when defaecation takes place.

aorta: the main *artery* of the body. In a mammal, it leaves the left *ventricle* of the heart and curves over to go through the abdomen. The arteries that supply all the other organs of the body (with the exception of the *lungs*) branch off the aorta. The wall of the aorta contains baroreceptors that monitor blood pressure and *chemoreceptors* that monitor changes in concentration of carbon dioxide and oxygen in the blood.

aortic body: an area in the wall of the *aorta* that contains *chemoreceptors*. These receptors respond to changes in the concentration of carbon dioxide in the blood. Sensory nerves take impulses from the aortic body to the *medulla* in the *brain*. The aortic bodies have a similar function to the *carotid bodies*.

aphid: a small insect that feeds by sucking the sap from the *phloem* of plants. Also known as greenfly or blackfly, aphids may occur in large numbers. They cause damage by removing nutrients from the plant. This affects growth and plants with heavy infestations of aphids are often deformed or stunted as a result. Aphids are also important economically as they may transmit *viruses* from one plant to another.

apical dominance: when the terminal bud on a shoot suppresses the growth of lateral buds further back on the same shoot. The terminal bud on an actively growing shoot produces *auxins* and other *plant growth substances*. These substances inhibit the growth of the lateral buds. If the terminal bud is cut off as when, for example, a plant is pruned, the lateral buds grow and the plant bushes out.

apoplastic pathway: the pathway by which substances go through the *cell walls* and the intercellular spaces of a plant. Water moves through the roots into the *xylem*. It also moves out of the xylem and through the leaf tissue. There are three separate pathways along which it can travel. The apoplastic pathway described here, the *symplastic pathway* through the *cytoplasm* of the cells, and the *vacuolar* pathway which involves passing from vacuole to vacuole. The apoplastic pathway is probably the most important of the routes by which water moves through plant tissues. There are no barriers to movement so it is able to diffuse freely.

apoptosis: a process involving the death of cells that plays an important part in the development of animals and plants. Early in human development, for example, the fingers are joined together. As a result of apoptosis or programmed cell death, as the process is also known, the tissue between the digits breaks down and the fingers separate. Apoptosis is triggered by a range of cell signals. These signals influence the activity of the *enzymes* that control biochemical pathways involved in the death of cells. (See also *cell signalling*.)

aqueous humour: a watery solution found in the front part of the eye underneath the *cornea*. It is continually being formed by a structure known as the ciliary body and reabsorbed back into the blood. It has two important functions. The *cornea* and *lens* are transparent but they are composed of living cells that require a supply of nutrients. The aqueous humour provides these nutrients. It also helps to maintain the shape of the front part of the eye.

archegonium: the female sex organ in plants such as mosses, liverworts and ferns. It contains an egg cell or female *gamete*. The male equivalent is the *antheridium*.

arrhythmia: cardiac arrhythmia involves abnormal electrical activity in the heart. As a result, the heart may beat too slowly, too rapidly or irregularly.

artefact: something seen in a specimen looked at under a microscope that was not present in the living cell. In order to prepare specimens for microscopic examination, they have to be treated in a number of ways. The tissue usually has to be preserved, cut into sections and stained. These procedures involve substances which may alter the cell. Artefacts which may arise in preparation of the specimen mean that care should be taken in interpreting what is seen under the microscope. This is particularly important with *electron microscopes* because it is not possible to observe living cells.

arteriole: a vessel that takes blood from the smaller *arteries* to the *capillaries*. Arterioles are very small in diameter and, like all *blood vessels*, have a lining of epithelial cells. Their walls contain muscle fibres. Many arterioles also have rings of muscle where they join with the capillaries. These are called *sphincter muscles*. By contraction of the muscle fibres in the walls and the sphincter muscles, the blood supply to particular capillary networks can be regulated to meet the needs of the part of the body concerned.

artery: a *blood vessel* that takes blood from the heart towards the *capillaries*. In mammals, arteries usually contain blood that is rich in oxygen but there is one important exception to this. The blood going to the *lungs* along the *pulmonary artery* has come from the right side of the heart. This is the blood that has returned from the tissues and is therefore low in oxygen. Like all blood vessels, arteries have a lining composed of epithelial cells. Their walls are thick and contain a large amount of elastic tissue and muscle. When the *ventricles* of the heart contract, blood at high pressure is forced into the arteries and causes the elastic tissue in the walls to stretch. Between ventricle contractions, the elastic tissue recoils. This stretching and recoiling of the elastic tissue in the artery walls helps to even out blood flow.

The activity of different organs in the body is always changing. During a period of strenuous exercise, for example, the muscles require as much blood as possible; less is required by the digestive system. The walls of smaller arteries contain a lot of muscle fibres. These may contract and make the diameter of the artery smaller. This helps to regulate the blood supply to a particular organ.

Arthropoda: the animal phylum containing insects, spiders and crustaceans such as crabs and woodlice. There are more arthropods both in terms of individuals and species than any other animal or plant phylum. Members of the Arthropoda share the following features:
- they possess an *exoskeleton* made of *chitin*. This hard outside layer has to be moulted before an arthropod can increase in size. Growth is therefore discontinuous
- they have jointed limbs or appendages. These are used for a variety of purposes including locomotion and feeding.

artificial fertiliser: a mixture of inorganic substances that provides the nutrients necessary for the growth of crops. Artificial or chemical *fertilisers* are made either from naturally occurring rocks or by industrial processes. The main plant nutrients found in these fertilisers are nitrogen, phosphorus and potassium (NPK).

artificial immunity: protection from *infectious disease* resulting from the deliberate exposure to an *antigen* such as occurs as a result of vaccination.

artificial selection: selective breeding carried out by humans to alter the characteristics of a *population*. The process of artificial *selection* has led to many of the characteristic features of domestic animals and crop plants. High milk yield in cattle, short stem length in modern varieties of wheat and fine wool in merino sheep have all arisen by artificial selection. The principles underlying artificial selection are exactly the same as those involved in natural selection. In both cases, individuals with certain characteristics have a greater probability of surviving and passing on their genes.

Ascaris: a *genus* of parasitic *roundworms* that live in the intestines of many vertebrates. *Ascaris lumbricoides* is found in humans. It is a large white worm, up to 40 cm in length. It has a worldwide distribution and probably affects over 600 million people.

asepsis: reducing or eliminating *pathogens*. Asepsis is important immediately before and during surgery. Procedures such as shaving and washing the skin of the patient with an *antiseptic*, and using sterile instruments, help to eliminate the risk of infection from pathogenic *microorganisms*.

asexual: reproduction that does not involve the fusion of *gametes*. One of the most important features of asexual *reproduction* is that it involves processes that give rise to the production of genetically identical offspring. Many *microorganisms*, such as bacteria and fungi, can reproduce in this way and asexual reproduction is also widespread in plants. Among animals it is less common and is generally found only in more primitive groups such as the *Cnidaria*. The production of buds by yeast cells, of *spores* by mosses and ferns and the splitting of an amoeba to form two separate individuals are all examples of asexual reproduction.

In *flowering* plants, asexual reproduction is also known as vegetative propagation. It frequently involves the development of fairly large structures such as bulbs in plants like daffodils and onions, or special types of underground stem as in potatoes and irises. Not only are these structures involved in asexual reproduction, they may also help the plant to survive extreme climatic conditions. With the first frosts, for example, the aerial parts of a plant such as an iris die. The underground stem, however, contains the nutrients necessary for the plant to survive the winter and to produce a number of new plants the following spring.

The fact that asexual reproduction produces genetically identical offspring is used commercially where it has considerable economic importance. Suppose a new variety of rose has been developed. Letting it reproduce sexually by pollinating the *flowers* and growing the resulting seeds would result in genetically different offspring. This would be of little use to the plant breeder. If more plants are produced by asexual reproduction, the offspring will all have the desired characteristic as they will be genetically identical. The process by which large numbers of genetically identical plants can be produced from cells taken from the growth region of a single parent is known as *micropropagation*.

association area: part of the *brain* involved with analysing and interpreting sensory information in the light of experience. The association areas form part of the *cerebrum* in the forebrain. They:
- link sensory information with previous knowledge, allowing us to recognise information
- link sensory information with information from other receptors
- interpret the information in a particular context.

Think about an image formed on the *retina* of the eye. You could get the same size image from a small object close to the eye or a much larger object at a distance. It is the visual association area that links all the sensory clues with information stored in the memory to interpret the image.

assimilation: the process by which substances absorbed from the *alimentary canal* after *digestion* are taken into the cells of the body and built up into useful substances.

asthma: a condition in which the flow of air to the gas exchange surface of the *lungs* is restricted. Some of the cells in the lining of the bronchi become sensitive to substances such as smoke, pollen or other atmospheric pollutants. Further exposure to these substances brings about an immune response which triggers an asthmatic attack. The smooth muscle in the walls of the bronchi contracts. Together with the production of large amounts of mucus, this leads to a narrowing of the airways and limits the amount of air that can reach the *alveoli*.

atheroma: fatty deposits formed in the walls of *arteries*. Atheroma is a common condition, seen in many adults, and is associated with a number of risk factors. These risk factors include diets high in **cholesterol** and refined sugars, smoking and lack of exercise. At first, atheroma may not produce any obvious signs of ill-health but, later in life, it may give rise to more serious diseases of the circulatory system such as an **aneurysm** or a **myocardial infarction**.

atherosclerosis: a disease of the arteries caused by **atheroma** or fatty deposits in the walls. These can either block an **artery** directly or increase the chance of it being blocked by a blood clot. When atherosclerosis affects the **coronary arteries**, it may lead to a heart attack or **myocardial infarction**.

atom: the smallest part of an **element**. It cannot be broken down further by chemical means. An atom is made up of a nucleus and **electrons**. The nucleus contains two sorts of particle: protons which have a positive charge and neutrons which, as their name suggests, have no charge. Negatively charged electrons orbit the nucleus. **Atoms** of the same or of different elements may combine to form **molecules**.

ATP: adenosine triphosphate, a substance found in all living cells, that is involved in the transfer of energy. A molecule of ATP is made up from a molecule of **adenine** joined to the five-carbon sugar **ribose** and three phosphate groups. When ATP is broken down, the third phosphate group is lost and a considerable amount of energy is released. The reverse reaction can also take place: **ADP** can join with a phosphate group to produce ATP. In this case, energy is required. These reactions are summarised in the diagram.

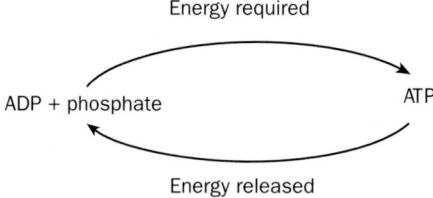

Energy required

ADP + phosphate   ATP

Energy released

Most of the ATP within a cell is produced using energy released during the process of **respiration**. A small amount is also made in **photosynthesis**. Some of the ways in which ATP is used in living organisms are:

- ATP releases energy. This energy may be used in many different chemical reactions, for **active transport** and for movement.
- ATP supplies phosphate to various chemical compounds. This is important in some of the biochemical reactions in respiration and photosynthesis.
- ATP is the molecule from which **cyclic AMP** is produced. Cyclic AMP is an important messenger molecule in cells.

ATPase: an **enzyme** catalysing the breakdown of **ATP** to **ADP** and phosphate. ATPase activity is associated with many cell functions. Pumps, for example, that actively transport sodium and potassium ions across **plasma membranes**, involve ATPase. (See also **active transport**.)

ATP synthase: an **enzyme** that catalyses the synthesis of **ATP** from **ADP** and phosphate. Some ATP synthases are found in the membranes of **mitochondria** and **chloroplasts**. Protons are produced in **respiration** and **photosynthesis**. When these protons flow back through a pore in the ATP synthase molecule, they cause the enzyme to spin. This results in ADP and phosphate bonding and forming ATP.

atrial systole: the stage in the *cardiac cycle* where the walls of the atria contract and force blood into the *ventricles*.

atrioventricular node (AVN): a small area of muscle in the wall of the heart between the *atria* and the *ventricles*. It plays an important part in coordinating the beating of the heart. At the start of each heart beat, a wave of electrical activity spreads from the *sinoatrial node (SAN)* over the walls of the atria. This brings about contraction of the atria. The muscle fibres in the atria are completely separate from those in the ventricle except in one small area: the atrioventricular node. It is only through this node that electrical activity can pass from the atria to the ventricles. There is a short delay here before electrical activity spreads to the base of the ventricles. This is important in allowing emptying of the atria to be completed before the ventricles start to contract.

atrioventricular valve: one of the valves between the *atria* and the *ventricles* in the heart. These valves are made of fibrous tissue and are opened and closed by blood pressure. During the part of the *cardiac cycle* in which the pressure in the *atrium* is higher than that in the ventricle, the valve is open and blood is able to flow through into the ventricle. During *ventricular systole*, the muscle in the wall of the ventricle starts to contract. The pressure of the blood in the ventricles is now greater than the pressure in the atria. As a result, the valve shuts. This prevents backflow into the atria and enables blood to be pumped out through the arteries leaving the heart. When a *stethoscope* is placed against the chest wall, sounds can be heard. The first of these heart sounds is due to the atrioventricular valves closing. The valve on the left side of the heart has two flaps of fibrous tissue and is called the *bicuspid valve*; that on the right side has three flaps and is called the tricuspid valve.

atrium: a chamber of the heart (plural atria). It receives blood returning to the heart. Fish have a *single circulation* and consequently the heart of a fish has only one atrium. Mammals, on the other hand, have a *double circulation*. Their hearts have two atria. The right atrium receives blood returning from the organs of the body, while the left atrium receives oxygenated blood which the pulmonary veins bring back from the *lungs*.

attenuation: the growth of bacteria or viruses in a culture so that they lose their ability to cause disease. Pathogens such as bacteria have a number of virulence genes. These genes play an important part in the bacterium completing its life cycle inside the cell of a host. They are not necessary, however, when the same bacterium is grown in an artificial culture. Consequently, over a period of time, the virulence genes in cultured bacteria tend to become ineffective through the normal processes of mutation. Attenuated pathogens are therefore harmless and are often used for making *vaccines*.

Australopithecus: a *genus* of extinct African *hominids*. They were similar to members of the genus *Homo*, because they could walk upright. There were, however, slight differences in the shape of the pelvis and upper leg *bones*. In relation to their body size, they had smaller *brains*. There were a number of species of *Australopithecus*, which could be roughly divided into two distinct groups. One group was rather lightly built; the other much more robust.

autoimmune disease: a disease caused by the body producing an *immune reaction* against its own tissues. Damage may either be caused by *antibodies* or by *T cells*. Autoimmune diseases are more common in older people. Some of these diseases, such as pernicious *anaemia*, affect specific organs or systems. In this disease, antibodies are made against a molecule in the gut necessary for the uptake of vitamin B12. Other examples, such as rheumatoid arthritis which causes painful swelling of the joints in many elderly people, have a more general effect on the body.

autonomic nervous system: the part of the nervous system that supplies the muscles and *glands* which are not under conscious control. The autonomic nervous system is divided into two sections. The *sympathetic nervous system* plays an important part when the body reacts to stress. The *parasympathetic nervous system* is more important when the body is at rest.

autoradiography: a technique in which radioactive substances are taken up by cells or tissues and concentrated inside them. The tissue is then placed on photographic film. The film is sensitive to these substances and shows up their position. Scientists have used autoradiography to investigate many biochemical and physiological processes. These processes include investigation of the *light-independent reaction* of *photosynthesis* and the passage of substances through *phloem*. Autoradiography is also used in *genetic fingerprinting* where it shows up pieces DNA that have bound to a radioactive probe.

autosome: a *chromosome* that is not a *sex chromosome*. In a human body cell, there are 23 pairs of chromosomes. One pair of these is a pair of sex chromosomes; the other 22 pairs are autosomes.

autotroph: an organism that has *autotrophic nutrition*.

autotrophic nutrition: a method of nutrition in which an organism builds up the organic molecules that it requires from simple inorganic molecules such as carbon dioxide and water. In order to do this, an energy source is necessary. In photoautotrophs, this energy source is light. In other organisms, however, the energy comes from another chemical reaction. These organisms are *chemoautotrophs*. Autotrophs are *producers* and are at the base of all *food chains*.

auxin: a *plant growth substance* that has a number of different functions. One of the most important of these is cell enlargement. The flow chart explains how it is thought that cell enlargement is brought about.

Presence of auxin leads to the active transport of H^+ ions into the cell wall

↓

The resulting change in pH leads to the breakdown of chemical bonds in the cell wall. As a result, the cell wall softens

↓

The cell can take in more water by osmosis. It can enlarge as the cell wall no longer restrains it

However, auxins have other effects and it is thought that they are also involved in switching on a number of different genes associated with growth.

Synthetic auxins have important commercial applications:

- They may be used as weedkillers. Since some of them are toxic to many *weeds* but are not toxic to grasses and cereals, they may be used to control weeds in cereal crops or on lawns.

- They can be used to stimulate the formation of roots on cuttings of plants such as geraniums.
- They help to prevent *fruit* drop. Fruit drop is a common problem in orchards. Large amounts of fruit fall from trees before it is ripe.

AVN: see *atrioventricular node (AVN)*.

axil: the angle between the upper surface of a leaf and the stem to which it is joined. It contains a bud, known as an axillary bud. One of the features of a potato tuber which confirms that it is a modified stem and not a root is that it has tiny scale leaves on its surface, each of which has a bud in its axil.

axon: a long process in a *neurone*. An axon carries *nerve impulses* away from the cell body.

Azotobacter: a *genus* of bacteria that live in the soil. These bacteria contain the enzyme nitrogenase. With the aid of this *enzyme*, they are able to convert nitrogen in the soil into ammonium compounds. They play an important part in the *nitrogen cycle* because they are able to fix nitrogen. (See also *nitrogen fixation*.)

A–Z Online

Log on to A–Z Online to search the database of terms, print revision lists and much more. Go to **www.philipallan.co.uk/a-zonline** to get started.

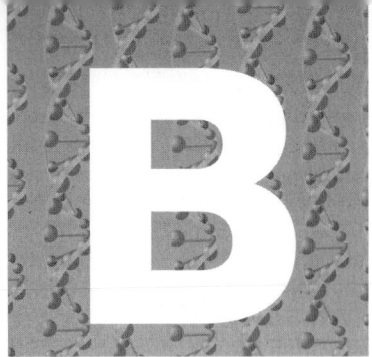

B cell: a *lymphocyte* that, when it comes into contact with the appropriate *antigen*, develops into an *antibody*-secreting cell. B cells have receptors on their surfaces that will only bind to specific antigens. Once a B cell has encountered this antigen, it divides rapidly and produces a large number of identical cells called a *clone*. This clone contains two sorts of cells: *plasma cells* and *memory cells*.

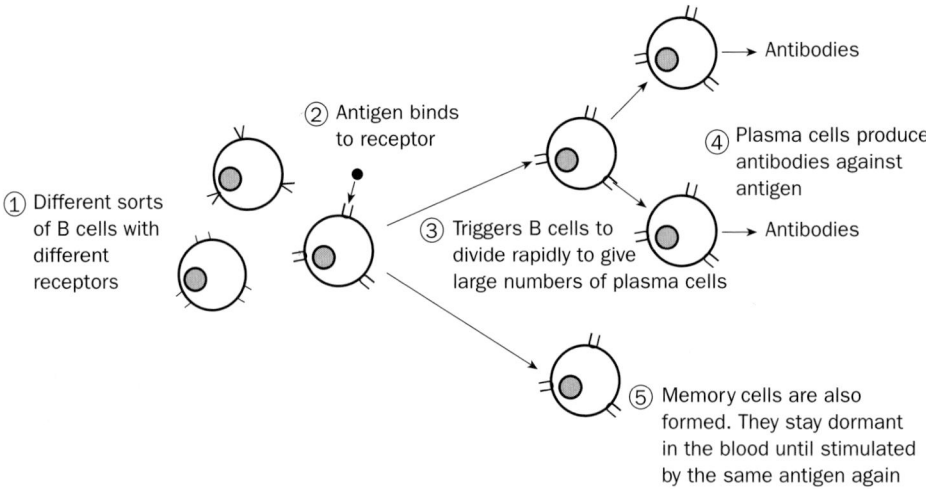

① Different sorts of B cells with different receptors

② Antigen binds to receptor

③ Triggers B cells to divide rapidly to give large numbers of plasma cells

④ Plasma cells produce antibodies against antigen

⑤ Memory cells are also formed. They stay dormant in the blood until stimulated by the same antigen again

Antibodies

The response of B cells to infection

B lymphocyte: see *B cell*.

B memory cell: see *memory cell*.

bacillus: a rod-shaped bacterial cell. Examples of bacilli include *Lactobacillus bulgaricans*, which is found in yoghurt, and the various species of *Salmonella*, which produce food poisoning.

bacteria: an important group of *microorganisms* in the kingdom *Prokaryotae*. The cells of bacteria (singular bacterium) are small, do not have nuclei or organelles such as *mitochondria*, and their *DNA* is in circular strands in the *cytoplasm*. Although some bacteria are harmful and are responsible for diseases such as *cholera* and plague, others form the basis of much of modern *biotechnology*. In addition, they play a vital role in ecological processes. *Nutrient cycles* rely on the action of bacteria in making the elements locked up in complex organic molecules available to plants once again.

A number of features are used to identify bacteria. The shape of an individual cell is important and enables it to be classified as a *bacillus*, a *coccus*, a *spirillum* or a *vibrio*.

Bacilli or **rods**

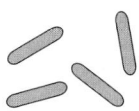

Cocci

May occur
singly

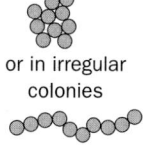

or in irregular
colonies

or in chains

Salmonella typhi causes typhoid
fever
Bacillus subtilis is used in
genetic engineering

Staphylococcus aureus causes boils

Streptococcus lactus is used in
making cheese

Spirilla

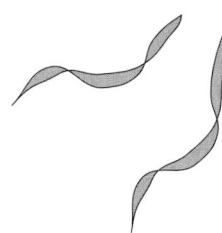

Vibrios or **curved rods**

Treponema pallidum causes
syphilis

Vibrio cholerae causes cholera

Identifying bacteria by their shapes

With their small size and the enormous number of species involved, shape alone is inadequate as a means of identifying individual bacteria. The molecules that they contain and the biochemical processes which occur within their cells are also important and allow techniques such as *Gram staining* to be used to distinguish between bacteria that look very similar under a microscope.

bacterial chromosome: see *chromosome*.

bactericidal: used to describe a substance that kills bacteria. Killing is essential for *sterilisation* so *disinfectants* and *antiseptics* are bactericidal compounds. There is a wide range of substances available, but their effectiveness depends on the *microorganism* involved. The *spores* formed by certain species of bacteria are extremely resistant to substances that would kill other forms of life. Many drugs used to combat disease, however, are *bacteriostatic* and simply prevent bacterial reproduction.

bacteriophage: see *phage*.

bacteriostatic: used to describe a substance that prevents bacterial reproduction. Many *antibiotics* are bacteriostatic as they interfere with the mechanism of protein synthesis. They prevent bacteria multiplying while the body's natural defences destroy those already present.

balanced diet: a diet that contains all the basic nutrients in the proportions necessary to maintain health. The exact composition of a balanced diet will vary throughout a person's life. Such factors as age, sex, pregnancy and *lactation* will all have an influence. To be balanced, consideration needs to be given to:

- *carbohydrates*. Sufficient carbohydrate needs to be taken in order to meet the energy requirements of the body. Excess can lead to being overweight. A *correlation* exists between an excessive intake of carbohydrates and a number of conditions. These include heart disease, *diabetes* and dental caries
- *lipids*. It is necessary to eat lipids, but there is concern over diets that contain high amounts of animal *fats*. There is a link between an excessive intake of animal fats and heart disease
- *protein*. People who are actively growing, and pregnant or lactating women, need increased protein in their diets. It is essential that the protein in the diet includes the necessary amounts of *essential amino acids*. This is particularly true for vegans, who eat no food of animal origin
- mineral ions. These are required in varying amounts and, again, the amount actually needed depends on the specific circumstances of the individual concerned. Women, for example, tend to require a higher intake of *iron* to meet the loss from the body through menstruation (see *menstrual cycle*). Pregnancy and lactation demand a higher level of *calcium*
- *vitamins*. These are required in small amounts for a variety of purposes. There is no evidence that consumption of extra amounts of vitamins has a beneficial effect. Most suitably varied diets will contain sufficient vitamins to meet a person's requirements adequately.

basal metabolic rate (BMR): the *metabolic rate* of a person who is at complete rest. It is a measurement of the amount of energy required for vital activities such as the action of the heart and the muscles associated with breathing. It should be measured some time after a meal, so as to reduce as far as possible the effect of food in the *stomach*. It also needs to be determined at a comfortable environmental temperature since processes such as shivering have a considerable effect on metabolism. It is usual to give figures for basal metabolic rate in energy units per square metre of body surface per hour. This takes into account variation in body size. There are a number of factors that affect basal metabolic rate:

- age. In humans, basal metabolic rate has a maximum value at about 1 year old. It then falls more or less continuously for the rest of a person's life
- sex. At all ages, females have a slightly lower basal metabolic rate than males
- state of nutrition. People who are undernourished tend to have lower basal metabolic rates.

base: in biology, the term usually refers to one of the components of a *nucleic acid* molecule. A molecule of DNA or RNA is made up of a large number of *nucleotides*. Each of these consists of a five-carbon sugar, a phosphate and a nitrogen-containing base. There are five different bases and one of these is found in each nucleotide. Some of the properties of these bases are shown in the table.

Property	Base				
	Adenine	Cytosine	Guanine	Thymine	Uracil
Type of base	Purine	Pyrimidine	Purine	Pyrimidine	Pyrimidine
Nucleic acid in which base is found	DNA and RNA	DNA and RNA	DNA and RNA	DNA only	RNA only
Base with which it pairs	Thymine (DNA) Uracil (RNA)	Guanine	Cytosine	Adenine	Adenine

base pair: a pair of complementary *bases* in a molecule of *DNA* or in *tRNA*. The term may also refer to a pair of complementary bases formed during the processes of *DNA replication* and protein synthesis. In the two *polynucleotide strands* in a molecule of DNA, the base *adenine* in one strand always pairs with *thymine* in the other strand. In a similar way, *guanine* always pairs with *cytosine*. Adenine and guanine are *purines*; *thymine* and cytosine are *pyrimidines,* so a larger purine always pairs with a smaller pyrimidine and the two polynucleotide chains remain a constant distance apart. *Hydrogen bonds* form between these base pairs. The principle of complementary base pairing is important in explaining *DNA replication*, *transcription* and *translation*.

basement membrane: a thin layer of material situated between *epithelium* and *connective tissue*. In the *renal capsule* in the *kidney*, the basement membrane acts as a filter. It lets smaller molecules and ions through but prevents large protein molecules from entering the *glomerular filtrate*.

batch culture: a method of growing *microorganisms* that involves separating the products at the end of the process. The microorganisms involved are put in a fermenter along with the necessary medium. Nothing is then added or removed during *fermentation*. At the end of the process, the required product is extracted and purified. This procedure may be contrasted with continuous culture, where the *substrate* is added and the products removed while fermentation is going on. Batch culture has some advantages over continuous culture:

- If contamination occurs, it only results in the loss of a single batch. Because it is relatively easy to set up a batch culture, this results in minimal financial loss.
- Batch culture is a relatively short-term process and does not involve setting aside a fermenter for a long period of time. It is much easier to switch the use of fermenters and make different products at different times. This is more versatile.
- It is much easier to maintain the correct environmental conditions while a single batch is cultured than over the long period of time required by continuous culture.

BCG vaccine: a vaccine used to protect against *tuberculosis (TB)*. It uses a strain of the bacterium *Mycobacterium bovis* that has lost its ability to cause disease.

Benedict's test: a *biochemical test* used to show the presence of *reducing sugars*. If the specimen is hydrolysed first by boiling with an acid, it can be used to test for a *non-reducing sugar*.

benign tumour: see *tumour*.

beta-antagonist: see *beta-blocker*.

beta-blocker: a drug that can be used to treat *angina* and high blood pressure. The membrane surrounding a cell, the *cell-surface membrane*, has a number of protein molecules in it. Some of these act as receptors. They have a particular shape into which another molecule such as a *hormone* fits. Heart muscle has beta-*adrenergic* receptors, which are the receptors into which *adrenaline* molecules fit. A beta-blocker or beta antagonist has a similar molecular shape to that of adrenaline. It will also fit into the beta-adrenergic receptor. When it does so, it blocks the receptor so adrenaline will not affect the heart.

Beta-blockers were first developed to combat angina, a condition characterised by severe chest pains that accompany exercise. When a person normally undertakes greater physical activity, adrenaline is secreted. This increases the activity of heart muscle with the result that more oxygen is required. In a person with heart disease, the supply of oxygen cannot be increased enough, so angina results. By blocking the action of the adrenaline, this condition should not arise. As well as preventing angina, beta-blockers can also be used to treat high blood pressure or *hypertension*.

α-glucose (beta glucose): a form or *isomer* of *glucose* in which the –H and –OH groups on carbon 1 are arranged as shown in the diagram.

β-glucose

Molecules of β-glucose link by *glycosidic bonds* formed by *condensation* to form *cellulose*.

β-pleated sheet (beta pleated sheet): a type of *secondary structure* found in *protein* molecules. In a β-pleated sheet, parts of the *polypeptide* chain are parallel to each other. They are held together by *hydrogen bonds* and form flat sheets. The protein fibroin is the main component of silk. It has molecules that form β-pleated sheets. Because the *amino acids* in these sheets have been pulled into an extended form, silk cannot be stretched.

bicuspid valve: the valve between the left *atrium* and the left *ventricle* in the heart of a mammal. (See also *atrioventricular valve*.)

bilateral symmetry: used to describe organisms that can only be cut in one plane to produce a mirror image. This is because their organs are arranged similarly on either side of a line that goes down the centre. Most bilaterally symmetrical animals are motile, that is, they can move. Bilateral symmetry is thought to be an adaptation for movement allowing the equal contraction of muscles on either side of the body. The other important feature of bilaterally symmetrical animals is that they tend to have a distinct head region. It is the front end of a moving animal that is most likely to encounter food and the different stimuli which occur when it enters a new *environment*. The feeding apparatus and sense organs therefore tend to be concentrated at the front end. Organisms which have more than one plane of symmetry are said to show *radial symmetry*.

bile: a solution produced in the *liver*, which empties into the *small intestine* through the bile duct. It is a slightly alkaline liquid containing:
- water and various inorganic ions
- *bile salts*. These play an important part in the *digestion* of *fat*
- *bile pigments*. These are excretory products formed from the breakdown of *haemoglobin*.

Bile does not contain any digestive *enzymes*. Between meals, bile is stored in the *gall bladder*.

bile pigment: a substance formed by the breakdown of *haemoglobin*. *Red blood cells* do not live for long. After about 120 days in the circulation, they die and the haemoglobin they contain is broken down. The diagram opposite shows that the haem group is split off from the globin part of the molecule. The *iron* from the haem group is used to make more haemoglobin while the rest of the group is converted into *bile* pigments that are excreted in the bile.

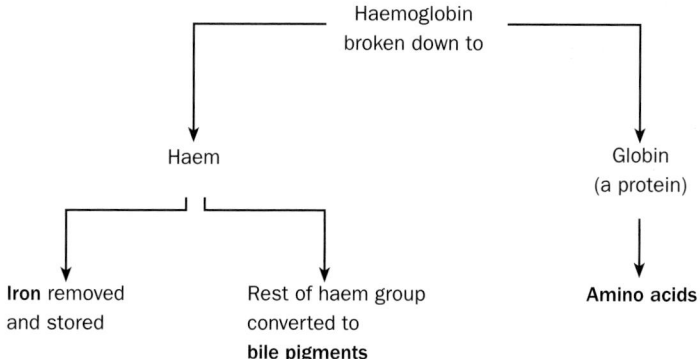

The formation of bile pigments from haemoglobin

bile salts: a component of *bile*. Bile salts are involved in the *digestion* and *absorption* of *fat*. As food passes along the *small intestine*, it is mixed and churned by muscles in the intestine wall. As a result, any fat it contains is emulsified or broken down into tiny droplets. Bile salts keep this fat in its emulsified form. Without bile salts, the droplets would rapidly run back together again. The emulsification of fat increases the surface area for the action of *lipase enzymes*. Without bile, digestion is much less efficient and there is a considerable increase in the amount of undigested fat found in the *faeces*. Most of the bile salts are reabsorbed by the small intestine and transported back to the *liver* where they enter the bile again.

binary fission: a form of *asexual reproduction* taking place in single-celled organisms. It is the way that most bacteria reproduce. The parent cell divides to form two identical daughter cells. Given favourable conditions, such as a plentiful supply of nutrients, binary fission can lead to a rapid increase in numbers.

binocular vision: vision in which both eyes face forward. It enables distance to be judged extremely accurately. Binocular vision is found in humans and other *primates*. It is thought to be an adaptation associated with living in trees where it is important to be able to judge the distance between branches as accurately as possible.

binomial system: the system of scientifically naming living organisms. The scientific name of a *species* consists of two words. The scientific name of a lion, for example, is *Panthera leo* and that of a tiger is *Panthera tigris*. The first word is the name of the *genus* to which the organism belongs, so lions and tigers both belong to the genus *Panthera*. The second word gives the name of the species. Lions belong to the species *leo* and tigers belong to the species *tigris*.

bioaccumulation: the way in which some substances are concentrated in the bodies of animals – the higher the position of the animal in a *food chain*, the greater the concentration of the substance. In the early 1950s, the substance DDT was widely used to control insect pests. It proved to be very effective but, after a while, it was noticed that there was a decrease in the number of predators, such as sparrowhawks, peregrine falcons and other birds of prey. DDT was present in the bodies of insects. These insects were eaten by small birds. The birds did not break the insecticide down but stored it in their body *fat*. A bird of prey, like a sparrowhawk, eats large numbers of smaller birds. Each time, a little more DDT enters its body and accumulates in its tissues. Although eating a single contaminated bird would have had little effect, eating many of them led to the build-up of lethal concentrations of insecticide. The table overleaf shows the results of one study into the effect of bioaccumulation on the concentration of organochlorine insecticides like DDT.

Species	Food	Concentration of organochlorine insecticides in the body tissues (parts per million)
Sparrow hawk	Small birds	3.8
Wood pigeon	Plant material	0.3

DDT is probably the best-known example of a substance that became concentrated by bioaccumulation, but there are other examples. In Japan in the 1950s mercury was discharged in factory waste. It became concentrated by the various organisms in the food chain, reaching high levels in shellfish. When these shellfish were eaten by the local people, they suffered the effects of severe mercury poisoning.

biochemical oxygen demand (BOD): the amount of oxygen removed from a sample of water in a given time. It is a measure of how polluted the sample is. If a sample of water contains a lot of organic matter, it will support large numbers of bacteria. The more bacteria there are present, the more oxygen they will remove from the water and, therefore, the higher the biochemical oxygen demand. The BOD of a water sample is determined by incubating it at a temperature of 20°C for 5 days and measuring the amount of oxygen used.

biochemical tests: a series of simple tests that can be used to identify *proteins*, *lipids* and *carbohydrates*. These tests are summarised in the table.

Substance	Test	Brief details of test	Positive result
Protein	Biuret test	• Add sodium hydroxide solution to the test sample • Add a few drops of dilute copper sulphate solution	Solution turns mauve
Lipid	*Emulsion test*	• Dissolve the test sample by shaking with ethanol • Pour the resulting solution into water in a test tube	White emulsion formed
Carbohydrates: *Reducing sugars*	*Benedict's test*	• Heat test sample with Benedict's reagent	Orange-red precipitate formed
Non-reducing sugars		• Heat test sample with Benedict's reagent to confirm that there is no reducing sugar present • Hydrolyse by heating with dilute hydrochloric acid • Neutralise by adding sodium hydrogencarbonate • Heat test sample with Benedict's reagent	Orange-red precipitate formed
Starch		• Add *iodine* solution	Turns blue-black
Cellulose		• Add Schultze's solution	Turns purple

biodegradable: something may be described as biodegradable if it can be broken down by the bacteria and fungi normally responsible for the processes of decay. Biodegradable *pesticides*, for example, are converted by soil *microorganisms* into less harmful substances and biodegradable plastics can be disposed of more easily when they are no longer required. One advantage of using *enzymes* in industrial processes is that they are biodegradable, because they are *proteins*. *Decomposers* involved in the *nitrogen cycle* are able to break them down and release the nitrogen they contain as ammonia, just as they would with naturally occurring proteins.

biodiversity: refers to the variety of living organisms in a particular area. Biodiversity has three main components:

- Species diversity. This is a measure of the number of individuals and the number of species in a *community*. (See also *index of diversity*.)
- Genetic diversity. There is genetic variation between members of a species. In animals such as tigers which have been reduced to a number of small isolated populations, the genetic diversity is often very low. This can result in the accumulation of harmful recessive *alleles* resulting from the breeding of closely related animals. Modern varieties of domesticated animal and cultivated plants have been bred for high productivity. They may also show low genetic diversity.
- *Habitat* biodiversity. A wide range of different habitats in an area allows a much greater species diversity. *Conservation* involves managing an area so as to maintain habitat diversity.

biofuel: a fuel produced from recently dead biological material. Biofuels include:

- *biomass* fuels such as woodchips, wood pellets and straw. They are commonly used for cooking and heating
- ethanol. This is produced from crops such as sugar cane, sorghum and sugar beet which have a high sugar content. The sugar is fermented by yeast and ethanol is produced
- biodiesel, which is produced from crops with a high oil content. Biodiesel can be used in diesel engines
- biogas, which is produced by anaerobic *digestion* of organic waste. This produces methane, which may be used as a fuel.

biological control: the use of natural predators or parasites to control pest populations. In their natural *ecosystems*, populations of organisms are kept more or less constant in size by their many *predators* and *parasites*. When an organism is introduced into a completely new area, these predators and parasites are often left behind. The introduced organisms can then multiply unchecked to become serious pests. This has happened with animals such as the rabbit in Australia, and plants such as ragwort in New Zealand. By finding and introducing a suitable predator or parasite, numbers can be brought under control again. The virus disease, myxomatosis, was used against rabbits and the cinnabar moth against ragwort. In the UK, the parasitic wasp, *Encarsia*, has been used successfully in controlling whitefly in large greenhouses. Biological control has some advantages over chemical control. Although not wiping out the pest completely, it reduces its numbers to a level where economic damage is no longer significant. If the parasite or predator has been chosen carefully:

- it is very specific, affecting only the pest species
- it only needs to be introduced once rather than used repeatedly like pesticides. It is therefore usually much cheaper

- it does not pollute the environment
- pests do not usually become resistant to organisms that are used for biological control.

biomass: a measure of the amount of living material present such as the biomass of plants in a rainforest or of worms in the soil. Biomass is expressed in units such as g m^{-2}. These units reflect both the mass and the size of the sample. Because the amount of water in living organisms is very variable, samples are often dried, in which case the term dry biomass is used. It is not essential, however, to use dry biomass. There are some circumstances where it is far more sensible and convenient to consider fresh or live biomass. *Pyramids of biomass* may be expressed in terms of the biomass at each *trophic level*.

biosensor: an instrument that uses biological molecules to detect and measure the concentration of certain substances. It usually consists of a membrane to which a layer of *enzymes*, *antibodies* or receptor molecules is attached. Binding of the biological molecule and the test substance produces an electrical signal which can be read by a suitable means. Biosensors have some advantages over chemical tests:

- Because the biological molecules in biosensors are very specific, they will only bind to the test substance. Chemical tests often give positive results with a range of similar substances.
- They can provide quantitative measurements, such as blood *glucose* concentration, without the need for time-consuming laboratory procedures.
- They can easily monitor changes over a period of time.

biosphere: all the living organisms found on Earth together with the non-living part of their *environment*. The biosphere includes the land, the oceans and the lower part of the atmosphere.

biotechnology: making use of *microorganisms* or biochemical reactions to produce useful products or to carry out useful processes. Although biotechnology was used in the making of beer and bread by the ancient Egyptians, there has been an enormous increase in the number and variety of processes in which it is now involved. Medically important substances such as *insulin*, factor VIII and *growth hormone* are produced with the aid of *genetic engineering*; *enzymes* are involved in a variety of applications ranging from the manufacture of 'stone-washed' jeans to soft-centred chocolates, and microorganisms play an important role in the manufacture of many food products.

biotic: an ecological factor concerned with the activities of living organisms. Predation, food availability and *competition* are all examples of biotic factors that affect the distribution of various organisms. Rainfall, soil pH and temperature, on the other hand, are concerned with the non-living part of the *environment* and are *abiotic* factors. Biotic factors are usually *density-dependent factors*, that is, they are factors whose effects on a population are relatively greater at higher population densities than at lower ones.

biotic potential: the maximum rate of reproduction of a population when the organisms concerned are living in optimum environmental conditions.

bipedal: able to walk on two legs. With the exception of birds, most land-dwelling vertebrates are quadrupeds – they use all four legs to walk. Humans, however, are fully bipedal. One of the most important ways of deciding whether fossil *bones* are human is to look for evidence of bipedalism. This is reflected in the structure of many parts of the skeleton.

- The *foramen magnum* is the opening in the skull where the *spinal cord* leaves the *brain*. It is where the vertebral column meets the skull. In bipedal animals it lies underneath the skull; in quadrupeds it is farther towards the back.

- The pelvis is short and broad instead of long and narrow.
- The femur or upper leg bone has a distinctive twist which brings the knee joint directly under the pelvic girdle.

bipolar cell: a type of *neurone* or nerve cell found in the *retina* of the *eye*. A bipolar cell has a cell body containing a nucleus. From this cell body there are two long, thin branches. One of these is a *dendron*, which brings *nerve impulses* from rod or *cone cells* to the cell body. The other is an *axon* that transmits impulses away from the cell body to other neurones. The *acuity* of the eye (its ability to distinguish between objects close together) depends on each cone cell having a connection with a single bipolar cell. Stimulation of cone cells, even if they are close together, therefore results in separate impulses in individual bipolar cells. This ultimately leads to impulses in different neurones in the *optic nerve*. The sensitivity of the eye, however, depends on the fact that several *rod cells* connect to a single bipolar cell. In this way, a number of weak stimuli can lead to the generation of a nerve impulse enabling a person to see a very faint image.

birth rate: the number of children born each year. It is usually given per thousand of the population.

Biuret reaction: a *biochemical test* used to show the presence of protein.

bivalent: paired homologous *chromosomes*. During the first stage of *meiosis*, homologous chromosomes come together so that they lie side by side. They shorten in length and get thicker. Each of the chromosomes forming a bivalent consists of a pair of *chromatids*. The formation of bivalents only occurs in meiosis; it does not happen in *mitosis*.

blastocyst: a stage in the development of humans and other mammals. The zygote that results from *fertilisation* of the egg cell divides and develops into a hollow ball. The wall in one part of this ball is thicker and forms a mass of cells. These cells eventually become the *embryo* while the *placenta* develops from the remaining cells. The blastocyst is really equivalent to the *blastula*, which is a characteristic stage in the development of all animals.

blastula: a stage in the development of an animal. When an egg cell has been fertilised by a sperm, a *zygote* is formed. The zygote is a single cell. It divides and divides again to produce a hollow ball of cells called a blastula. The blastula will eventually develop into an *embryo*. The presence of a blastula is a characteristic shared by all members of the animal *kingdom*.

blood cells: one of the different types of cell found in the blood. In a mammal, there are two main types of blood cell. These are *red blood cells*, or erythrocytes, whose main function is the transport of respiratory gases. White blood cells, or *leucocytes*, are larger than red blood cells and there are fewer of them. They are formed from cells in the *bone* marrow and they are concerned with *immunity*.

blood clotting: the mechanism by which a blood clot forms when tissue is damaged. A protein called *fibrinogen* is present in blood *plasma*. When the body is wounded, soluble *fibrinogen* is converted to insoluble *fibrin*. This forms a mesh over the surface of the wound which traps red blood cells and forms a clot. Clots stop further blood from escaping and also help to prevent the entry of pathogens. A complex mechanism controls the process of blood clotting. Fibrinogen can only be converted to fibrin in the presence of the enzyme *thrombin*. Thrombin is normally present in the blood in an inactive form known as prothrombin. Prothrombin can be converted to thrombin in the presence of a number of substances or factors, some of which are only released at the site of tissue damage. This is summarised in the diagram overleaf.

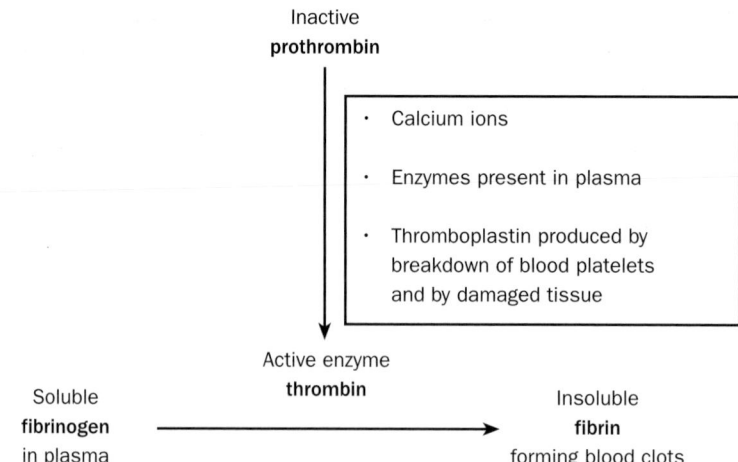

The blood clotting process

In addition to this mechanism, blood is normally prevented from clotting inside the body by **anticoagulants** produced in the body.

blood glucose pool: the total amount of **glucose** in the blood at any one time. Accurate control of blood glucose is important. If the concentration falls too low, the **central nervous system** ceases to function correctly. If it rises too high, there will be a loss of glucose from the body in the **urine**. **Hormones** including **insulin** and **glucagon** play an important part in keeping the concentration of glucose in the blood constant. Although the overall concentration stays within narrow limits, glucose is always being added to and removed from the blood glucose pool. **Digestion** and **absorption** of **carbohydrates** and conversion from the body's stores of **glycogen** and **fats** tend to increase the blood glucose concentration, while such processes as **respiration** decrease it. This is summarised in the diagram.

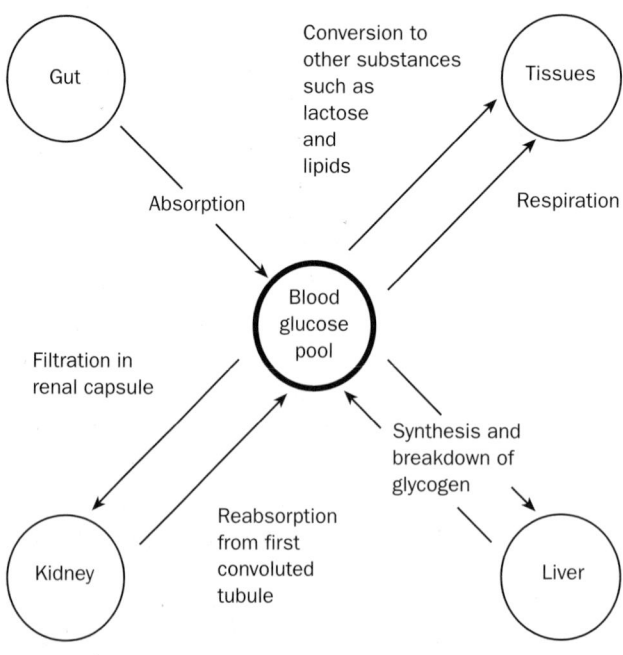

The blood glucose pool

blood group: one of the types into which a person's blood may be separated. *Red blood cells* have protein molecules in their *cell-surface membranes*. Some of these *proteins* act as *antigens* and determine blood group. There are many different systems of grouping blood, but the ABO system is the best known. There are four blood groups in this system: A, B, AB and O. These are determined by the presence of the relevant antigen. People who are blood group A, therefore, have antigen A on their red blood *cell* membranes; those who are group B have antigen B; group AB individuals have both antigen A and antigen B, whereas group O have neither of these antigens. The antigens determine whether or not one person's blood may be safely given to another during *blood transfusion*. The antigens are inherited and their presence is determined by a single gene with three *alleles*: I^A, I^B and I^O. The alleles I^A and I^B are *codominant*, whereas the allele I^O is *recessive* to both I^A and I^B. The table shows how these blood groups are genetically determined.

Blood group	Antigens on plasma membrane of red blood cells	Possible genotypes
A	A	$I^A I^A$ or $I^A I^O$
B	B	$I^B I^B$ or $I^B I^O$
AB	A and B	$I^A I^B$
O	None	$I^O I^O$

blood system: the transport system found in many animals. Blood systems have a number of characteristics:

- They contain blood, a fluid whose composition varies from animal to animal. Blood transports a number of different substances. It often contains a *respiratory pigment* such as *haemoglobin*, which makes it effective in transporting oxygen.
- They are mass-flow systems that provide a rapid means of transport from one part of the body to another.
- Substances are exchanged between the blood and the surrounding cells by means of *diffusion* and *active transport*.
- A pump is necessary to move the blood within the system. This may be a *blood vessel* with contractile walls or a specialised structure known as a *heart*.

Animals such as insects have an open system in which blood is pumped out by the heart into large blood spaces. It flows slowly through these and bathes the surrounding tissues directly. Fish and mammals, on the other hand, have a closed circulatory system in which the blood flows through and stays in a series of *blood vessels*. This has a major advantage because it allows for much better control of the supply of blood to particular organs or regions of the body. The blood system in fish is a *single circulation*, passing through the heart only once in its passage round the body. A mammal has a *double circulation*. The blood passes through the heart twice in its passage round the body.

blood transfusion: the transfer of blood from one person to another. The ability to do this safely depends on the *blood groups* of the people concerned. *Red blood cells* have protein molecules in their *cell-surface membranes*. Some of these *proteins* act as *antigens* and determine blood group. For example, the red blood cells belonging to people who are group A have antigen A on their surfaces while those belonging to people with group B blood have antigen B.

In addition to this, people may also have **antibodies** present in their plasma. Someone with group A will have a specific **antibody** against antigen B. We will call this antibody b. Similarly, a person with group B blood will have antibody a. Because of this, group A blood can obviously be safely given to someone else with blood group A. Both have antigen A but neither has antibody a. However, if blood of group B is given to a person with group A blood, there is a problem. We would be giving blood with antigen B to someone with antibody b. The antigen would combine with the antibody causing the red blood cells to stick to each other. This is called **agglutination** and can lead to blockage of **blood vessels** and possibly the death of the person concerned.

blood vessel: part of the system that transports blood round the body. In a mammal, blood is pumped away from the heart along large **arteries**. These branch into smaller arteries and eventually lead into still smaller **arterioles**. Blood passes from the arterioles into the **capillaries** which are very small in diameter. Capillaries form a network of fine tubes in almost every organ in the body. They are vital to the blood system as it is here that exchange of substances between the blood and the cells of the body takes place. From the capillaries, blood returns to the heart via the venules and, finally, the **veins**.

There are differences in the pressure of the blood and the way in which it flows in the various parts of the blood system. The structures of the different sorts of blood vessel are related to this. The table shows some of these features.

Feature	Artery	Capillary	Vein
Pressure of blood	High	Lower than in the arteries	Very low
Blood flow	Pulsatile. The flow is not even; it surges	Even blood flow	Even blood flow
Structure of walls	Thick wall containing a lot of elastic tissue and muscle	Wall made up of lining cells only. No muscular or elastic tissue	Wall not as thick as that of an artery. Little elastic tissue or muscle
Presence of valves in wall	Not present	Not present	Present

BMR: see **basal metabolic rate (BMR)**.

BOD: see **biochemical oxygen demand (BOD)**.

body mass index (BMI): a method of relating a person's body mass to his or her height. If body mass is measured in kilograms and height in metres, it is calculated from the formula:

$$BMI = \frac{mass}{height^2}$$

The importance of this index is that it lets us compare the body mass of people of different height. A person who has had or is recovering from a serious illness may be underweight and will have a low body mass index, while someone who is obese will have a high body mass index. The table opposite shows some typical values.

BMI	Interpretation
Under 20.0	Underweight
20.0–24.9	Normal
25.0–29.9	Overweight
Over 30.0	Obese

A man is 1.8 m tall and weighs 80 kg. Calculate his body mass index.

$$\text{Body mass index} = \frac{\text{mass}}{\text{height}^2} = \frac{80}{1.8^2} = \frac{80}{3.24} = 24.7$$

Bohr effect: this is what happens to the *oxygen dissociation curve* for *haemoglobin* when there is an increase in the concentration of carbon dioxide. The greater the concentration of carbon dioxide in the blood, the further the curve is pushed to the right. This means that, at high concentrations of carbon dioxide, haemoglobin will give up a greater amount of the oxygen that it is carrying. This is particularly important during exercise. As activity increases, so does the rate of *respiration* of the muscles. They will, therefore, produce greater amounts of carbon dioxide and this will result in the release of greater amounts of oxygen from the blood.

bone: a supporting tissue in the skeletons of mammals and some other vertebrates. Bone is made up of cells embedded in a matrix of *connective tissue*. This matrix contains organic material that is largely the protein *collagen*. This is impregnated with calcium phosphate and other inorganic salts. The arrangement of the materials within the bone matrix gives it great strength and enables it to withstand the considerable forces which act on it. Bone cells are found throughout this matrix in small spaces called lacunae. Tiny canals filled with *cytoplasm* connect these lacunae to each other, while *blood vessels* supply the living bone tissue with the oxygen and nutrients it needs and remove waste products. The activity of the various types of bone cell enables bone to be reabsorbed or added to. This property enables an individual bone to be repaired, to grow or to change to meet different forces which act on it at various stages in an animal's development.

bone density test: a method of detecting *osteoporosis*. An *X-ray* beam is used to measure the *calcium* content of *bone* in areas of interest. These are usually the hips and lower backbone. From the results, a doctor can predict the probability of a fracture occurring and prescribe suitable preventive treatment.

Bowman's capsule: see *renal capsule*.

brachycardia: a heart rate of less than 60 beats per minute. Brachycardia may occur in healthy people such as athletes. It is also associated with reduced activity of the *thyroid gland* and *heart block*.

brain: the organ responsible for coordinating the activities of the nervous system. The brain of a mammal is complex. It is divided into three main regions, each of which is made up of a number of parts with their own specific functions. The main functions of some of the major parts of the brain are given overleaf.

- The forebrain
  - The *cerebrum*, or cerebral hemispheres, controls the body's voluntary behaviour. It is responsible for analysing and interpreting much of the information from sense organs such as the eyes, as well as for controlling many motor activities, learning and memory.
  - The *hypothalamus*. This is an area on the floor of the forebrain that has a number of important functions. These include the control of many aspects of *homeostasis*, integration of the nervous and hormonal systems and the coordination of the *autonomic nervous system*.
- The midbrain
- The hindbrain
  - The *cerebellum* controls posture and balance as well as coordinating the overall smooth movement of the body.
  - The *medulla* oblongata contains centres that coordinate and control breathing, heart rate and blood pressure.

bronchiole: one of the small airways in the lung which connect the larger bronchi with the *alveoli* or air sacs. The structure of bronchioles is different from that of *bronchi* as there is no *cartilage* in the walls of the bronchioles.

bronchitis: a disease of the gas-exchange system. In bronchitis, the main airways in the *lungs* – the bronchi – become inflamed. The onset of acute bronchitis (see *acute disease*), caused by bacterial or viral infection, is quite sudden. Chronic bronchitis, however, is a long-term condition associated with atmospheric pollution and cigarette smoking. The early stages of chronic bronchitis involve the production of considerable amounts of *mucus* from the enlarged mucus-producing *glands* that line the bronchi. This is accompanied by frequent bouts of coughing. As the disease progresses, there is a gradual narrowing of these airways. The resulting breathlessness may condemn patients to increasing disability. The condition is often associated with *emphysema*.

bronchis: one of the larger tubes that take air into and out of the *lungs*. Air passes from the mouth and nose into the lungs through the *trachea* or windpipe. The trachea divides into two main bronchi (singular bronchus). These then divide into other small bronchi. The epithelial cells that line the walls of the bronchi, like those lining the trachea, have *cilia* on their surface. Particles in the air are trapped in *mucus* secreted by *goblet cells* in the airways of the lung. The cilia beat continuously and the mucus and the particles that have been trapped are carried upwards through the bronchi and trachea into the back of the throat. There are rings of *cartilage* in the bronchial walls that prevent the bronchi from collapsing as air is breathed in.

Bryophyta: the plant phylum that contains mosses and liverworts. Members of the Bryophyta share the following features:

- Like all plants, members of this phylum show an *alternation of generations*. In the Bryophyta, the *dominant* generation is the *gametophyte*.
- They do not have true roots or leaves.
- They do not have vascular tissue. There is no *xylem* or *phloem* present, so bryophytes rely on *diffusion* to move substances from one place to another.

buffer: a solution or substance that can absorb hydrogen ions. As a result of this, the *pH* of a buffer solution changes very little when small amounts of either an acid or an alkali are added to it. Buffer solutions are used in investigations where it is necessary to maintain a solution at a particular pH value. They also occur naturally. *Haemoglobin*, *plasma proteins* and phosphates are all important buffers that help to keep the pH of mammalian blood constant.

bundle of His: specialised muscle fibres in the heart that go from the *atrioventricular node* to the base of the heart. They play an important part in coordinating the beating of the heart. A wave of electrical activity passes rapidly down them to the base of the *ventricles*. It then spreads upwards to all parts of both ventricles along finer branches known as *Purkyne tissue*. This electrical activity stimulates the heart muscle to contract. Because of the arrangement described, the ventricle muscle starts contracting from the bottom. This squeezes blood upwards into the arteries leaving the heart.

Do you need revision help and advice?

Go to pages 306–314 for a range of revision appendices that include plenty of exam advice and tips.

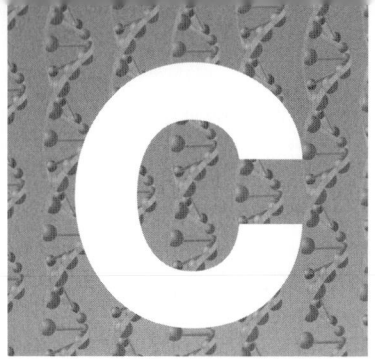

C₃ pathway: part of the biochemical pathway of *photosynthesis*. In the *light-independent reaction*, carbon dioxide is taken into the plant and a molecule is formed that contains three carbon *atoms*. This is part of a cycle of chemical reactions known as the *Calvin cycle* or, because the first molecule produced contains three carbon atoms, the C₃ pathway. The reaction in which carbon dioxide combines with ribulose bisphosphate to produce *glycerate 3-phosphate (GP)* is not efficient. Some plants such as maize and sugar cane use a different method of trapping carbon dioxide. This is known as the *C₄ pathway*.

C₄ pathway: a photosynthetic pathway in which carbon dioxide taken into the plant is first used to form a molecule that contains four carbon *atoms*. In the C₃ pathway, the reaction in which carbon dioxide combines with ribulose bisphosphate to produce the three-carbon molecule *glycerate 3-phosphate (GP)* is slow, particularly if the concentration of carbon dioxide falls much below its normal concentration. In hot, dry conditions the *stomata* of plants often close during the daylight and the carbon dioxide concentration in the leaf falls as it is used for *photosynthesis*. The C₄ pathway involves a carbon dioxide concentrating system. Carbon dioxide is taken into the *mesophyll* cells of the plant and converted to a molecule that contains four carbon atoms. This reaction can take place in low concentrations of carbon dioxide. It can then be released into the relatively few cells in which the C₃ pathway takes place, producing high concentrations of carbon dioxide in these cells. The disadvantage of this pathway is that it requires rather more energy in the form of *ATP*. It is therefore only effective in conditions of high light intensity. Plants that have the C₄ pathway include a number of tropical species such as sugar cane and maize.

calcium: an important nutrient required by both animals and plants. Calcium has a number of functions:

- Calcium pectate is an important component of the *cell walls* of plant cells. Calcium salts are also important in skeletal structures such as the *bones* of mammals and the shells of molluscs.
- Calcium is essential for a variety of processes in a mammal, including *blood clotting*, muscle contraction and transmission across *synapses* in the nervous system.
- A *second messenger* is a molecule found inside a cell which responds to the presence of *hormones* outside the cell. It works by activating a particular enzyme inside the cell. Calcium may act as a second messenger.

Calvin cycle: a biochemical cycle that forms part of the *light-independent reaction* of *photosynthesis*. The main stages in the process are summarised in the diagram.

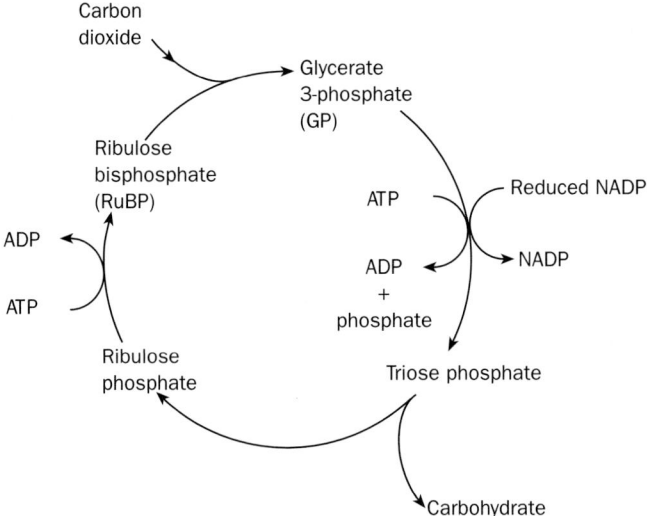

The Calvin cycle

1 Carbon dioxide enters the cycle, combining with the five-carbon sugar ribulose bisphosphate (RuBP) to form two three-carbon molecules of **glycerate 3-phosphate (GP)**.

2 GP is converted to triose phosphate. This reaction requires **ATP** and reduced **NADP**, two substances formed in the **light-dependent reaction**.

3 Some of the triose phosphate is used to produce **glucose** and other substances. This is not part of the Calvin cycle.

4 The rest of the triose phosphate goes through a series of reactions in which RuBP is again produced. ATP is also required for this process.

cambium: a layer of unspecialised cells between the **xylem** and the **phloem** in a plant. Cambium cells can divide by **mitosis** and continually replace themselves. The daughter cells they produce when they divide differentiate into more xylem vessels and phloem sieve tubes (see **sieve tube element**). Cambium cells are therefore described as **stem cells**.

cancer: see **tumour**.

capillary: a very small **blood vessel**. The wall of a capillary consists of a single layer of epithelial cells. There is no muscle and no elastic tissue, both of which are found in the walls of larger **arteries** and **veins**. Capillaries play an important role in the body because it is as blood passes along capillaries that substances are exchanged. There is a large number of capillaries, so most cells in the body are close to one of these tiny blood vessels.

The blood pressure at the arterial end of a capillary is high and forces water and small soluble molecules out through the walls forming the **tissue fluid** that surrounds the body cells. Much of this tissue fluid flows back into the capillary at its venous end because the **water potential** of the tissue fluid is higher than that of the blood **plasma** at this point. There is, therefore, a continuous circulation of fluid out of the capillaries and back into them. This takes useful substances to the cells and returns waste products to the blood.

Blood flows into capillaries from the **arterioles**. Arteriole walls contain muscle fibres that may contract and reduce the diameter of the vessel. In this way, the blood supply to the capillary system in a particular organ is always being adjusted to meet the needs of the organ.

capillary action: the ability of water to move upwards through a fine tube. *Xylem* vessels in plants have diameters of between 20 and 400 µm, so we can think of xylem as being made up of a series of fine tubes that extend through a plant from its roots to its leaves. Water therefore can move up through the xylem by capillary action. Although capillary action helps to explain how water gets to the leaves of a plant, the maximum height that could be achieved by this process is only about 1 m. This is certainly not enough to get the water to the top of a tree. Therefore, there must be other mechanisms that are responsible for the movement of water. One of these is *cohesion–tension*.

capsid: the protein coat that surrounds a *virus* particle. The main function of the capsid is to protect the *nucleic acid* molecule contained inside the virus. As well as this, some viruses have specific receptor molecules in their protein coats. These enable recognition of, and attachment to, host cells.

capsule: a layer that may be found outside the *cell wall* in a bacterial cell. Capsules are not found in all *bacteria*, or even in all bacteria of the same species. Where they occur, they are usually made of polysaccharides although some species of bacteria have protein capsules. The presence of a capsule is one of the features that helps to decide whether a particular strain of bacterium is harmful or not. Bacteria that are enclosed in capsules are often able to resist the action of *phagocytes* in an animal's blood and are, therefore, not as easily destroyed.

capture–recapture: see *mark–release–recapture*.

carbohydrate: a substance that contains the chemical elements carbon, hydrogen and oxygen. Carbohydrates get their name because the hydrogen and oxygen are in a 2 : 1 ratio, the same as in water. They are either sugar molecules or are built up from sugar molecules joined by means of *condensation*. Carbohydrates whose molecules contain a single sugar unit are *monosaccharides*; those with two sugar units are *disaccharides* and those with many sugar units are *polysaccharides*. The table shows some features of some biologically important carbohydrates.

Example	Type of carbohydrate	Role in living organisms
Glucose	*Monosaccharide*	Important as a source of energy in *respiration*
Ribose		A five-carbon sugar found in *nucleic acids*
Sucrose	*Disaccharide*	The form in which sugars are transported in the *phloem* of plants
Lactose		The main carbohydrate in milk
Maltose		A product of *digestion* of *starch*
Glycogen	*Polysaccharide*	The storage carbohydrate in animals
Cellulose		Important structural molecule in plant *cell walls*
Starch		The storage carbohydrate in plants

carbon cycle: the way in which the element carbon cycles in an *ecosystem*. The basic principles are exactly the same as for other *nutrient cycles* and the diagram is based on this.

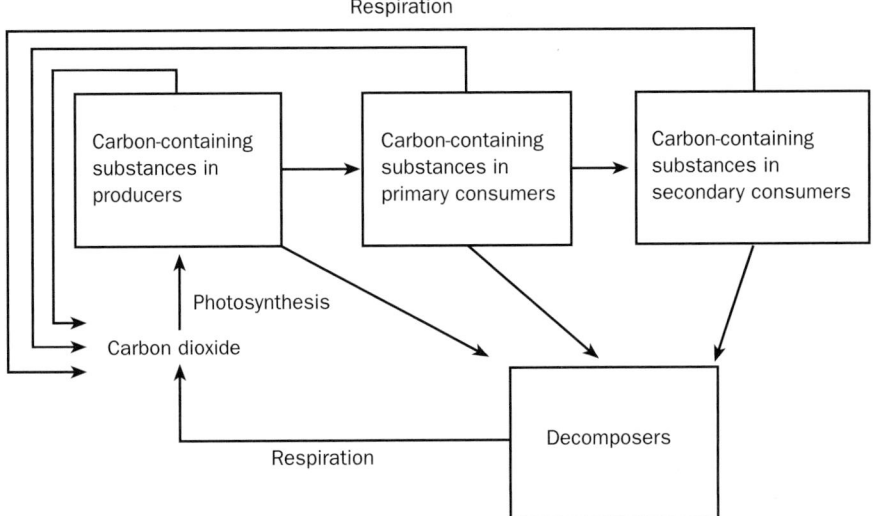

The carbon cycle

The particular points to note in the biological part of the cycle are:

- The inorganic form of carbon taken up by the plants is carbon dioxide. This is absorbed by the leaves and used in **photosynthesis**. Carbon is not taken up through the roots like other elements.
- Carbon dioxide is released directly as a result of **respiration** by all the other organisms involved in this cycle, including the **decomposers**.
- A lot of carbon has been locked up in coal, oil and limestone formed many millions of years ago, as well as in forests that have existed for hundreds, if not thousands, of years. The burning of fossil fuels and the clearing of forests have led to the release of this carbon dioxide and an increase in its concentration in the atmosphere. This is regarded as one of the main contributors to the **greenhouse effect** and **global warming**.

carbon fixation: the conversion of an inorganic source of carbon to an organic substance. In the **light-independent reaction** of **photosynthesis**, carbon fixation involves combining carbon dioxide with ribulose bisphosphate to produce two molecules of **glycerate 3-phosphate (GP)**.

carbon sink: a stage in the **carbon cycle** involving a net removal of carbon dioxide from the atmosphere. One of the most important carbon sinks is provided by the huge number of tiny algae, which form the plankton floating in the surface waters of the oceans. As these algae photosynthesise, they convert carbon dioxide dissolved in the water into various organic compounds. The seas in many parts of the world are heavily polluted and there are concerns that pollution could have an adverse effect on planktonic algae.

carbonic anhydrase: an enzyme that catalyses the reaction in which carbon dioxide and water combine to produce carbonic acid. The carbonic acid then dissociates to form hydrogen ions and hydrogencarbonate ions.

$$CO_2 + H_2O \rightarrow H_2CO_3 \rightarrow H^+ + HCO_3^-$$

In the blood *plasma*, there is no carbonic anhydrase and this reaction takes place very slowly. However, carbonic anhydrase is present in the **red blood cells** so the reaction takes place about 10,000 times quicker here. As a result, there is a higher concentration of hydrogencarbonate ions in the red blood cells than in the blood plasma. The hydrogencarbonate ions diffuse into the plasma and are transported to the *lungs*. The reverse series of reactions occur in the lungs and the carbonic anhydrase catalyses the breakdown of carbonic acid to form carbon dioxide and water.

carcinogen: a substance or other agent that increases the probability of a person developing cancer (see *tumour*). Examples of carcinogens include ultraviolet light and substances such as those found in the tar in cigarette smoke. Carcinogens cause damage to *DNA*. Cells in which DNA has been damaged, particularly in people who have inherited a tendency to develop cancer, may go on to become cancerous.

cardiac accelerator nerve: a *nerve* that increases heart rate. The *cardiovascular control centre* in the *medulla* of the *brain* controls heart rate. The cardiac accelerator nerve is part of the *sympathetic nervous system*. *Nerve impulses* pass along the cardiac accelerator nerve from the cardiovascular control centre to the *sinoatrial node (SAN)*. This leads to an increase in heart rate.

cardiac acceleratory centre: part of the *cardiovascular control centre* in the *brain*. It sends *nerve impulses* down the *cardiac accelerator nerve* to the *sinoatrial node (SAN)*. This speeds up the heart rate.

cardiac arrest: the sudden failure of the heart to pump blood. There is no pulse, breathing stops and the person loses consciousness. Cardiac arrest commonly occurs when the heart stops beating or when the muscle in the *ventricles* contracts erratically and fails to pump blood into the arteries (*ventricular fibrillation*). Prompt medical attention is essential because *brain* damage will result if some blood supply is not restored within 2 or 3 minutes.

cardiac cycle: the sequence of events that makes up a heart beat. It is a continuous cycle but it may be divided into three main parts.

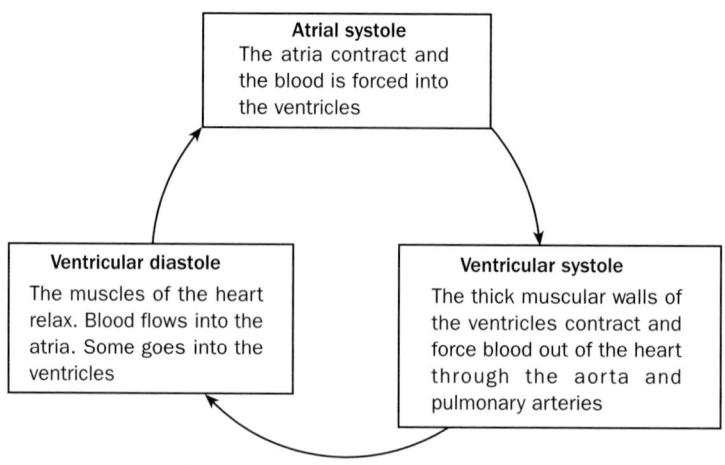

The cardiac cycle

The heart beat is controlled by a small area of muscle in the wall of the right *atrium* called the *sinoatrial node (SAN)*. From here, a *nerve impulse* spreads over the surface of the heart, bringing about the events described in the diagram. Two nerves from the *brain*, the *cardiac accelerator nerve* and the *vagus nerve*, are able to send impulses to the sinoatrial node and alter the heart rate.

cardiac inhibitory centre: part of the *cardiovascular control centre* in the *brain*. It sends *nerve impulses* down the *vagus nerve* to the *sinoatrial node (SAN)*. This slows down the heart rate.

cardiac muscle: the type of muscle found in the heart. Its microscopic structure is similar to *skeletal muscle* and it has the same characteristic striations. The individual muscle fibres, however, branch frequently to form a network that spreads through the walls of the *atria* and *ventricles*. The heart is able to beat continuously throughout the life of the animal. This requires a lot of energy. This energy is supplied by *aerobic respiration*. Heart muscle, therefore, has an abundant blood supply and a large number of *mitochondria*.

If the nerves supplying a skeletal muscle are cut, the muscle will not contract. Cardiac muscle shows an important difference. It will continue to beat rhythmically even after all the nerves supplying it have been cut. Because it can beat without being stimulated by nerves, cardiac muscle is said to be *myogenic*.

cardiac output: the total volume of blood pumped out by the heart in one minute. There are two things that contribute to cardiac output. The first of these is *stroke volume*. This is the volume of blood pumped out each time the heart beats. The second is the heart rate, the number of beats per minute. The relationship between cardiac output, stroke volume and heart rate is given by the formula:

$$\text{cardiac output} = \text{stroke volume} \times \text{heart rate}$$

An increase in physical activity requires an increase in cardiac output. More oxygen has to be supplied and more waste products removed. This increase in cardiac output is achieved by increasing both the stroke volume and the heart rate. In some animals, fish for example, it is accomplished mainly by increasing the stroke volume. The heart rate increases very little. In mammals and birds, however, an increase in cardiac output is mainly due to a faster heart rate.

cardiopulmonary resuscitation (CPR): emergency treatment aimed at restarting the heart and breathing after *cardiac arrest*. It involves three procedures:

1 The patient's airway is cleared of mucus or vomit. His or her neck is extended and the chin is raised.
2 Direct mouth-to-mouth breathing is used to restart breathing.
3 Cardiac massage is given. Both hands are placed on the patient's lower breastbone. They are pushed down hard between 60 and 100 times a minute.

cardiovascular control centre: part of the hindbrain responsible for regulating heart rate. There are two centres concerned. These are the *cardiac acceleratory centre*, which increases heart rate, and the *cardiac inhibitory centre*, which slows it. Both centres are found in the part of the hindbrain known as the *medulla* oblongata. These two centres do not work independently. Their activities are integrated and they work together to match the heart rate to the needs of the body. The flow chart overleaf shows how these centres influence the rate at which the heart beats.

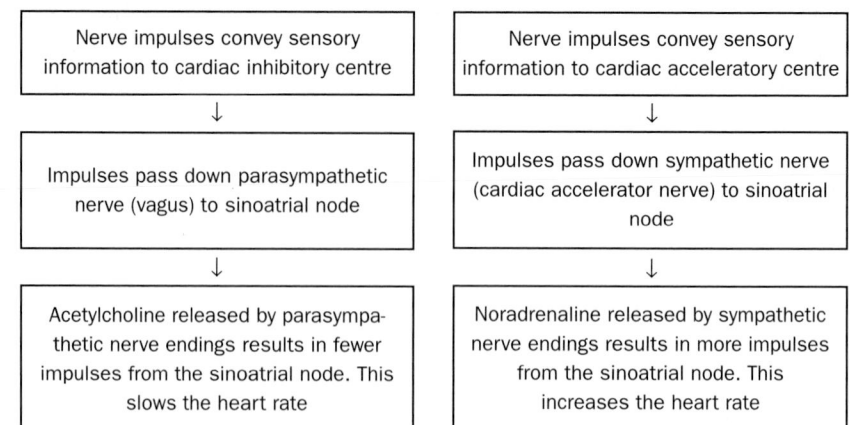

The cardiovascular control centre and the control of the heart rate

carotid body: a group of *chemoreceptors* found in the walls of the carotid arteries. They detect changes in the carbon dioxide concentration in the blood. Sensory information from the carotid bodies goes to the centres in the *brain* that regulate breathing.

carrier protein: a protein found in the *plasma membranes* that helps to transport substances across the membrane. Substances can pass through plasma membranes in different ways. One of these is *facilitated diffusion*. This is a process that requires carrier *proteins*. Carrier proteins have specific shapes. In particular, they have receptor sites into which the substance being transported fits. This brings about a change in the shape of the carrier that results in the transported molecule passing through the membrane. A similar process occurs with *active transport*, the main difference being that, in this case, the change in shape of the carrier protein requires energy from *ATP*.

carrying capacity: the maximum stable *population* that a particular *environment* can support. The wildebeest is an antelope that lives in the Serengeti National Park in Tanzania. In 1960, an outbreak of disease reduced its numbers to approximately 10 animals per km². Its population then rose and levelled out at between 50 and 60 animals per km². This figure represents the carrying capacity.

cartilage: a hard, flexible supporting tissue important in skeletons. In animals such as sharks, the skeleton is made completely of cartilage. There is no *bone* present at all. In humans and other mammals, there is a lot of cartilage present in the young animal, but some of this is replaced by bone as the animal grows. Cartilage has various functions. It is more compressible than bone so cartilage found at the ends of bones and between the vertebrae enables the body to withstand the shocks and jarring which accompany movement. Its flexibility is ideally adapted to supporting such structures as the nose, the larynx and the trachea.

Cartilage is similar to bone in that it consists of cells embedded in a matrix. This matrix contains a large number of fibres of the protein *collagen*, but it is not impregnated with inorganic salts. Cartilage-forming cells or chondrocytes are found in little groups in this matrix, enclosed in spaces called lacunae. Substances needed by these cells diffuse through the matrix. There are no *blood vessels* found in cartilage.

cascade effect: the way in which tiny amounts of a *hormone* can cause a target cell to produce very large amounts of a particular product. A single molecule of a hormone such as **adrenaline** binds to a receptor molecule on the **cell-surface membrane**. This produces many molecules of **cyclic AMP**, which acts as a messenger inside the cell. Each molecule of cyclic AMP activates a large number of enzyme molecules and each enzyme molecule affects large numbers of **substrate** molecules. In this way, a single molecule of adrenaline can trigger the production of many thousands of **glucose** molecules in a **liver** cell.

Casparian strip: a band of a waterproof material that runs round the walls of the endo-dermal cells in a root. The **endodermis** in a plant is a ring of cells between the outer part of the root and the **vascular tissue** in the centre. The Casparian strip is made of a sub-stance called **suberin**. It prevents water and dissolved substances going through the **cell walls** and intercellular spaces into the vascular tissue. As a result, these substances pass through the **cytoplasm** of the endodermal cells which can control their movement into the **xylem**.

catabolic reaction: a chemical reaction in living organisms, which involves the breakdown of larger molecules and the release of energy. Examples of catabolic reactions include **respi-ration** and the breakdown of storage molecules such as **starch** and **glycogen**. Metabolism involves a combination of catabolic reactions and **anabolic reactions**. Anabolic reactions are reactions in which large molecules are synthesised from smaller ones.

cataract: the clouding of the **lens** of the eye. It results in blurred vision. The common-est type of cataract is that associated with old age, but cataracts may also arise as a result of **diabetes** or from injury or prolonged exposure to infra-red radiation. Cataracts may be treated surgically and the diseased lens removed. Glasses, contact lenses or an artificial lens inserted in the eye may be used to compensate for the loss of the original lens.

cDNA: see **complementary DNA (cDNA)**.

cell: the basic unit from which living organisms are built up. A cell consists of a mass of **cytoplasm** surrounded by a **cell-surface membrane**. The body of an individual plant or animal contains many different types of cell. Each type is specialised and adapted for a particular function. The process by which cells become specialised in this way is called **differentiation**. A group of similar cells that carry out a particular function is known as a **tissue**. Different tissues make up an **organ**, while several organs combine to form a **system**.

The development of **electron microscopes** has allowed cell structure to be investigated in detail. The cytoplasm that makes up most of the cell has been shown to contain a large number of different **organelles**, each of which has a specific function.

There are two basic types of cell found in living organisms. **Prokaryotic** cells are found in bacteria. They are characterised by the absence of a nucleus and have no membrane-bound organelles. **Eukaryotic** cells are found in animals and plants. They have nuclei as well as large numbers of different organelles. Some, like **mitochondria**, **chloroplasts** and **lysosomes**, are surrounded by membranes.

cell cycle: the cycle of cell growth and cell division that occurs in animals and plants. The cell cycle consists of three main stages:

1 interphase – the cell grows and increases in size; new **proteins** are synthesised and new cell organelles are made; DNA replication takes place
2 mitosis – the genetic material divides
3 **cytokinesis** – the **cytoplasm** of the cell and its organelles divide more or less equally between the two daughter cells.

The length of an individual cell cycle, even in the same organism, is variable. Many things combine to determine its precise length. It depends, for example, on temperature and nutrient supply.

The cells of many higher organisms are generally only able to go through a limited number of cell cycles. Once they have become specialised, they are unable to divide any more. Tumour cells, however, carry on dividing indefinitely. Knowledge of the mechanism which controls the cell cycle may lead to the discovery of ways in which tumour growth may be limited.

cell fractionation: the process in which cells are broken up and the different types of *organelle* they contain are separated from each other. The flow chart summarises the main steps in the procedure.

> Tissue homogenised. It is broken up and suspended in a buffer solution to keep the pH constant. This solution has the same water potential as the original tissue and is kept cold

↓

> The mixture is filtered to remove any material which has not been properly broken up

↓

> The filtrate is put in a centrifuge and spun at a low speed. The larger organelles such as nuclei and chloroplasts fall to the bottom to form a pellet. These can be removed and suspended in a fresh solution if they are required

↓

> The fluid supernatant can be put back in the centrifuge and spun at a higher speed. Smaller organelles such as mitochondria now separate and form a pellet

Cell fractionation

It is possible to use cell fractionation to get a suspension containing a single type of organelle. Various techniques can then be used to investigate the structure and functions of the organelle concerned.

cell membrane: see *plasma membrane*.

cell sap: the liquid contained inside the *vacuole* of a cell. It contains water in which there are dissolved substances such as sugars and mineral ions. These substances are usually at a much higher concentration than they are in the surrounding *cytoplasm*. This means that there is a *water potential* gradient and water moves into the vacuole by *osmosis*. This is important in maintaining the cell in a turgid state and providing a plant with support. Cell sap may also contain *pigments* such as the anthocyanins which are responsible for producing the blue or red colours of many *flowers*.

cell signalling: the ways in which cells communicate. Cell signalling controls the activities of cells and coordinates their actions. There are three main ways in which cells communicate:

- by direct contact. When heart muscle contracts, a **nerve impulse** spreads directly from cell to cell in the **cardiac muscle**
- by chemical means over short distances. When a nerve impulse reaches a **synapse**, **neurotransmitters** diffuse across the synaptic cleft and trigger an impulse in the postsynaptic neurone
- by chemical means over longer distances. **Insulin** released from cells in the **pancreas** affects the uptake of **glucose** by **liver cells**.

Errors in processing of these signals can lead to the development of diseases such as **diabetes** and cancer. More research on cell signalling might lead to new ways of treating disease. (See also **tumour**.)

cell-surface membrane: the **plasma membrane** found on the outside of a **cell**.

cell wall: a rigid layer that surrounds individual cells in plants, bacteria and fungi. Animal cells do not have cell walls. The most abundant component of plant cell walls is **cellulose**. Cellulose molecules form long parallel chains that are linked into bundles called microfibrils. These cellulose microfibrils are cemented together in a matrix containing a variety of other substances. The resulting structure is very strong and resistant to both compression and tensile (pulling) forces. It is, however, freely permeable, allowing water and other substances to pass readily through it. In the cell walls of **sclerenchyma** and **xylem**, this wall is further strengthened by the addition of **lignin**. Lignin cements and binds the cellulose microfibrils together.

The cell wall provides support for the cell. In addition, it plays a vital part in allowing turgor pressure to build up in the cell. This helps to support the plant. The cell walls of bacteria and fungi share many of the functions of the cell walls of plants, although there are differences in their chemical structure. The main molecule in bacterial cell walls is peptidoglycan, whereas in fungal cell walls it is the nitrogen-containing polysaccharide **chitin**.

cellulose: a **polysaccharide** that is one of the main components of plant **cell walls**. The diagram shows how molecules of a form of **glucose** called β-glucose are joined by **condensation** to give long, straight chains of cellulose.

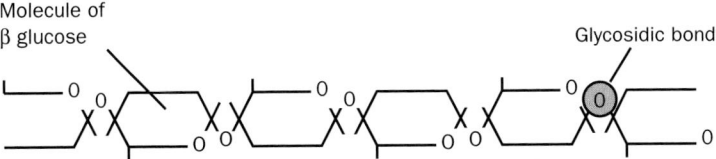

The structure of cellulose

These chains are linked into bundles by **hydrogen bonds**. It is because cellulose is arranged in these bundles that it can form strong but permeable cell walls in plants. An additional feature of cellulose is that, although it is a common **carbohydrate**, it is difficult to digest. Mammals that live to a large extent on plant material usually have partnerships with **microorganisms**. The mammal provides the microorganisms with cellulose from its diet. In turn, the microorganisms produce **enzymes** that break the cellulose down into smaller molecules which can be absorbed by the mammal concerned. This is an example of **mutualism**.

central nervous system: the *brain* and *spinal cord*. The central nervous system is responsible for coordinating and controlling the activity of the nervous system.

centrifuge: a piece of apparatus that is able to separate out particles of different size and density by spinning them at high speed. *Cell fractionation* is a process in which cells are broken up and the resulting suspension is centrifuged. The different organelles have different densities and settle to the bottom of the centrifuge tube at particular spinning speeds. A centrifuge of the type normally found in school or college laboratories generates a force about one thousand times that of gravity (1000 *g*). This is enough to separate larger organelles such as nuclei and *chloroplasts*. Ultracentrifuges which generate forces up to 10 000 *g* can separate much smaller organelles.

centriole: a structure found in an animal cell that is associated with the separation of *chromosomes* during *mitosis*. They are small, hollow cylinders. Each contains a ring of *microtubules*. The centrioles separate from each other during the early stages of mitosis. One goes to one pole and the other goes to the other pole.

centromere: the region on a *chromosome* that holds the sister *chromatids* together during the early stages of cell division. The centromere is also the region to which the *spindle* fibres attach. Shortening of the spindle fibres pulls the sister chromatids apart during *anaphase*.

cerebellum: part of the hindbrain that controls posture and balance as well as coordinating the overall smooth movement of the body.

cerebral hemispheres: see *cerebrum*.

cerebrovascular accident (CVA): this is a stroke. The term refers to the effects produced by a serious interruption to the blood supply of the *brain*. It may be caused by a blood clot or by the rupture of an *artery* wall. The effects vary according to the part of the brain involved. Damage to the right side of the *cerebellum*, for example, may produce loss of feeling in or paralysis of the left side of the body.

cerebrum: part of the forebrain, made up of the two cerebral hemispheres, responsible for control of voluntary behaviour. In humans, this part of the *brain* is large and covers most of the midbrain and hindbrain. Studies of the brain suggest that different parts of the cerebrum have different functions. These may be grouped together and involve:

- sensory areas – those parts of the cerebrum associated with receiving sensory information. They include areas involving specific senses, such as vision and hearing, as well as those concerned with sensory information from the general body surface. The size of the area is related to the number of receptors involved
- *association areas* – the parts of the cerebrum responsible for interpreting sensory information in the light of experience. They are obviously associated with memory and learning
- motor areas – these are the areas from where the *nerve impulses* that go to the muscles originate.

CFC: chlorofluorocarbon, a substance used as an aerosol propellant, in refrigeration and in containers used in fast-food packaging. It is biologically important as it is mainly responsible for the thinning of the ozone layer.

chance: luck. Suppose measurements were made of the lengths of two sets of bean seeds and the mean values calculated. The means would probably differ slightly from each other. There could be two possible explanations for this difference. It might be that it resulted from the beans having been obtained from different genetic varieties or from plants grown in different environments. On the other hand, the difference might simply be due to chance, a matter of luck that there were more small beans in one sample than in the other. In analysing the results of investigations like this, we can use statistical tests to determine the **probability** of differences being due to chance. We can then decide whether the results are biologically significant or not.

chemiosmosis: the way in which **ATP** is produced in **mitochondria** and **chloroplasts**. **Electrons** are passed from molecule to molecule along **electron transport chains** in the membranes of the organelles. At each transfer, a small amount of energy is released. The energy released is used to pump the protons through the membrane in which the carrier molecules are found. This produces a proton gradient with more protons on one side of the membrane than on the other. The proton gradient acts as a store of potential energy. As protons return through the membrane they release this energy, which is then used to produce ATP.

chemoautotroph: an organism that obtains energy from chemical reactions. It uses this energy to produce its organic compounds. All autotrophs make the organic substances which they require from simple inorganic ones such as carbon dioxide and water. They require a source of energy to do this. **Photoautotrophs** use light as their source of energy. Chemoautotrophs obtain the energy they require from chemical reactions. Some nitrifying bacteria convert ammonia into nitrites. Others produce nitrates from nitrites. Both of these reactions involve oxidation. The reactions release the energy that enables the organism to build up its **carbohydrates** and other organic compounds from carbon dioxide and water. Other chemoautotrophic organisms are found in the ocean depths. They are associated with volcanic vents and also use the energy released from chemical reactions to synthesise organic molecules. Since no light can penetrate this far, these chemoautotrophs are the **producers** on which a variety of marine worms and other organisms depend.

chemoreceptor: a receptor that detects changes in its chemical environment. Chemoreceptors in the walls of the carotid arteries detect changes in the carbon dioxide concentration in the blood. Sensory information from these receptors goes to the centres in the **brain** that regulate heart rate and breathing.

chemotherapy: the use of drugs and other chemical substances to treat disease. The term is often used to refer to the treatment of cancer. There are many different drugs used in the chemotherapeutic treatment of cancer. Some of their sites of action are shown in the diagram. (See also **tumour**.)

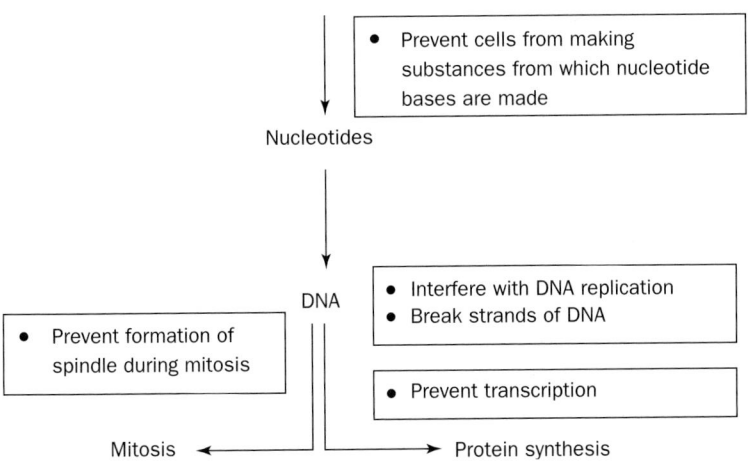

The site of action of some anticancer drugs

chiasma: (plural chiasmata) the place where *chromatids* cross over each other in a cell which is undergoing *meiosis*. During the first stage of meiosis, *chromosomes* come together in their homologous pairs, each chromosome consisting of a pair of chromatids. There are, therefore, two pairs of sister chromatids. Chiasmata occur at the points where these chromatids break and rejoin in the process known as *crossing over*. The number of chiasma on a pair of chromosomes varies. Generally speaking, the longer the chromosome, the greater the number of chiasmata that form.

chitin: a nitrogen-containing *polysaccharide*. Chitin is found in the *exoskeletons* of insects and in the *cell walls* of fungi. The basic units of the molecule are linked together by *condensation* to make up long chains. *Hydrogen bonds* link the chains together and help to make chitin rigid and strong.

chloride shift: the *diffusion* of chloride ions across the *cell-surface membrane* of a *red blood cell*. Respiring tissues produce carbon dioxide. This enters a red blood cell and combines with water to produce carbonic acid. The carbonic acid dissociates and forms hydrogen ions and hydrogencarbonate ions. An enzyme called *carbonic anhydrase* is present inside red blood cells. This enzyme catalyses the reaction in which the hydrogencarbonate ions are formed. As a result there is a higher concentration of hydrogencarbonate ions in the red blood cells than in the blood *plasma*. The hydrogencarbonate ions therefore diffuse into the plasma. This is balanced by the chloride shift.

chlorophyll: the green pigment found in organisms that photosynthesise. It is responsible for the capture of light energy. There are a number of slightly different sorts of chlorophyll molecule which differ from each other in the details of their chemical structure but they all share certain features:

- they have a head region with a magnesium atom at its centre. This is the part of the molecule that absorbs light energy
- slight differences in the chemical groups attached to the head region produce different chlorophyll molecules. These different chlorophyll molecules are able to absorb light with slightly different wavelengths
- they have a tail region that is soluble in *lipids*. Because of this, the tail region is the part that is anchored in the membranes of the *chloroplast*.

Chlorophyta: a group of algae that contain *chlorophyll* pigments similar to those found in plants. Some are single cells. Others live in colonies formed of large numbers of these cells arranged in filaments or jelly-like balls.

chloroplast: a *chlorophyll*-containing *organelle* found in the cells of plants and algae. Chloroplasts are where *photosynthesis* takes place. The chloroplasts found in plant cells are small, flattened discs, each about five micrometres in diameter. They are found in the *cytoplasm* but can change their position in response to differences in light intensity. Chloroplasts are surrounded by an outer envelope consisting of two *plasma membranes*. Inside there are further membranes. These run through the chloroplast and are stacked in places to form structures rather like piles of coins. These membrane stacks are the *grana*. The membranes inside the chloroplast contain molecules of chlorophyll and they are where the *light-dependent reaction* of photosynthesis takes place. Light energy is trapped by the chlorophyll molecules and used to produce chemical energy in the form of *ATP*. The light energy also reduces the coenzyme *NADP*.

Surrounding these membranes is the rest of the chloroplast. This is called the **stroma**. It contains the **enzymes** associated with the **light-independent reaction** of photosynthesis. In this reaction, sugars are produced from carbon dioxide, using the ATP and reduced NADP formed in the light-dependent reaction.

cholera: an **infectious disease** of the **small intestine** caused by the bacterium **Vibrio** cholerae. The disease starts with severe muscle and **stomach** cramps. Vomiting and fever soon follow, and then the victim develops diarrhoea. This leads to a huge loss of body fluid. Without treatment, people with cholera often die. This is usually because they have lost so much fluid that their circulatory system fails.

cholesterol: a **lipid** that plays an important part in living organisms. Some of the choles-terol required by the body is taken in with food and some is synthesised in the **liver**. Cho-lesterol is an important component of **plasma membranes**. It is also a precursor of **bile salts** and steroid **hormones** such as **testosterone** and **progesterone**. High concentrations of cholesterol in the blood, however, are associated with **atheroma** and **coronary heart disease**. Its concentration is therefore often monitored in older people and drugs such as **statins** may be given to reduce blood cholesterol.

cholinergic: a **synapse** in which the **neurotransmitter** is **acetylcholine (ACh)**. In mammals and other vertebrate animals, cholinergic synapses are common. They are found throughout the **central nervous system**. Other examples of cholinergic synapses are the neuromuscular junc-tions between **motor neurones** and the muscles they supply. In the autonomic nervous system, some of the synapses in the sympathetic part and all of those in the parasympathetic part are cholinergic. (See also **parasympathetic nervous system** and **sympathetic nervous system**.)

chordae tendinae: string-like **tendons** in the heart that are attached to the flaps of the **atrioventricular valves**. They prevent blood from flowing back into the **atria** when the pres-sure in the **ventricles** rises and closes the valves.

Chordata: the animal **phylum** to which humans and other mammals belong. Fish, amphibia, reptiles and birds are also chordates. Some of the features shared by members of this phylum are:

- at some stage in their life cycles, they have **gill** slits. These are obvious in fish and can also be seen clearly in the tadpoles of frogs and toads. In the other groups, they are only visible while the animal is an **embryo**
- there is a post-anal tail. An earthworm is not a chordate; its **anus** is right at the end of the body and there is no tail behind it. A snake may look worm-like but, if you examine it carefully, you will see that the anus is about two-thirds of the way back along the body. The rest of the animal is tail. Snakes are members of the Chordata
- the nerve cord forms a hollow tube just under the dorsal surface. In animals that are not chordates, the nerve cord is solid and is on the ventral side
- there is a cartilaginous rod-like structure called a notochord below the nerve cord. This is the feature that gives the phylum its name of Chordata.

chorionic villus sampling (CVS): a method of obtaining cells from an **embryo** for **genetic screening**. The technique can be carried out as early as eight weeks into the pregnancy. **Ultrasound** is first used to find the position of the developing embryo. Then a fine tube is inserted through the vagina and cervix. Some cells are collected through this tube. These cells are taken from the chorionic villi, structures that form part of the developing **placenta**.

chromatid: one of the two strands of genetic material that make up a *chromosome*. When chromosomes appear during the early stages of cell division, each can be seen to consist of a pair of chromatids. These are held together by a *centromere*. The two chromatids are then pulled apart to the opposite poles of the cell during the later stages of cell division. In *mitosis*, the sister chromatids are genetically identical. In *meiosis*, they may differ genetically because of *crossing over*.

chromatin: the *DNA* present in the nucleus of a non-dividing cell. It is only when a cell is dividing that the DNA and *proteins* contained in its nucleus can be seen to be packaged into *chromosomes*. When the cell is in interphase, these substances are much more spread out and form chromatin. Chromatin consists of two forms: *euchromatin* and *heterochromatin*.

chromatography: a technique used to separate substances in a mixture. A solution of the substances is made in a suitable solvent. A small drop of this solution is then loaded on to a strip of absorbent chromatography paper. The paper is suspended so that its end dips into a solvent. The solvent moves up the paper by *capillary action* taking the substances that were loaded on as it goes. These substances separate out at various levels. The places where they end up depend on their solubility in the moving solvent and the ease with which they are absorbed by the paper. It is possible to identify the individual components by calculating their R_F *values*.

chromosome: a thread-like structure found in the *nucleus* on which the genetic material of the cell is organised. It may be thought of as a package of *genes*. A chromosome consists of a tightly coiled length of *DNA*, closely associated with a small amount of *RNA* and a number of *proteins*. These proteins have a variety of different functions. One group, the *histones*, are involved in the packaging of the DNA.

In non-dividing cells, the chromosomes are long and thin and cannot be seen as separate structures. The DNA and proteins from which they are made, however, can be stained and form *chromatin*. During cell division, the chromosomes become shorter and thicker and can be identified as distinct structures. In a human body cell, there are 23 *homologous* pairs of chromosomes. The members of each pair have the same distinctive appearance. This enables them to be identified and the chromosomes of one person compared with those of another. Twenty-two of these pairs are found in the body cells of all humans and these are known as *autosomes*. In addition, there are two *sex chromosomes*. These are referred to by the letters X and Y.

In *prokaryotic* cells, such as those of bacteria, there is less DNA present. It forms a loop that is sometimes called a bacterial chromosome. Since this structure is not associated with histone packaging proteins, it is not a true chromosome.

chromosome mutation: see *mutation*.

chronic disease: a disease that develops gradually. It is long lasting and changes in the condition often only take place slowly. Chronic disease may be compared with *acute disease*.

chronic obstructive pulmonary disease (COPD): a long-term disease in which the flow of air to and from the *lungs* is obstructed. Chronic *bronchitis* and *emphysema* are two distinct conditions, but patients frequently have both at the same time. When this happens, doctors find it difficult to determine the relative effects of each condition and use chronic obstructive pulmonary disease to refer to the combined effects of both chronic bronchitis and emphysema.

cilia: tiny, hair-like structures found on the surface of certain cells. They have a distinctive internal structure. Each has a ring of nine pairs of tubules near the outside and a single pair in the centre. These tubules enable the cilia to beat. Cilia are found in a number of places:

- Single-celled organisms like paramecium have rows of cilia covering the cell surface. The organism moves through the water by beating these cilia in a rhythmic, coordinated way.
- Filter-feeding worms and molluscs have tentacles or *gills* covered in ciliated cells. These maintain a constant flow of mucus, which traps small food particles and draws them towards the mouth.
- Epithelial cells line the larger airways in the gas exchange system of a mammal. Cilia on the surface of these cells waft a mixture of mucus and trapped particles away from the gas-exchange surface.

ciliated epithelium: epithelial *tissue* lines the inside of an organ. Epithelial cells that line the larger airways in the gas exchange system of a mammal have **cilia** on their surface. These cilia waft a mixture of mucus and trapped particles away from the gas-exchange surface.

ciliary muscles: the muscles in the eye that are responsible for altering the shape of the *lens*. The way in which these muscles work is shown in the diagram.

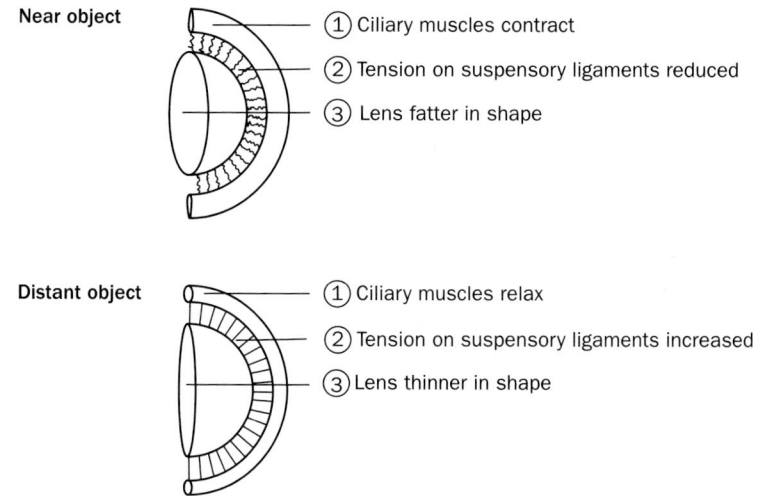

Focusing on near and distant objects

While you are reading this book, you are focusing on the print. This is very near your eye, so your ciliary muscles have contracted. This reduces the tension in the suspensory *ligaments* attaching the muscle to the lens. As a result, the lens becomes fatter in shape.

When you look at an object in the distance, the ciliary muscles relax. This increases tension in the suspensory ligaments and the lens is pulled into a thinner shape. Change in shape of the lens allows both close and distant objects to be seen clearly, a process called *accommodation*.

circadian rhythm: the internal 24-hour rhythm that governs the behaviour of many organisms. The changes that take place during the course of a single day have a considerable effect on animals. In hot, dry climates, for example, it may be a considerable advantage to be nocturnal, hunting for food during the cooler hours of the night and resting in the shade during the day. Organisms have a natural biological clock that enables them to adapt to conditions like this. Experimental evidence suggests that this is an internal mechanism, but that it has to be maintained by making constant reference to slight changes in environmental conditions. Lizards, for example, when kept under constant environmental conditions, still show a 24-hour pattern of activity. However, unless there are slight differences in the environmental conditions, this pattern starts to drift. Circadian rhythms are found widely in living organisms. Examples include leaf movements and growth patterns in plants, the vertical *migration* of small planktonic organisms in the sea and patterns of *hormone* secretion in animals.

CITES (Convention on International Trade in Endangered Species): a voluntary international agreement controlling trade in endangered species and products obtained from them. It aims to prevent international trade endangering the survival of species of animals and plants. Approximately 5000 species of animals and 28 000 species of plants are protected by this convention.

class: a level of *classification* below that of *kingdom* and phylum.

classical conditioning: a form of learning. An unrelated, neutral stimulus produces the response originally associated with another, more appropriate, stimulus. The early work on classical conditioning was carried out by a Russian scientist called Pavlov. He fed dogs and then measured the amount of *saliva* they produced. The stimulus of the presence of food in the mouth produced the response of an increase in the amount of saliva. Pavlov found that when he presented a neutral stimulus such as a flashing light at the same time as the original stimulus, the dog eventually responded to the neutral stimulus alone.

classification: the way in which living organisms are divided into groups. No one knows for certain how many different species of living organisms there are, but it is probably around 15 million. Biologists sort out and classify all these organisms using similarities and differences between them. The system that we use has two particular characteristics:

- It is a hierarchical system. This means that we keep breaking everything down into smaller and smaller groups. All living organisms belong to a *kingdom*; each kingdom is divided into a number of phyla; each phylum is broken down into classes and so on down to genera, which are divided into individual species.
- It is phylogenetic. There are all sorts of ways of classifying a group of living organisms – whether they are edible or not, their colour and size – but none of these would help us to understand the evolutionary relationships between the different members. The system used is based on important biological differences so it shows these relationships clearly.

The table shows how biologists classify a brown rat, **Rattus norvegicus**.

Kingdom	Animalia	Rats are clearly members of the animal kingdom
Phylum	Chordata	They possess the characteristics of the phylum, Chordata
Class	Mammalia	They have hair, sweat **glands** and suckle their young
Order	Rodentia	Gnawing mammals
Family	Muridae	The group containing rats and mice
Genus	*Rattus*	The scientific name of a species has two words. The first word gives the genus
Species	*norvegicus*	and the second, the species to which the organism belongs

climatic factor: an **abiotic** factor that influences the distribution of organisms. Climatic factors depend on climate and include temperature, light intensity and availability of water.

climax community: the different species of organisms that make up the final stage in an ecological succession. Over most of Britain, the climax community is some form of **deciduous** woodland.

Clinistix™: a test strip used to test for **glucose**. It consists of a plastic strip. The tip of this strip is impregnated with the enzyme glucose oxidase. When glucose is present, this enzyme catalyses a reaction that causes the tip of the strip to turn purple.

clone: a group of genetically identical organisms all produced from the same parent by **asexual reproduction**. The ability to produce clones is important in plant breeding. Cloning allows varieties that have arisen by **mutation** or as the result of deliberate crosses to be multiplied. This gives large numbers of identical plants for commercial purposes. All named varieties of roses and apples, for example, are clones and came from a single parent plant. Cloning is also important in **genetic engineering**.

close season: a part of the year in which a particular species of animal may not be caught or harvested. The stocks of almost all the world's commercially important fish have been seriously depleted by overfishing. To maintain existing stocks and allow them to recover, various **conservation** measures need to be put in place. One of these is to use close seasons and stop fishing during the main breeding season of the species concerned. This measure is difficult to enforce, but it is used with freshwater species.

closed circulatory system: see **blood system**.

Cnidaria: the animal phylum containing hydra, jellyfish, sea anenomes and corals. Members of the Cnidaria share the following features:
- They are **diploblastic** and radially symmetrical (see **radial symmetry**).
- They catch their prey with special stinging cells called cnidocysts.
- Their digestive cavity or **enteron** has only one opening. The enteron also acts as a **hydrostatic skeleton**.

coccus: a bacterium that is more or less spherical in shape. Cocci range from approximately 0.5 μm to 1.2 μm in diameter. They include a number of species of *Streptococcus*. These bacteria are important in cheese-making as well as being the bacteria responsible for boils and some of the infections that may follow surgical operations. Other cocci include species of *Staphylococcus* such as methicillin-resistant *Staphylococcus aureus* (**MRSA**).

codominant: two *alleles* are said to be codominant when both are expressed in the *phenotype* of the *heterozygote*. In showing genetic crosses, a different system is used to represent codominant alleles from that used to represent a pair of alleles in which one is *dominant* and one *recessive*. In cattle, one of the genes responsible for coat colour has two codominant alleles. The gene may be represented by the letter C. The two alleles are C^R, which is responsible for producing red hairs, and C^W, which is responsible for producing white hairs. The heterozygote will have the *genotype* C^RC^W. Since these alleles are codominant, both will be expressed and the animal will have a mixture of red and white hairs. This produces an overall colour called roan.

codon: a sequence of three bases on an *mRNA* molecule that codes for an amino acid (see *genetic code*). There are also three codons that do not code for any amino acid. They act as stop signals and indicate the end of a particular protein. They act rather like a full stop at the end of a sentence.

coeliac disease: a condition in which the *villi* in the *small intestine* are damaged and the person fails to digest and absorb food properly. The damage is caused by the immune system responding to gluten. This is a protein found in cereals such as wheat and oats. The condition may be treated successfully with a life-long gluten-free diet.

coelom: a cavity that develops in a *triploblastic* animal. It separates the muscles of the body wall from the various internal organs, so locomotion is independent of movements of the internal organs. The coelom can also transport molecules and cells from some parts of an animal to others. In soft-bodied animals such as worms belonging to the phylum *Annelida*, the coelom acts as a fluid-filled or *hydrostatic skeleton*.

coelomate: an animal that has a *coelom*.

coenzyme: a molecule, other than the *substrate*, that is needed for an *enzyme* to work. It acts like a shuttle and transports *atoms* or molecules from one enzyme-controlled reaction to another. Some important coenzymes are involved in the processes of *respiration* and *photosynthesis*. In many of the stages of *aerobic respiration* the coenzyme, *NAD*, is reduced. The reduced NAD is then fed into the *electron transport chain*. Many of the coenzymes important in respiration are made from *vitamins*. NAD, for example, is produced from nicotinic acid, also known as vitamin B3.

cofactor: a substance, other than the *substrate*, that is essential for the action of a particular *enzyme*. Unlike a *coenzyme*, which shuttles *atoms* or molecules between chemical reactions, a cofactor stays firmly bound to the enzyme molecule as it catalyses the reaction. Metal ions are common cofactors. In humans these include iron, manganese, copper and zinc.

cohesion–tension: one of the ways in which water moves in the *xylem* from the roots up to the leaves of a plant. The main steps in the process are:
- water is lost from the leaves during *transpiration*. The water molecules evaporate and move from the saturated air inside the leaf to the drier air outside

- because of the formation of **hydrogen bonds**, water molecules stick to each other. This property is called cohesion. As water is lost by evaporation from the leaf, more is pulled up because the water molecules cling to each other
- since transpiration is pulling the water column in the **xylem** upwards and gravity is tending to pull it down, the water column is stretched. It is under tension
- water molecules are also attracted to and cling to the walls of the xylem. This is known as adhesion and helps in pulling the water upwards.

coliform bacteria: a group of bacteria whose presence in a water sample indicates that the sample has been contaminated with **faeces**. It would not be practicable to test drinking water for all the many potentially harmful **microorganisms** that might be present, so coliform bacteria are used as **indicator organisms**. Coliform bacteria are good indicators because they live in the intestines of humans and other mammals. If these bacteria are present, it is likely that the water has been contaminated by faeces and might also contain pathogens.

collagen: a fibrous **protein** found in animals. Almost a third of all the protein found in the human body is collagen. Each collagen molecule consists of three **polypeptide** chains that are wound tightly round each other to form a helix. This gives the molecule a structure that is very resistant to stretching. This is an important property in tendons, for example, that join muscle and **bone**. When collagen is boiled it is converted to gelatin.

collecting duct: the last part of a **nephron** or **kidney** tubule. The part of the nephron known as the **loop of Henle** sets up a gradient of salt concentration in the **medulla** of the kidney. This means that the **water potential** of the fluid in the collecting duct is always higher than that in the medulla. Water is therefore removed by **osmosis** all the way down the collecting duct. This is important in producing concentrated **urine**.

Antidiuretic hormone (ADH) is a **hormone** that increases the permeability of the cells that line the collecting duct. When the body is dehydrated, this hormone is secreted. It makes the cells lining the collecting duct more permeable to water by increasing the number of channels in their **cell-surface membranes** through which water passes. More water is reabsorbed and smaller quantities of more concentrated urine are produced.

collenchyma: supporting tissue found in plants. Collenchyma cells are slightly elongated and have **cell walls** with extra **cellulose** thickening at the corners. Collenchyma tissue is found either in strands or in a continuous cylinder round the outside of young stems. This enables it to resist the compression and stretching that occur as the plant is blown from side to side by the wind. Like **parenchyma** cells, the cells of collenchyma have living contents. Their walls can still be stretched, so they are particularly important in the support of young, growing stems.

colon: part of the **alimentary canal** between the **small intestine** and the **rectum**. This is the part in which **faeces** are produced and stored before they are removed from the body. The main function of the rectum is the **absorption** of water. The colon contains many bacteria. In some herbivorous mammals such as rabbits and horses, these bacteria play an important part in the **digestion** of **cellulose**.

colostrum: a secretion produced by the mammary **glands** of the mother immediately after birth. It is different from milk because it contains more protein and much less fat and sugar. It provides the young mammal with **antibodies**, which provide **passive immunity** against some diseases. These antibodies are thought to be particularly important in preventing gut infections that might otherwise prove dangerous to a newborn animal.

commensalism: a relationship between two different species of organisms where one gains but the other suffers no direct harm. There is no physiological link between the organisms concerned. House sparrows may be described as having a commensal relationship with humans. The sparrow consumes considerable amounts of human food waste. This causes no direct harm to humans, however. More different organisms have probably established commensal relationships with humans than with any other species.

community: all the living organisms present in an *ecosystem*. These organisms have different feeding habits and occupy different *trophic levels*. Organisms within these trophic levels are related to each other by *food chains* and *food webs*.

companion cell: a type of cell found in the *phloem* of a plant. Companion cells may be recognised by their densely stained *cytoplasm* and conspicuous nucleus. They lie side by side with *sieve tube elements*.

competition: a relationship between different organisms that require the same resources. In most ecological situations, resources are limited. There is, for example, only a limited amount of nitrate in the soil, or of light energy striking the surface of the ground. Some organisms gain these resources at the expense of others. These will be successful in terms of competition and will be more likely to survive. *Intraspecific competition* is competition between members of the same species. *Interspecific competition* is competition between members of different species.

competitive inhibition: when the rate of an enzyme-catalysed reaction is slowed down or stopped by a substance whose molecules are similar in shape to those of the *substrate*. An *enzyme* molecule has an *active site* formed by a group of amino acid molecules. The active site has a particular shape into which the substrate molecule normally fits to form an *enzyme–substrate complex*. This complex then breaks down to give the products of the reaction. A competitive inhibitor has molecules that are similar in shape to those of the substrate. They fit into the enzyme's active site and prevent it from being occupied by substrate molecules. Because there are now fewer enzyme molecules available, the rate of reaction slows down. Competitive inhibitors, therefore, compete with the substrate for the active site of the enzyme.

complementary DNA (cDNA): a single strand of *DNA* that is produced from *mRNA*. Biologists use complementary DNA in *genetic engineering*. Each *somatic cell* in an adult organism contains a complete copy of all the organism's DNA. Specialised cells, however, only transcribe some of this DNA, so they only contain mRNA transcribed from some of the genes in the nucleus. mRNA is isolated and mixed with free DNA nucleotides and an enzyme called *reverse transcriptase*. This enzyme catalyses a process that is the reverse of *transcription* and makes a molecule of complementary DNA.

complementary therapy: methods of healing that doctors and other members of the medical profession do not consider to be standard. Complementary therapies include acupuncture, the use of herbal remedies and homeopathy. Many of these therapies have not been tested scientifically, but they may be tried by people with *chronic diseases* that have failed to respond to standard medical treatment.

complete metamorphosis: the process in which an organism goes through a dramatic change of form between the *larva* and the adult. Complete metamorphosis is a feature of the life cycles of insects such as flies, butterflies and moths, and beetles. The larva moults into a pupa. Inside the pupa, most of the existing larval tissues are broken down. New tissues which make up the adult then develop. The pupa moults and the adult insect emerges.

compost: organic material that has been broken down by microorganisms. The compost that is produced may be added to the soil as a conditioner and *natural fertiliser*. The main steps in making compost are as follows:

1 Suitable organic material such as grass clippings and vegetable waste are collected into a heap.
2 Soil *microorganisms* colonise this heap. They are called mesophils because they live at temperatures between 10°C and 45°C. As they multiply and respire the compost heap heats up.
3 The mesophils are replaced by thermophils. Thermophils are microorganisms that live at a much higher temperature. They convert the organic material into useful compost.
4 The last stage is the curing stage. The compost matures and is suitable for adding to the soil.

computed tomography (CT): the use of an *X-ray* scanner to record the appearance of slices through a patient's body. This information is then used by a computer to produce a cross-sectional image in which the soft tissues of the body show clearly. It is particularly useful in investigating diseases of the *central nervous system*.

condensation: a chemical reaction involving the joining together of two molecules with the removal of a molecule of water. Condensation is important in the formation of biologically important *polymers*. It is the way in which *amino acids* join to produce *polypeptides*, *glucose* molecules form *starch* and *glycogen,* and individual nucleotides combine to produce DNA and RNA. Molecules that are formed in this way can usually be broken down by the addition of water molecules, a process called *hydrolysis*.

conditioned reflex: see *classical conditioning*.

cone cell: a colour-sensitive receptor cell found in the *retina* of the eye. Cone cells contain pigments that are sensitive to light of different colours. In the cone cells of the human eye, there are three different pigments. One is most sensitive to blue light, one to green light and one to red light. Blue light only stimulates cones that contain the blue-sensitive pigment. This means that the *brain* will interpret the object concerned as being blue. Orange objects appear orange because they reflect both green and red light. If an orange object is viewed, the light reflected from this will stimulate the cones containing the green-sensitive pigment as well as those containing the red-sensitive pigment. The brain will interpret this object as being orange. Cone cells are tightly packed into the *fovea*. There are about 50 000 of them to each mm^2. This distribution of cone cells and the fact that each one is connected individually to a *sensory neurone* in the retina give the eye its ability to see fine detail. Each cone cell is able to produce a *nerve impulse* in a separate sensory neurone.

conjugated protein: a *protein* to which a non-protein chemical group called a prosthetic group is attached. *Haemoglobin* is a conjugated protein.

conjugation: a process by which *DNA* can be transferred from one mature bacterial cell to another. A thin tube forms between two cells. Some of the DNA in the donor cell moves through this tube into the recipient cell. *Antibiotic resistance* may spread from one species of bacterium to another when DNA is transferred in this way.

connective tissue: a type of *tissue* found in mammals and other vertebrates. Connective tissue often contains cells that produce fibres of *proteins* such as *elastin* and *collagen*. These cells and the fibres they produce are contained in a non-cellular matrix. Connective tissue has many different functions. It binds other tissues together and it forms *tendons*, *ligaments*, *bone* or blood. Many of these tissues have functions associated with support. Connective tissue also plays an important part in defending the body against infection.

conservation: a way of maintaining the *biodiversity* of *ecosystems*. There are two views of conservation. At one extreme, there are people who feel that nature should be left to take its course. They argue that natural environments should be preserved and protected from all human influence. At the other extreme is the idea that wildlife should be actively managed so that it can be exploited on a sustainable basis. An example of these two approaches is the conservation of African elephants. Before poaching became a serious problem in Kenya and some other East African countries, total protection of relatively small areas meant that the elephant population exceeded the *carrying capacity* of the land. As a result, serious damage was done by the animals to their *environment* and large numbers died. In parts of southern Africa, however, surplus animals were deliberately killed in such a way that a sustainable yield of meat, skins and ivory could be produced. (See also *genetic conservation*.)

consumer: a *heterotroph*. All *food webs* and *food chains* depend on *producers*. Producers are organisms such as plants that produce organic molecules from simple inorganic ones. Consumers eat producers. *Primary consumers* feed directly on producers, secondary consumers feed on primary consumers and tertiary consumers feed on secondary consumers. In the simple food chain:

nettle plant → large nettle aphid → two-spot ladybird

the large nettle *aphid* is a primary consumer and the two-spot ladybird is a secondary consumer. However, it is not always easy to describe the *trophic level* to which an organism belongs as precisely as this. Many animals feed at more than one trophic level. A bush cricket, for example, feeds directly on nettle plants so it is a primary consumer. It also feeds on other animals, which makes it a secondary or even a tertiary consumer.

continuous variation: *variation* in which there is a complete range of measurements from one extreme to the other. Individuals do not fall into discrete categories. The graph shows the milk yields of a large number of cows. There is a complete range from the cow with the lowest yield to the one with the highest. A pattern of continuous variation like the one shown often reflects a characteristic that is controlled by a number of different genes. The *environment* may also play a significant part in determining the final pattern of variation.

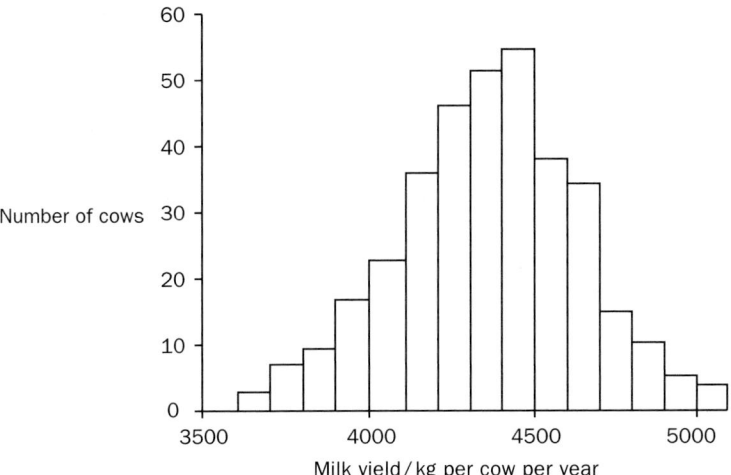

Milk yields in cows

contractile vacuole: an *organelle* found in the *cytoplasm* of some single-celled organisms. It is involved in the removal of excess water from the cell. Amoeba is a small organism which lives in freshwater. Its cytoplasm contains a higher concentration of dissolved substances than is present in the surrounding water. As a result, the cytoplasm has a lower *water potential* and water moves into the organism by *osmosis*. The contractile vacuole is involved in removing the excess water. It gradually fills with water. This is an active process and involves *ATP*. Eventually, when it is full, it empties its contents out through the *cell-surface membrane*. The contractile vacuole may also be involved in the *excretion* of waste products.

COPD: see *chronic obstructive pulmonary disease (COPD)*.

cornea: the transparent area at the front of the eye. The cornea helps to focus light on the *retina*. Light rays are bent when they pass from one medium to another with a different density. This bending is called refraction. In the eye, refraction occurs mainly when the light rays pass from the air into the cornea. However, the angle through which rays of light are bent by the cornea is always the same. This creates a potential problem because light rays from objects which are close to the eye need to be bent more if they are to be focused. The *lens* is able to change shape so it can change the amount of refraction, allowing both close and distant objects to be focused on the retina.

The cornea consists of living cells which must be supplied with nutrients such as *glucose*. This is done by the liquid which lies behind it, the *aqueous humour*. Some medical conditions result in the cornea becoming cloudy and sight being lost. They can be treated by a corneal transplant. A piece of cornea can be transplanted from one person to another without being rejected because the cornea does not have a blood supply.

corolla: a part of a *flower*. The corolla consists of a number of petals. In insect-pollinated flowers it has two important functions. Its bright colours attract insects and it also serves as a suitable landing platform for them. Many different insects visit flowers and the colour and shape of the corolla is often related to the species that pollinate them. Generally speaking, bees are attracted to flowers that are blue or purple, flies to those that are yellow or white and butterflies to red and pink-coloured flowers. Flowers that have an open structure are more likely to attract flies and small beetles, while those that are tube-shaped are often pollinated by bees. The corolla is very small in wind-pollinated plants.

coronary artery: a *blood vessel* supplying oxygenated blood to the muscle of the heart wall.

coronary bypass: a piece of *vein* that is used to provide an alternative pathway for blood to flow round a blockage in a *coronary artery*. The vein is usually taken from the person's leg. One end is grafted into the *aorta*. The other end is grafted into the coronary artery below the blockage. The blockage is bypassed and, as a result, oxygenated blood reaches all of the heart muscle.

coronary heart disease: a disease affecting the coronary arteries that supply the muscle of the heart. When one of these arteries is blocked, oxygen cannot get to the area of heart muscle that it supplies. The muscle therefore dies. This causes a heart attack or *myocardial infarction*. Two of the main reasons why blockages occur in the coronary arteries are:

- *atheroma*. This is a fatty deposit formed in the *artery* wall. Eventually, it narrows the *lumen* of the artery so less blood can pass through
- thrombosis. The presence of a blood clot in one of the coronary arteries. This is often associated with atheroma. A blood clot may form when the surface of one of the atheroma plaques breaks away. The clot blocks the lumen of the artery.

corpus luteum: a group of cells in the ovary of a mammal. The corpus luteum develops from an *ovarian follicle* after *ovulation* has taken place. It produces the *hormone proges-terone*. During the *menstrual cycle*, progesterone is responsible for maintaining the lining of the uterus. If *fertilisation* has not taken place, the corpus luteum gradually disintegrates and the concentration of progesterone falls. As a result, the lining of the uterus starts to break down and is lost from the body during menstruation. If pregnancy does occur, however, the corpus luteum remains active and continues to produce progesterone for about three months. After this, progesterone production is taken over by the *placenta*.

correlation: two variables are correlated if a change in one of them is reflected by a change in the other. A simple method by which we can find out if a correlation occurs is to plot the two variables on a graph and draw the line of best fit. These graphs are called scatter graphs. Care must be taken in interpreting scatter graphs. Two things may be correlated, but this does not mean that one causes the other.

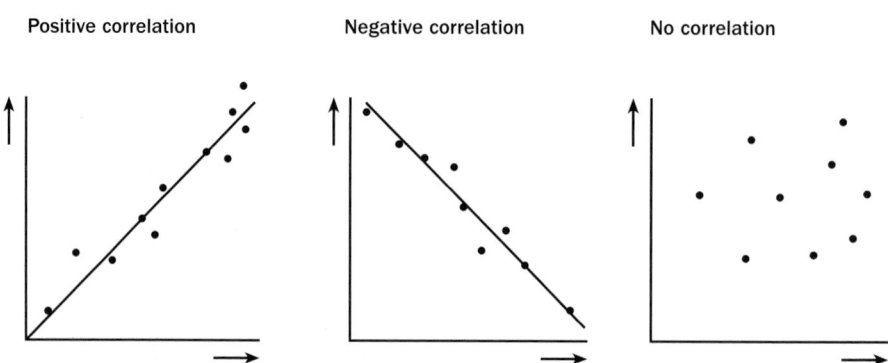

Scatter graphs showing correlation

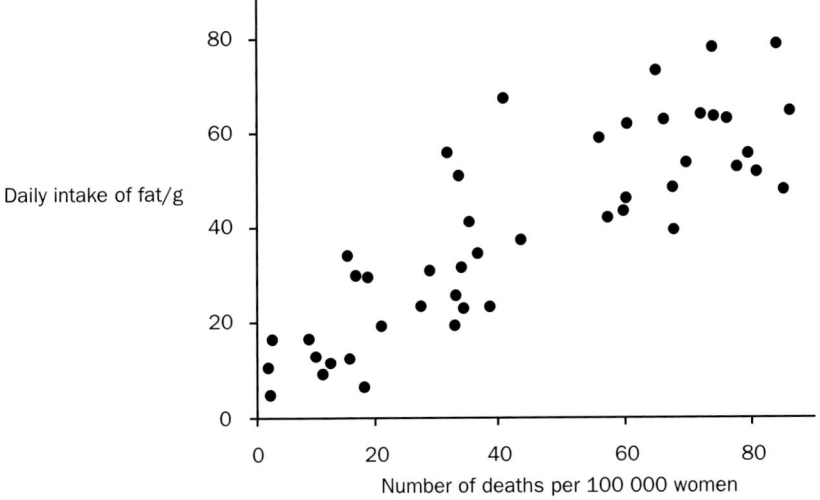

Scatter graph showing correlation between mean dietary fat intake and breast cancer

The scatter diagram above shows the relationship between mean daily intake of fat and the number of cases of breast cancer in various countries. It shows a clear positive correlation. We must not jump to conclusions, however. It does not mean that eating fat causes breast cancer. There may be any number of other possible explanations. For example, we could argue that women who eat more fat come from more affluent areas. They may therefore drink more alcohol or smoke more cigarettes. Either of these might be risk factors for breast cancer.

cortex: in a plant, the cortex is the tissue between the *epidermis* and the *xylem* and *phloem*.

co-transport: the way in which *glucose* and sodium ions are absorbed from the *small intestine*. There are *carrier proteins* called co-transport proteins in the *cell-surface membranes* of epithelial cells lining the small intestine. Each time a glucose molecule is transported into a cell through one of these proteins, so is a sodium ion. This is a form of *facilitated diffusion*. *Oral rehydration solutions (ORH)* are used to rehydrate a person who has become dehydrated because of diarrhoea. These solutions stimulate the *absorption* of glucose and sodium ions through the co-transport proteins. This lowers the *water potential* inside the epithelial cells and water is reabsorbed from the small intestine.

cotyledon: a leaf from a plant embryo found inside a *seed*. In some plants, when seeds germinate, the cotyledons stay below the surface of the soil and supply nutrients to the developing plant. In other plants they emerge above ground level, turn green and photosynthesise. The number of cotyledons present is an important feature in classifying *flowering* plants. The seeds of *monocotyledons* have one cotyledon in each seed, while those of the *dicotyledons* have two.

counter-current system: a system in which two fluids flow in opposite directions. This increases the efficiency with which substances are exchanged. In a fish, water with a high concentration of dissolved oxygen flows over the surface of the *gills*. Oxygen diffuses from this water into the blood flowing through the *capillaries*. If both water and blood flowed in the same direction, equilibrium would rapidly be reached and no more oxygen would be taken up by the blood. By flowing in the opposite direction, however, the water always has a higher oxygen concentration than the blood in the nearby *capillary*. Thus a concentration gradient is maintained along the entire exchange surface. This makes the uptake of oxygen much more efficient. A similar system is found in the *placenta* where the blood of the mother and the blood of the fetus generally flow in opposite directions. Counter-current systems between blood entering and leaving limbs and organs are involved in maintaining body temperature.

courtship behaviour: the behaviour that comes before, and is associated with, mating. Many different patterns of courtship behaviour exist. In some species there is little or no preliminary interaction between the individuals involved. In others, courtship involves long, complex sequences of behaviour. Courtship has a number of different functions:

- Courtship behaviour attracts a mate, possibly from a considerable distance away. Female silk moths, for example, produce substances called *pheromones*. Male silk moths respond to these and can use them to locate females from a distance of several kilometres.
- Courtship behaviour enables an animal to identify a member of its own correct species. The males of different species of firefly produce different patterns of flashing lights. Females will only respond to the correct pattern.
- Courtship behaviour enables males and females to synchronise their reproductive behaviour. With sticklebacks, the behaviour of one individual stimulates a response from the other. This leads to *fertilisation* of eggs laid by the female in a nest built by the male.
- Courtship behaviour may keep male and female together while the young are reared. Storks greet each other when food is brought to the nest with a noisy bill-clattering display.

CPR: see *cardiopulmonary resuscitation (CPR)*.

crenation: the shrinking and shrivelling of an animal cell that results when water is lost by *osmosis*.

cristae: the folded inner membranes of *mitochondria*. In *aerobic respiration*, the *electron transport chain* is associated with these membranes. The cristae contain the series of protein molecules along which *electrons* are transferred. They also contain *enzymes* through which protons pass and generate *ATP*.

crop rotation: changing the crop grown in a particular field each year. This has a number of advantages:

- Many important agricultural pests only affect particular crops. Their eggs and *spores* often remain in the soil after the crop has been harvested and infect next year's crop if the same plant is grown. Changing the crop regularly limits the build-up of pests.
- Different crops remove different proportions of mineral ions from the soil. By varying the crop, the mineral resources in the soil may be exploited to the full.
- The bacteria present in the *root nodules* of *leguminous plants* (such as clover) fix nitrogen. When these plants decay, their nitrogen-containing compounds are broken down by soil bacteria. This increases the soil nitrate content for the following crop.

cross fertilisation: *fertilisation* in which male and female *gametes* come from different organisms. This should be compared with *self-fertilisation* where both gametes come from the same organism.

cross pollination: *pollination* in which the pollen comes from a different plant. This should be compared with self pollination where the pollen comes from the same plant. Many plants have evolved mechanisms that increase the probability of cross pollination:

- They may have male and female *flowers* on separate plants.
- The organs of one sex in an individual flower develop before the organs of the other sex. In some species, male organs develop before female organs. In other species, the female organs develop first.
- The pollen is incompatible with the female organs of the same plant, so the pollen tube is unable to grow down the *style*.

cross-sectional study: a method of collecting information, such as about growth, by measuring individuals of different ages. Each individual is only measured once. A *longitudinal study* is where the same individual is measured at different ages. Cross-sectional studies of growth have some advantages over longitudinal studies:

- They can be carried out more rapidly as there is no need to wait for individuals to age. This, of course, makes them less expensive.
- Large samples can be measured as there is no need to keep track of the individuals in the study. Because of this, they provide a lot of information about the range of variation at a particular age.

However, there are some disadvantages as well:

- Cross-sectional studies cannot follow the growth patterns of individuals.
- Because the data obtained from a cross-sectional study is presented as the average size at a particular age, the growth curve that results can be misleading. If data is collected for humans, for example, it is difficult to see the growth spurt associated with puberty. This shows more clearly in longitudinal studies.

crossing over: this is where *chromatids* break and rejoin during cell division. During the first stage of *meiosis*, the *chromosomes* come together in their homologous pairs, each chromosome consisting of a pair of chromatids. There are, therefore, two pairs of sister chromatids. These chromatids may break and rejoin with other chromatids. This results in the genes on one part of a chromatid becoming attached to the genes which were present on part of another chromatid. If one of the chromatids concerned originally came from the mother while the other came from the father, there would be a new mixture of genetic material. Crossing over is important as it produces genetic variation.

cruciate ligaments: a pair of *ligaments* inside the knee joint. Sports injuries, particularly to football players, often involve damage to these ligaments.

crypt of Lierberkühn: a gland in the wall of the *small intestine*. It secretes *mucus* and some *enzymes*. It also contains *Paneth cells*. Cells at the base of the crypt divide by *mitosis*. This replaces the cells that are continually worn away from the villi.

CT: see *computed tomography (CT)*.

cultural evolution: the changes that have taken place in human societies over the course of time. The large *brains* and upright posture that are features of modern humans have evolved gradually. Evidence from fossils can reveal a lot about these physical changes. At the same time, however, there have been changes in the structure of human communities. Human communities have progressed from small groups that depended on hunting and scavenging for meat and gathering wild plants, to the complex societies that exist today.

There were a number of significant steps in our cultural evolution. Some of the more important ones are summarised in the table.

Step	Evidence first found	Evidence
Making of tools	About 2.5 million years ago	• Presence of marks and scars on the surface of the tool where it was deliberately shaped • Marks on accompanying **bones** suggesting deliberate cutting or hitting • Presence of stones of different material from surrounding rocks
Control of fire	Good evidence only from about 0.5 million years ago	• Ash deposits found associated with signs of human occupation
Domestication of plants	Between 12 000 BP and 9000 BP	• Evidence poor • Presence of cereals in which the grains do not fall easily from the stalks
Domestication of animals	Between 9000 BP and 8000 BP	• Discovery of relatively large numbers of bones from young and female animals • Evidence from animal bones showing a reduction in body size • Much more variation in features such as horn length and shape. These are features by which individual animals may be recognised

cuticle: a layer found on the surface of leaves and on the outside of many animals. In plants
it consists of wax. This is impermeable to water and helps to prevent water loss. *Xerophytes*
are plants that are adapted to live in dry conditions. They have very thick cuticles. In insects,
the cuticle is a more complex structure consisting of three layers:

• a very thin outer waxy layer known as the epicuticle. This layer is waterproof and, like
  the cuticle on a leaf, helps to limit water loss
• the exocuticle, which is extremely hard and strong. It is made of *chitin* and *protein*
• a slightly softer inner layer, the endocuticle. This consists mainly of chitin.

The cuticle completely surrounds the insect and is unable to stretch. Because of this,
insect growth takes place in a series of steps. An insect moults its cuticle and growth only
takes place before the new cuticle has fully hardened.

CVA: see *cerebrovascular accident (CVA)*.

cyanobacteria (blue-green algae): a group of *prokaryotic* organisms related to
bacteria. They are interesting to biologists in a number of ways:

- They are among the oldest organisms known. Stromatolites are a mixture of sediment and cyanobacteria. Some fossil stromatolites have been estimated to be over 2 billion years old.
- Some cyanobacteria are important nitrogen-fixing organisms (see **nitrogen fixation**).
- In many nutrient-rich lakes and rivers, cyanobacteria may be present in enormous numbers. The death of these organisms leads to a reduction in the amount of dissolved oxygen (see **eutrophication**). Some of the cyanobacteria involved produce toxins, which are poisonous to animals drinking the water.

cyclic AMP: a messenger molecule found inside cells. It is formed when a **hormone** binds with a receptor site on the **cell-surface membrane**. This causes an enzyme to be released that converts **ATP** to cyclic AMP. Cyclic AMP then activates the **enzymes** that control the biochemical pathways associated with the action of the hormone.

cyclic photophosphorylation: photophosphorylation is the process in which energy from light is used to produce **ATP** in the **light-dependent reaction** of **photosynthesis**. In cyclic photophosphorylation, the **electrons** that drive this process pass along an **electron transport chain** and return to the **chlorophyll** molecule from which they came.

cystic fibrosis: an inherited condition that affects **cell-surface membranes**. One of the **proteins** present on a cell-surface membrane is the CFTR protein. In healthy people, this protein transports chloride ions out of the cell. The cells lining the airways of the **lungs** are normally coated in mucus. The chloride ions enter this mucus and lower its **water potential**. As a result, water moves into the mucus by osmosis. The mucus is now sticky enough to trap particles that have been breathed in, but thin enough to be moved by **cilia**.

People with cystic fibrosis have a faulty CFTR protein. As a result, they produce thick, sticky mucus. This has several effects on the body. It can accumulate in the lungs making the person susceptible to lung infections, and it can also prevent the secretion of pancreatic **enzymes**. At present, physiotherapy is used to get rid of the mucus that collects in the lungs and extra pancreatic enzymes are provided so that food can be digested efficiently. It is hoped that **gene therapy** may be used in the future. Researchers are investigating ways of introducing the gene for making the CFTR into the cells of affected people.

Cystic fibrosis is a **recessive** condition and the gene coding for the protein concerned has been identified on **chromosome** number 7. Because there is a one in four chance of a second child with cystic fibrosis being born to parents who have already had one affected child, it is important that such parents should receive **genetic counselling**.

cytochrome: a molecule that forms part of the **electron transport chain**. A cytochrome is a protein which is combined with another chemical group containing iron or copper. Cytochromes are found on the **plasma membranes** inside **mitochondria** and **chloroplasts**.

cytokinesis: the division of the **cytoplasm** of a cell during cell division. The process usually starts in late **anaphase**. In animal cells the **cell-surface membrane** around the equator of the cell starts to fold inwards, a process which is controlled by protein filaments in the cytoplasm. Eventually the folds meet and fuse, and two daughter cells are produced.

cytokinin: a **plant growth substance** that has several different functions. One of the most important of these is to stimulate cell division. Cytokinins are produced in the growing root tips of plants and are transported upwards to the stems and leaves.

Micropropagation of plants requires the isolation of plant cells and their growth in a suitable medium to produce a mass of similar cells called callus. If cytokinin and another plant growth substance, *auxin*, are added to the callus, roots and shoots will form. Roots form if a lot more cytokinin is added than auxin; shoots form if the amount of cytokinin is only a little more than the amount of auxin. Use of these two plant growth substances in the correct ratios is important in producing new plants by this technique.

cytoplasm: the contents of a cell outside its *nucleus*. The cytoplasm consists of a solution containing ions and smaller molecules, such as simple sugars and *amino acids*, in which larger insoluble molecules are suspended. It also contains a network of protein filaments, which provide the cell with its shape and make movement possible. *Organelles* such as *mitochondria* and *ribosomes* are suspended in the cytoplasm.

cytosine: one of the *nucleotide bases* found in *nucleic acid* molecules. Cytosine is a pyrimidine. This means that it has a single ring of *atoms* in each of its molecules. When two polynucleotide chains come together, cytosine always bonds with *guanine*. The atoms of these two bases are arranged in such a way that three *hydrogen bonds* are able to form between them.

cytoskeleton: a network of filaments and tiny tubes found in the *cytoplasm* of a cell. The cytoskeleton plays an important part in maintaining the shape and structure of the cell. It is also involved in transporting substances from one part of a cell to another and in enabling cell movement.

cytosol: the fluid which forms part of the *cytoplasm* of a cell. It consists of a solution of ions and small molecules as well as various larger, insoluble molecules.

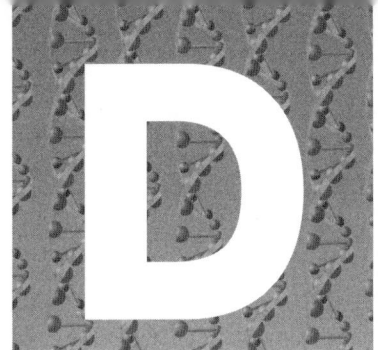

deamination: the removal of the amino (NH_2) group from an *amino acid*. If more protein is eaten than is required, the body is unable to store the excess *amino acids* that result. They are broken down in the *liver*. The amino group is removed from the molecule and converted into ammonia. Ammonia is very toxic. It can only be diluted sufficiently to be excreted safely in animals that live in water. In mammals, it is converted into a much less poisonous compound, *urea*, in a cycle of biochemical reactions called the ornithine cycle. The remainder of the amino acid molecule is either respired or converted into *carbohydrate* or *fat*.

decarboxylation: a chemical reaction in which carbon dioxide is removed. Decarboxylation occurs in the biochemical pathway of *aerobic respiration*. In the *link reaction*, the conversion of pyruvate to acetylcoenzyme A involves decarboxylation. Decarboxylation also takes place in the *Krebs cycle* where a six-carbon compound is converted to a five-carbon compound and then to a four-carbon compound.

deciduous: a plant that loses its leaves for part of the year. This process is controlled by *auxins*. A layer of cells called the abscission layer forms between the leaf and the stem. These cells are sensitive to auxins. In autumn the amount of auxin produced by the leaf falls. The cells in the abscission layer elongate and cause the leaf to drop off the plant.

decline phase: the final stage in a *population growth curve* in which there is a fall in the number of living organisms present. The reasons for this are usually either a lack of nutrients or a build-up of toxic waste products.

decomposer: a *microorganism* that breaks down the organic compounds in dead plants and animals and releases carbon dioxide and simple inorganic ions. As soon as a plant (for example) dies, its dead tissues are colonised by bacteria and fungi. Soluble sugars are rapidly broken down by the first microorganisms to arrive. Structural substances such as the *cellulose* and *lignin* that make up plant *cell walls* take much longer to decompose. Decomposers play an important part in breaking down dead plant material in *compost* heaps.

deep vein thrombosis (DVT): a condition that results when a blood clot forms and blocks one of the *veins* deep in the body, often in the lower leg. This produces symptoms of swelling and pain. People who have had surgery are most at risk from deep vein thrombosis but another *risk factor* is the immobility that occurs on long-haul flights. Deep vein thrombosis may be treated with *anticoagulants*. These prevent more blood clots forming. In more serious cases, clot-breaking drugs such as *streptokinase* may be given.

defibrillator: a device used to give an electric shock that may cause the heart to start beating normally again after its rhythm has been disrupted by *ventricular fibrillation*. The shock is given through electrodes placed on the chest.

deficiency disease: a disease caused by the lack of particular nutrients in the diet. Some examples of deficiency diseases are shown in the table.

Nutrient lacking	Deficiency disease	Explanation
Iron	*Anaemia*	Iron is a component of *haemoglobin*. Insufficient iron in the diet lowers the amount of haemoglobin present. This reduces the amount of oxygen that can be transported by the blood.
Iodine	Goitre	Iodine is a component of *hormones* produced by the *thyroid gland*. Inadequate iodine causes swelling of the neck produced by enlargement of the thyroid gland.
Vitamin A	'Night blindness'	Vitamin A is essential for formation of *rhodopsin*. This is a pigment found in *rod cells* in the eye. Lack of rhodopsin results in an inability to see in dim light.
Vitamin C	Scurvy	Essential for the formation of *collagen*, a protein important in *connective tissue*. The symptoms of scurvy reflect this and involve bleeding from the gums and the failure of wounds to heal.
Vitamin D	Rickets	Vitamin D controls the *absorption* of *calcium* from the gut and its metabolism. Without adequate calcium, bones fail to calcify and skeletal deformities result.

deflected succession: a type of ecological *succession* in which human activity prevents a *climax community* from being established. An area of grassland, for example, would normally give way to scrub and, if left long enough, to woodland. The woodland would be the climax community. If, however, grazing animals such as sheep are kept on the area, they will eat the young woody plants which will be unable to establish themselves. As long as the land is grazed by the sheep, succession will not progress beyond the grassland stage. This is a deflected succession.

degenerative disease: a disease caused by a decline in the efficiency of the body systems as a result of ageing. Examples of degenerative diseases include senile *dementia* resulting from degeneration of the *brain*, *cataracts*, which affect the *lens* in the eye, and *osteoarthritis*, which affects joints. As people live longer, so the importance of degenerative disease increases.

deletion: see *gene mutation*.

dementia: loss of mental ability as a result of a condition that affects the *brain*. People with dementia suffer from loss of memory. As the condition becomes more severe, this is accompanied by slowing down of the ability to solve problems, difficulties with speech and, in its most severe forms, abnormal patterns of behaviour. Although dementia is mainly a condition that affects the elderly, symptoms may develop in middle age. Two important causes of dementia are **Alzheimer's disease** and **stroke**.

demographic transition: a model used to describe the pattern of change in the human population as a country develops. It shows development from a mainly rural and illiterate society depending on farming to an urban, literate, industrial one. There are four main stages in the demographic transition:

1 High stationary phase: a stable population with a high **birth rate** and a high death rate.
2 Early expanding stage: increasing population with high birth rate and falling death rate.
3 Late expanding stage: increasing population with falling birth rate and low death rate.
4 Late stationary stage: stable population with low birth rate and low death rate.

dendrochronology: the study of growth rings in trees over a period of time. Many trees that grow in temperate regions produce one growth ring every year. This means that we have a pattern of ring formation for the entire period of a tree's life, which reflects the climate at the time. Warm, moist conditions favour growth. A ring formed in these conditions will be wide. A drought or a very cold year, on the other hand, will produce a narrow ring. By studying large samples and analysing the data with computers, we can draw inferences about past climates.

dendron: a long, thin process in a **neurone**. It carries **nerve impulses** towards the cell body.

denitrification: the conversion of nitrates to nitrogen gas by denitrifying bacteria. This is part of the **nitrogen cycle**. It is important in soils such as those that are waterlogged or poorly aerated. In these soils the amount of oxygen is low and denitrifying bacteria are common. Economically, these bacteria are important because they reduce the concentration of nitrate by changing it to nitrogen gas. Plants cannot use nitrogen gas directly.

density-dependent factor: a factor whose effect on a population is relatively greater at higher population densities than at lower ones. Take, as an example, caterpillars feeding on cabbage plants. If there is a high population density of caterpillars, they will eat a lot of the cabbage. The resulting lack of food will cause large numbers of caterpillars to die from starvation. On the other hand, if there is a low population density of caterpillars, food supply will have little effect on the caterpillar population. Food supply is said to be a density-dependent factor regulating population size. The higher the population density of caterpillars, the greater the effect on the food supply, and the greater the number of caterpillars that will die. Most **biotic** factors are density-dependent factors. Disease, predation and parasitism all have a relatively greater effect on populations at high population densities.

density-independent factor: a factor whose effect on a population is more or less the same, whatever the population density of the organism concerned. Many **abiotic** factors are density-independent. Heavy rain may flood the soil around a pond. It is likely to have a similar effect on the populations of small animals living there whatever their population densities.

deoxyribonucleic acid: see *DNA*.

deoxyribose: the five-carbon sugar or pentose found in *DNA*.

detritivore: an animal that feeds on pieces of partly broken down plant or animal tissue (detritus). Many soil organisms such as earthworms and woodlice are detritivores. Together with *decomposers*, they play an important part in the cycling of soil nutrients. Investigations have shown that breakdown by decomposers is much faster if detritivores are present. Since the detritivores break the tissue into smaller pieces, they expose a much greater surface area to attack and breakdown by decomposers.

diabetes: also called diabetes mellitus. A condition in which the concentration of *glucose* in the blood cannot be properly controlled. There are two forms of diabetes. In the insulin-dependent form (type 1 diabetes), the *pancreas* fails to secrete enough *insulin*. After a meal, glucose is absorbed from the *small intestine*. Not enough insulin is produced to convert the excess glucose to *glycogen* for storage in the *liver*. As a result, the blood glucose concentration rises. The concentration of glucose in the blood is now too high to be reabsorbed in the *kidneys*. Because of this, it appears in the *urine*. As little glycogen is stored, the body starts to break down *fats* and *proteins* to use as *respiratory substrates* and there is a rapid loss in body mass. The condition can be controlled effectively by carefully regulating the diet and insulin injections. The second form of diabetes (type 2 diabetes) is non-insulin dependent. This usually affects older people. They may continue to produce insulin, but it is no longer effective. This results in cells failing to take up glucose and a consequent rise in blood glucose concentration. The condition can often be controlled by careful regulation of dietary *carbohydrate*.

dialysis: a method of separating small molecules from larger ones. Suppose it is necessary to separate a mixture of *glucose* and *starch*. The solution containing both substances is put in a partially permeable bag and suspended in distilled water. The small glucose molecules are able to diffuse through the pores in the membrane of the bag into the surrounding distilled water. The starch molecules are too large and remain behind. Haemodialysis is used to treat *kidney* disease. It relies on this principle.

diaphragm: a thin sheet of muscle that separates the thorax from the abdomen. When a person breathes in, the diaphragm muscle contracts and flattens. This increases the volume of the thoracic cavity and decreases the pressure inside it. Air enters the *lungs* down the resulting pressure gradient. When a person breathes out, the diaphragm relaxes and returns to its original dome shape.

diastole: the stage in the *cardiac cycle* that involves relaxation of the heart muscle. During this stage, the *ventricles* fill with blood.

diastolic blood pressure: the blood pressure in the main arteries measured when the *ventricles* are filling (*diastole*). At this stage in the *cardiac cycle*, blood pressure in the arteries will be at its lowest. It is the pressure to which blood falls between heart beats.

dichotomous key: a method of identifying different species by comparing specific features. A dichotomous key consists of a series of paired statements. People using the key have to decide which statement refers to the specimen that they are trying to identify. This leads on to other pairs of statements and finally to the species concerned.

dicotyledon: a member of the group of *flowering* plants characterised by having two *cotyledons* in each of their seeds.

dietary fibre: a component of food that cannot be digested by *enzymes* in the course of its passage through the human gut. The most important components of dietary fibre are the non-starch *polysaccharides*. These are substances such as *cellulose* that make up plant *cell walls*. Cereals and vegetable foods are particularly good sources of dietary fibre. It is not possible to say that a lack of fibre in the diet causes a particular disease but a low fibre diet is a risk factor that has been linked with *diabetes*, heart disease and cancer of the *colon*. One of the main problems in interpreting evidence is that low fibre diets are often associated with other risk factors such as high fat intake.

dietary reference values: a set of figures that relates to the intake of particular nutrients in the human diet. The amount of a given nutrient needed by different people varies. Some people require larger than average amounts; some require less than average. The graph shows the estimated requirements for a particular nutrient in a large sample of people.

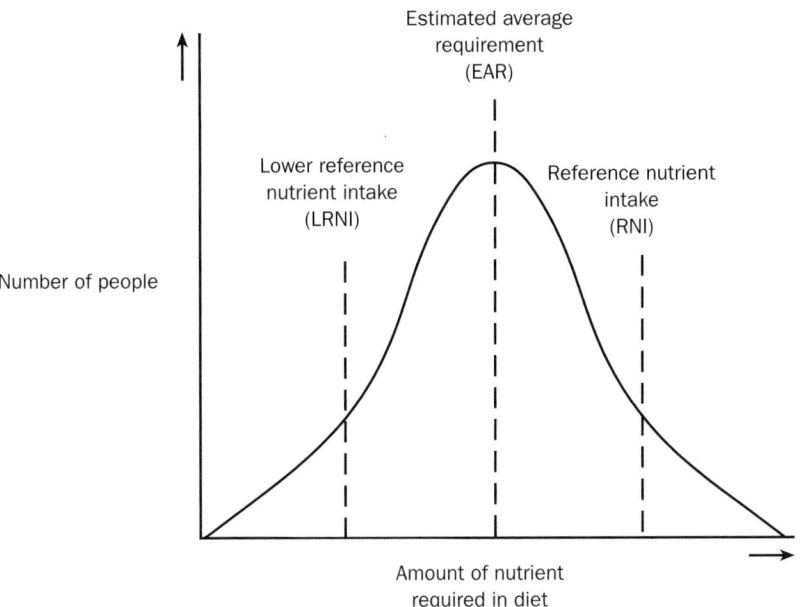

The graph is a bell-shaped or *normal distribution* curve. It allows us to fix several other points. These are:

- the Estimated Average Requirement (EAR). This is the mean value
- the Reference Nutrient Intake (RNI). This is the point that is defined mathematically as two *standard deviations* above the Estimated Average Requirement. Intakes of this amount will be adequate to meet the needs of most people
- the Lower Reference Nutrient Intake (LRNI) is defined as the point two standard deviations below the Estimated Average Requirement. An amount less than this might be adequate for some individuals in the population but, for many, it will almost certainly not be enough.

Unfortunately, the data on which such values are based are not very reliable. Because of this, care has to be taken both in interpreting them and in using them in practical situations.

differentiation: the process in which a cell becomes specialised for a particular function. In a plant root, for example, only the cells immediately behind the tip are able to divide. When a longitudinal section of a root is examined with a microscope, these cells can be identified quite easily, as various stages of *mitosis* can be seen. Once they have divided, the cells elongate and finally develop into tissues such as *xylem* and *phloem*. Differentiated cells, such as those in the xylem and phloem, no longer have the ability to divide. Undifferentiated cells like those in the root and shoot tips are used when large numbers of plants are produced from a single parent by *micropropagation*.

diffusion: the movement of *molecules* or *ions* from where they are in a high concentration to where they are in a lower concentration. In other words, particles of a substance diffuse down a concentration gradient. Since energy from *ATP* is not required for diffusion, diffusion is described as a passive process.

In a solid, the particles are fixed and cannot move relative to one another. In a liquid, the particles are free to move but are close together, so bump into each other and change direction. In a gas, the molecules travel much further before colliding with each other. In liquids and gases, the kinetic energy that the molecules possess results in them continually moving about. As this movement is random, an equilibrium is eventually reached where the molecules are spread out evenly.

Diffusion is important in the movement of substances through the *cell-surface membrane*. *Lipid*-soluble molecules dissolve in the *phospholipid* part of the membrane and diffuse readily. Molecules such as water, oxygen and carbon dioxide diffuse rapidly because of their small size. Ions, which are charged particles, and larger molecules such as *glucose* cannot pass through the phospholipid layer. Instead, they go through special channel *proteins*. The rate at which substances diffuse into and out of cells is governed by a number of different factors. These include the:

- area of the surface over which they diffuse
- difference in concentration on either side of the exchange surface
- thickness of the exchange surface.

Fick's law is a simple way of expressing this relationship.

digestion: the process in which the large insoluble molecules that make up the food of an organism are broken down to smaller, soluble molecules. The place where digestion takes place differs from organism to organism. Mammals, for example, have specialised guts, whereas amoeba engulfs its food by *phagocytosis* and then digests it once it is inside the cell. Fungi are *saprophytes*. They secrete *enzymes* on to the surface of their food, digest it and then absorb the soluble molecules that result.

In digestion, enzymes break down the individual food molecules. All digestive enzymes work in the same way. They are *hydrolases*, which means that they are able to break chemical bonds by the addition of water. This is *hydrolysis* and it is involved in the breaking down of *proteins* to *amino acids*, *disaccharides* and *polysaccharides* to *monosaccharides*, and *lipids* to *fatty acids* and glycerol.

dihybrid cross: a genetic cross between individuals in which genes at two different loci (see *locus*) are considered. It is important in explaining a cross such as this to set out the explanation clearly and logically. The example below shows you how to do this.

In peas, the height of the plant and the colour of the pods are determined by two genes. These genes are on different pairs of *chromosomes*. Pure-breeding, tall plants with green pods were crossed with pure-breeding dwarf plants with yellow pods. The F_1 plants that resulted from this cross were all tall with green pods. Show by means of a suitable genetic diagram the expected results of a cross between two of these F_1 plants.

If all the F_1 plants were tall and had green pods, then the *allele* for tall would be dominant to that for dwarf and the allele for green pods would be dominant to that for yellow pods.

These alleles may be represented by the following letters:

T	tall	G	green pods
t	dwarf	g	yellow pods

Parental phenotypes: tall, green pods   tall, green pods

Parental genotypes: TtGg   TtGg

Gametes: TG Tg tG tg   TG Tg tG tg

Offspring genotypes:

		Male gametes			
		TG	Tg	tG	tg
	TG	TTGG	TTGg	TtGG	TtGg
Female	Tg	TTGg	TTgg	TtGg	Ttgg
gametes	tG	TtGG	TtGg	ttGG	ttGg
	tg	TtGg	Ttgg	ttGg	Ttgg

Offspring phenotypes:

Tall, green pods:	9
Tall, yellow pods:	3
Dwarf, green pods:	3
Dwarf, yellow pods:	1

dilution plating: a method of counting the number of live bacteria in a culture. Suppose a sample of the culture was poured on an *agar* plate and incubated for a suitable time. Colonies of bacteria would appear on the surface of the agar. Each of the colonies would have come from a single bacterium. By counting the colonies, the number of live bacteria in the original sample could be found.

In practice, there would be so many bacteria present in the culture that it would not be possible to do this. It would be necessary to dilute the sample first. A 1 cm³ volume is taken from the culture and diluted by adding 9 cm³ of distilled water. After mixing thoroughly, 1 cm³ of the diluted suspension is removed and further diluted with distilled water. This process is continued, resulting in a series of dilutions, each one-tenth of the concentration of the previous one. A sample is taken from each dilution and poured on a separate agar plate. The plate on which the number of colonies is small enough to count but large enough to give accurate results is taken. The colonies are counted and the result is multiplied by the dilution factor to give the number of live bacteria in the original sample.

diploblastic: an animal body pattern in which there are two layers of cells. The outer layer is called the ***ectoderm*** and the inner layer is called the ***endoderm***. Together, the two layers surround a gut cavity or ***enteron***. Animals that belong to the phylum ***Cnidaria***, which includes jellyfish, sea anenomes and corals, have this basic body structure. Most other animal phyla are ***triploblastic*** and have a body pattern based on three layers of cells.

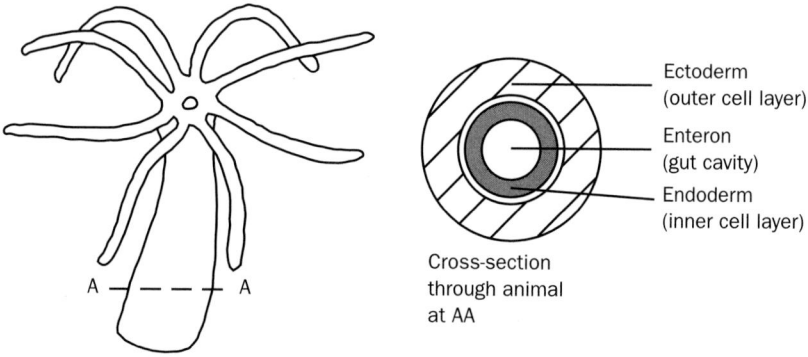

Ectoderm (outer cell layer)

Enteron (gut cavity)

Endoderm (inner cell layer)

Cross-section through animal at AA

A diploblastic animal

diploid: a term that can refer to cells, organisms or stages in the life cycle in which the nuclei contain two copies of each ***chromosome***. The diploid number of chromosomes is variable and differs from one species to another. In humans, this number is 46, in pea plants it is 14, and in fruitflies it is 8.

Of the 23 pairs of chromosomes that are found in each body cell in humans, 22 pairs are found in both males and females. They are called ***autosomes***. The remaining pair, the ***X chromosomes*** and ***Y chromosomes***, are the ***sex chromosomes***. When ***gametes*** are formed, the process of ***meiosis*** results in male and female gametes each having half the number of chromosomes present in a body cell. In other words, the gametes are ***haploid***.

directional selection: *selection* that operates against one extreme in a range of variation. The graph shows variation in hair length in house mice.

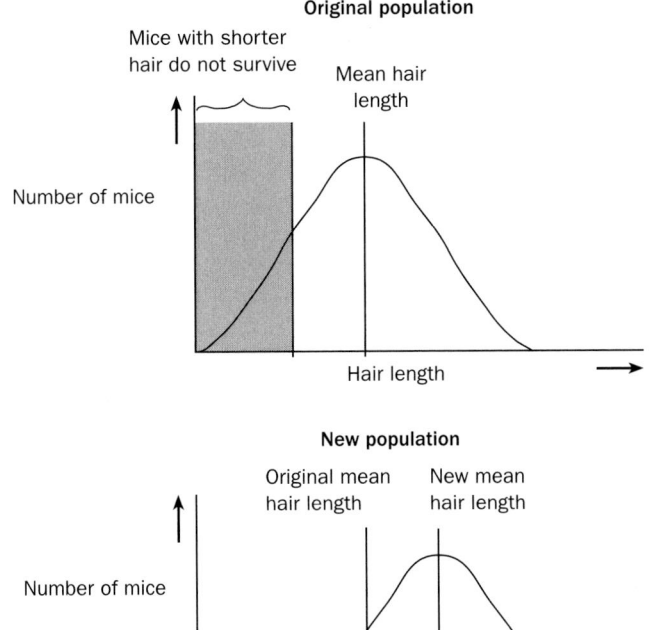

Original population

Mice with shorter hair do not survive

Mean hair length

Number of mice

Hair length

New population

Original mean hair length

New mean hair length

Number of mice

Hair length

Directional selection

House mice are common animals that live in many different places. Some populations of house mice are able to live in cold stores. In cold stores, there will be selection against mice with short hair. Mice with long hair will be at an advantage and will be selected for. Over a period of time, directional selection has resulted in an evolutionary change. The mean hair length of house mice living in cold stores is now longer than the mean hair length of house mice living in other places.

disaccharide: a *carbohydrate* whose molecules contain two sugar units. *Maltose* is an example of a disaccharide. It is made from two *glucose* molecules joined together by a reaction in which a molecule of water is removed. This is known as **condensation** and produces a bond between the two glucose molecules called a glycosidic bond. The reaction also goes in the reverse direction. If a molecule of water is added to a maltose molecule, two molecules of glucose are produced. This is *hydrolysis* and happens during the final stages in the *digestion* of maltose.

discontinuous variation: *variation* in which individuals fall into distinct categories. This is in contrast to **continuous variation** in which there is a complete range of measurements from one extreme to the other. The graph overleaf shows figures relating to the number of petals in a sample of celandine *flowers*.

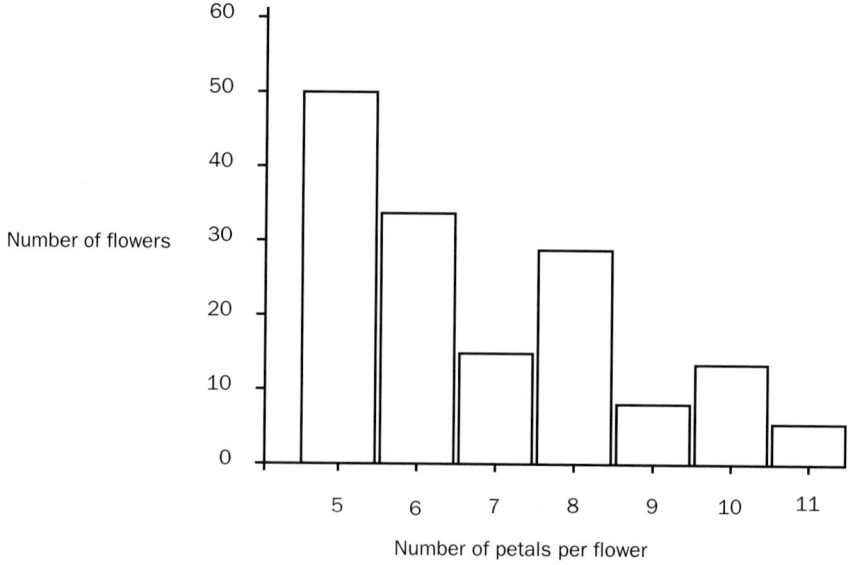

Discontinuous variation: the number of petals in celandine flowers

Discontinuous variation is based on features that are not measured. They are either present or they are not present. In this example, celandine flowers may have any number of petals from five to eleven, but they have either one number or another. There are no intermediate categories. Where there is a pattern of discontinuous variation like this, it often shows a characteristic controlled by a small number of genes or even a single gene. It is also likely that the **environment** only plays a small part in determining the final pattern of variation.

disease: a disorder that affects an organism. It has a specific cause and can be recognised by particular signs and symptoms. We can group human diseases into two broad categories. Some are caused by **microorganisms** such as bacteria and viruses. As these diseases can be passed from person to person, they are called **infectious diseases**. Non-infectious diseases, on the other hand, are not caused by microorganisms. They may result from the genes that we inherited from our parents or they may be linked with our lifestyle and associated with factors such as diet, smoking and lack of exercise.

disinfectant: a substance that is used to kill **microorganisms** that are outside the body. Disinfection kills most of the harmful bacteria but it does not necessarily kill all microorganisms. It is enough, for example, to disinfect a thermometer before taking a person's temperature. Surgical instruments, however, need to be sterilised before an operation. **Sterilisation** kills all microorganisms.

diversity: see **species diversity**.

diverticulosis: the presence of pouches in the wall of the **large intestine**. Pressure in the intestine results in the inner wall pushing out through the muscle layer to form a pouch. Diet may be partly responsible for the condition, as it is rare in parts of the world where the diet has a high fibre content. Diverticulosis is most frequent in people living in more developed countries where processing of food removes much of the fibre. When the pouches or diverticula become inflamed, the condition is known as diverticulitis.

division of labour: the adaptation of different parts of an organism to carry out different functions. Even within a simple one-celled organism like amoeba, different *organelles* have different roles. Regulation of the water content of the cell is carried out by a *contractile vacuole*; the *metabolic pathways* associated with *respiration* take place in the *mitochondria*, and *lysosomes* release the *enzymes* necessary for the *digestion* of food particles that have been taken into the cell. Larger organisms have many cells. These cells are grouped together to form *tissues*, the tissues form *organs* and the organs form *systems*. Each system is adapted to carry out a specific function within the organism.

One of the evolutionary trends that is apparent in the animal *kingdom* is that the more advanced the animal, the greater the division of labour. Hydra is a small animal belonging to the phylum *Cnidaria*. It has a sac-like body consisting of just two layers of cells that enclose a cavity, the *enteron*. The enteron does several things. It functions as a simple gut as food is taken in and the process of digestion is started here. Oxygen diffuses from the water in the enteron into the surrounding cells, so the enteron is also associated with gas exchange. Finally, it acts as a simple *hydrostatic skeleton*, providing support for the organism. Thus the enteron is a single part of the hydra, with a number of different functions. In a mouse, on the other hand, these functions are all carried out by different systems. There are separate systems for digestion, gas exchange and support.

DNA: an information-carrying molecule that forms the genetic material in living organisms. A molecule of DNA consists of two *polynucleotide* strands. One strand runs in one direction, the other runs in the opposite direction. Each strand has a sugar-phosphate backbone. One of four bases – *adenine, cytosine, guanine* or *thymine* – is attached to each sugar in this backbone. These bases are joined by *hydrogen bonds* formed between the adenine in one strand and the thymine in the other, and between the cytosine in one strand and the guanine in the other. The whole molecule is twisted to form a double helix with one complete turn of the spiral for every ten *base pairs*. In animals and plants, DNA is an important component of *chromosomes*, but it is also found in *mitochondria* and *chloroplasts*.

A sequence of three bases, called a triplet, codes for a specific amino acid. Genes are sections of a DNA molecule that code for particular polypeptides. The base sequence of a gene determines the amino acid sequence in a polypeptide.

During *protein synthesis*, another *nucleic acid, mRNA*, is produced which takes the gene code from the DNA in the nucleus of the cell to the *cytoplasm* where *amino acids* are assembled in the right order to produce the required polypeptide.

During the *cell cycle, DNA replication* occurs. A DNA molecule is copied accurately and an exact copy is made that is passed on to genetically identical daughter cells.

DNA helicase: an *enzyme* involved in *DNA replication*. Replication starts when the two polynucleotide strands that make up the *DNA* molecule start to unwind. The *hydrogen bonds* between complementary *base pairs* break and the strands move apart. DNA helicase is involved in breaking the hydrogen bonds when the two DNA polynucleotide strands separate.

DNA hybridisation: a method used to compare the *DNA* of different *species*. Biologists make a hybrid DNA molecule. This consists of a *polynucleotide* strand from one species joined to the complementary strand from another species. The more closely the two species are related, the greater the number of **base pairs** that will be complementary. The greater the number of complementary base pairs, the more **hydrogen bonds** that will form between them and the more strongly the two strands will be held together. More heat is required to separate strands joined together with many hydrogen bonds than is required to separate strands that are joined together with few hydrogen bonds. By measuring the temperatures at which different hybrid molecules of DNA are split apart, we can see how closely different species are related.

DNA ligase: see *ligase*.

DNA polymerase: an enzyme that catalyses the chemical reaction in which DNA *nucleotides* join together by **condensation** to produce a molecule of DNA. DNA polymerase plays an important part in **DNA replication**. Biologists also make use of DNA polymerase in the **polymerase chain reaction (PCR)**. This is a process used to make large amounts of identical DNA from very small samples.

DNA probe: a single strand of **DNA** used to identify a particular gene. A DNA probe has a base sequence that is complementary to the base sequence on part of one of the DNA chains in the **gene** to be identified. The DNA that is being tested is separated into its two chains. The separated DNA chains are then mixed with the probe. The probe binds to the complementary bases on one of the chains. Because the probe is usually made from *nucleotides* containing radioactive phosphorus, the site at which it binds may be identified.

DNA replication: the process by which a **DNA** molecule can produce two exact copies of itself. DNA replication is an important part of the **cell cycle**. It takes place in interphase. The flow chart summarises the basic steps involved in DNA replication.

The DNA molecule unwinds, the hydrogen bonds break and its two strands separate. This process involves enzymes and other proteins

$\downarrow$

Each of the original strands acts as a template for the formation of a new strand. The bases of free nucleotides in the cell line up against the complementary bases on each of the original DNA chains

$\downarrow$

The nucleotides are now joined together by enzymes forming two new DNA molecules, each formed from one of the original strands which provided a template for the formation of a new one. This is known as semi-conservative replication

DNA normally replicates very accurately and cells have a number of repair mechanisms that correct errors in replication. Nevertheless, uncorrected errors do occasionally happen and these are the causes of **gene mutations**.

There are some important differences between DNA replication and the process of *transcription* which takes place during *protein synthesis*. These are summarised in the table.

Feature	DNA replication	Transcription
Molecule or molecules produced	Two molecules of DNA	Single molecule of *mRNA*
Strands of DNA involved	Both strands of DNA, each producing a new DNA molecule	One strand of DNA, the coding or template strand, producing an mRNA molecule
Amount of DNA involved	Whole DNA molecule unwinds and separates	Only part of the DNA molecule unwinds and separates

domain: a term used in a recently devised system of classifying organisms. The three-domain system is based on differences in the nucleotides that make up ribosomal RNA (*rRNA*). Ribosomal RNA has the same function in all organisms. Because of this, its structure has changed little over time. Similarities and differences between rRNA are therefore good indicators of how closely different organisms are related. The three-domain system proposes three domains:
- archaea or primitive bacteria
- bacteria or typical bacteria
- eukarya, which contains all other organisms. The eukarya is divided into four *kingdoms*: *Protoctista*, *Fungi*, *Animalia* and *Plantae*.

dominant: an *allele* is described as dominant if it is always expressed in the *phenotype* of an organism. In peas, the allele for green pods, G, is dominant to that for yellow pods, g. Because the allele for green pods is dominant, pea plants with either the *genotype* Gg or the genotype GG will have green pods.

dopamine: a *neurotransmitter* found in the *brain*. It is associated with the pleasure system of the brain and is thought to provide feelings of enjoyment. Dopamine is also involved in the control of motor activity. People with *Parkinson's disease* do not synthesise sufficient dopamine. As a result they are unable to carry out smooth, controlled movements.

dorsal root: part of a *spinal nerve* that joins with the *spinal cord*. The spinal nerves contain both sensory and motor *neurones*. Just before each nerve joins with the spinal cord, it splits in two producing a dorsal root and a *ventral root*. The dorsal root contains the *sensory neurones* that carry *nerve impulses* from the receptors to the spinal cord. There is a small swelling called a *ganglion* present on the dorsal root. This contains the cell bodies of the sensory neurones.

double circulation: a blood system in which the blood passes through the heart twice in its passage round the body. In mammals, blood flows from the right side of the heart to the *lungs*. After it has become oxygenated, it returns to the left side of the heart from where it is pumped to the rest of the body. A complete circuit therefore involves passing through the right side and through the left side of the heart. The advantages of a circulation like this are that:
- oxygenated and deoxygenated blood are kept completely separate. They are not mixed. Because of this, the blood supplying the tissues is always saturated with oxygen

- blood is supplied to the tissues at a much higher pressure than is the case in a single circulatory system. This results in a very efficient circulation.

Down syndrome: a human condition that results from three copies of **chromosome** 21, one more than normal. During the first division of **meiosis**, the chromosomes belonging to pair number 21 fail to separate. This produces **gametes** containing an extra chromosome. When **fertilisation** takes place, a zygote will be produced with 47 chromosomes instead of the usual 46. Abnormalities arising from the failure of chromosomes to separate properly during meiosis become more common with age and there is a **correlation** between the age of the mother and the probability of having a child with Down syndrome.

Down syndrome is an example of a **chromosome mutation**. Chromosome mutations involve changes in large amounts of genetic material and many of them are either fatal or produce individuals who are severely disabled. However, although people with Down syndrome show a number of distinctive physical features, such as a characteristic facial shape and stature, and have limited mental development, they are able to lead relatively normal lives.

downstream processing: the name given to the steps that follow the biological part of a biotechnological process (see **biotechnology**). Downstream processing results in a pure concentrated product. There are three basic steps involved:
- separation of the crude product from the various other products, **substrates**, **enzymes** or **microorganisms** that may be present in the mixture
- concentration of the product to remove the large amount of water usually present
- purification of the crude extract to produce a single product.

Duchenne muscular dystrophy: an inherited condition that involves progressive wasting and weakening of muscles. The **allele** for this condition is **recessive** and is found on the **X chromosome**. It is therefore **sex-linked** and is almost completely confined to males.

ductus arteriosus: a **blood vessel** found in the fetus. It links the **pulmonary artery** directly to the **aorta**. While a fetus is developing inside the uterus, its oxygen supply comes from its mother via the **placenta**. Its **lungs** do not function. Blood entering the right **atrium** of the heart either passes through a hole called the **foramen ovale** into the left atrium and on into the aorta, or it goes from the pulmonary artery through the ductus arteriosus into the aorta. At birth, the foramen ovale and the ductus arteriosus both close. Blood then flows from the right side of the heart to the lungs.

duodenum: the first loop of the **small intestine**. When a ring of muscle called the pyloric sphincter relaxes, partly digested food flows from the **stomach** into the duodenum. Alkaline mucus is secreted by **glands** in the wall of the duodenum. This neutralises the partly digested food and provides the correct pH for **enzymes** to function. Two important secretions are added to the partly digested food in the duodenum. These are **bile** from the **gall bladder** and digestive enzymes secreted by the **pancreas**. The inner surface of the duodenum, like that of the rest of the small intestine, has many villi on its surface. These provide a large surface area for the **absorption** of the products of **digestion**.

DVT: see **deep vein thrombosis (DVT)**.

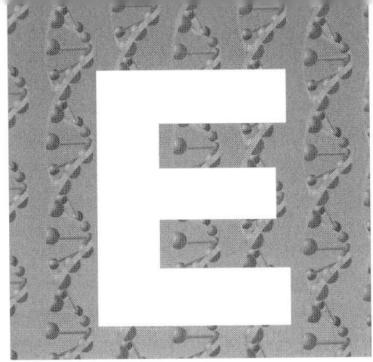

ECG: see *electrocardiogram (ECG)*.

Echinococcus granulosus: a species of tapeworm that normally lives in the intestines of dogs. If a human accidentally ingests eggs of this parasite, they may hatch and form large balloon-like cysts in organs such as the *liver*, *brain* or *lungs*.

ecological pyramids: see *pyramids of number, biomass and energy*.

ecosystem: an ecological unit that includes all the organisms living in a particular area, as well as the *abiotic* features of their *environment*. Ecosystems can be of very different sizes. The desert ecosystem of North Africa, for example, stretches for many thousands of kilometres. A bird's nest could just as well be considered as an ecosystem inhabited by large numbers of insects and *microorganisms*. Although ecosystems are usually thought of as being self-contained, they influence each other considerably. A migrating bird, for example, may fly between ecosystems in different continents, whereas pollution of farmland in Europe and North America has resulted in increased amounts of pesticides in the bodies of Antarctic penguins.

ecotourism: tourism that involves travel to areas where wildlife is the main attraction. Ideally, ecotourism should:

- conserve the *biodiversity* of the area concerned and promote its sustainable use
- minimise the impact on the environment, both directly and indirectly, such as by avoiding wastage associated with unnecessary luxury
- involve the local community, providing a positive experience for visitors and hosts.

ectoderm: the outer layer of cells in a *triploblastic* animal. Cells in this layer form the animal's *epidermis* and nervous system.

ectoparasite: an organism that lives on the surface of a host organism. The parasite gains a nutritional advantage from this relationship, but the host suffers a disadvantage. Head lice are ectoparasites.

ectotherm: an animal that makes use of the *environment* to regulate its body temperature. The core body temperatures of many ectotherms are therefore similar to those of the environments in which they live. Some ectotherms, however, are able to maintain their core body temperatures within very narrow limits. One of these is the crocodile. A crocodile regulates its temperature by moving between water and land. At night, when the air temperatures are low, the animal stays in the warmer water. Nevertheless, its body temperature gradually falls. With daylight, it moves up on to the shore and basks in the sun. As a result of this, its temperature rises again. During the middle of the day, when it can be very hot, it returns to the cooler water. A crocodile is able to maintain its own temperature between remarkably narrow limits by making use of differences in air and water temperatures.

edaphic factor: an *abiotic* factor that influences the distribution of organisms. Edaphic factors relate to the physical and chemical composition of the soil. They include soil pH and the availability of nutrients such as nitrates and phosphates.

effector: when stimulated by a *nerve impulse,* an effector brings about a response or a change. In animals, *muscles* and *glands* are effectors.

efferent: means coming from. In a *reflex arc*, for example, the efferent neurone transmits a *nerve impulse* from the *spinal cord*. In the *kidney*, the efferent *arteriole* takes blood from the *glomerulus*.

egestion: the removal of waste products from the body as *faeces*. The *lumen* of the gut forms a cavity that runs through the body, so digested food that is in the lumen of the gut is not actually inside the body. Only when this digested food has been absorbed does it enter the body itself. Much of the material present in the faeces of a mammal is undigested food and has, therefore, never actually been inside the body. Because of this, we should not refer to defaecation as *excretion*; we should use the word egestion instead.

elastin: a protein that forms long, flexible fibres. These fibres are found in the skin, in *ligaments* and in the walls of arteries where their elastic properties are important.

electrocardiogram (ECG): a record of the electrical events associated with the beating of the heart. Electrodes are attached to the skin on the chest and the limbs. These are connected to the recording apparatus. When the heart beats, a wave of electrical activity passes over its surface, producing an electrocardiogram such as that shown in the diagram.

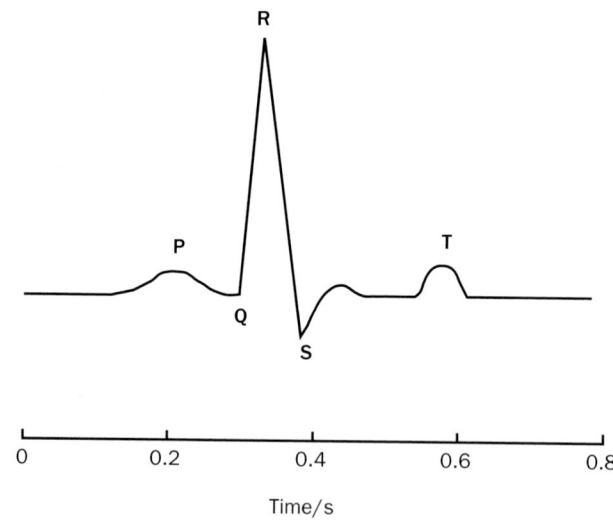

Time/s

An ECG from a healthy heart

- The part labelled P represents the spread of electrical activity from the *sinoatrial node (SAN)* over the surface of the atria.
- The part labelled QRS represents the passage of electrical activity over the *ventricles*.
- The part labelled T represents the electrical changes that occur during the filling of the heart.

Although electrocardiograms can be used to investigate heart function in healthy people, the information they contain is also used to monitor the hearts of patients recovering from heart attacks and to investigate other forms of heart disease.

electron: a negatively charged particle that orbits the nucleus of an *atom*. The possession of electrons is important in allowing atoms to form chemical bonds.

electron microscope: a type of microscope that uses a beam of *electrons* rather than a beam of visible light. Because the wavelength of electrons is much smaller than the wavelength of visible light, an electron microscope not only gives a high *magnification* but it also has high *resolution*. This means that details can be seen clearly. An electron microscope is similar to a light microscope in the way in which it works but, instead of using glass lenses to focus a beam of light, it uses magnets to focus a beam of electrons.

We cut very thin sections through the material we are going to examine. These sections, or specimens, are then put into a sealed chamber in the microscope. The air is sucked out of the chamber and a vacuum is produced. Technicians carry out many chemical processes in preparing the specimen and these processes may change its appearance considerably. Features that result from the techniques used in preparing the specimen are called *artefacts*. Care needs to be taken in interpreting electron micrographs because of the possible presence of artefacts. The main differences between a light microscope and an electron microscope are shown in the table.

Feature	Light microscope	Electron microscope
Source of illumination	Visible light	Beam of electrons
Method of focusing	Glass lenses	Magnets
Specimens that can be examined	Live or dead	Dead
Magnification	Student microscopes magnify up to about 400 times but the maximum magnification that can be achieved with a light microscope is about 1500 times	Up to 500 000 times
Resolution	About 0.2 μm	About 1 nm or 0.001 μm

There are two main types of electron microscope: the *transmission electron microscope* and the *scanning electron microscope*.

electron transport chain: a series of protein molecules along which *electrons* are passed. This releases energy that is used to produce *ATP*. In *respiration*, oxidation reactions take place in the *Krebs cycle*. Hydrogen is released in this process. It acts as a source of protons and electrons, as each hydrogen atom consists of one proton and one electron. The electrons are passed from molecule to molecule along the electron transport chain. At each transfer, a small amount of energy is released. This is used to pump the protons through the mitochondrial membrane in which the carrier molecules are found. This produces a proton gradient with more protons on one side of the membrane than on the other. The proton gradient acts as a store of potential energy. When the protons return through the membrane, they release this energy which is then used to produce ATP. The protons and electrons finally combine with oxygen to produce water. The carrier molecules that make up the electron transport chain in respiration are found on the inner membranes of the *mitochondria*.

A similar system produces ATP in the *light-dependent reaction* of *photosynthesis*. Here, *chlorophyll* provides the electrons, while the protons come from water molecules.

electrophoresis: a method of separating molecules based on differences in their electrical charge. The substances to be separated are placed in a buffer solution on a layer of gel or a sheet of filter paper. Electrodes are arranged so that a direct current is passed through this medium. As a result, each of the substances to be separated has a different charge and moves a different distance. The separated substances are found in bands whose position can be shown by using a suitable *stain*.

Genetic fingerprinting involves separating *DNA* fragments of different size. This is done by gel electrophoresis and relies on the fact that the distance a particular fragment of DNA moves is proportional to its mass.

ELISA technique: a sensitive method used by biochemists for measuring the concentration of a particular substance. It stands for enzyme-linked immunosorbent assay. The term can be explained by looking at these words in reverse order:

- An assay is a test used to measure the amount of a particular substance.
- Immunosorbent refers to the fact that the test is based on an immune reaction involving *antibodies*. Antibodies are specific and a particular *antibody* will only bind with a molecule of a specific substance. In this technique, an antibody is used which is specific to the molecules of the substance being measured. The antibodies form complexes with these molecules. The more molecules of the substance present, the greater the number of antibody complexes that are formed.
- The number of antibody complexes can then be determined by using an *enzyme*. This enzyme catalyses an easily measured reaction involving the test substance. Hence, the process is described as being enzyme-linked.

embryo: a young animal in the later stages of development. In humans, the term embryo is usually used to refer to the two months following conception. After two months, once it is becoming recognisable as a human, it is called a fetus.

embryo sac: a large cell found in the ovary of a *flowering* plant, which contains a number of nuclei including the female *gamete* or egg nucleus. Inside the ovule of an immature plant is a single cell. This cell divides by *meiosis* to produce four *haploid* daughter cells. Three of these cells disappear. The other one undergoes a number of mitotic divisions to produce the embryo sac, containing eight nuclei. At *fertilisation*, two of these nuclei – the polar nuclei – will fuse with one of the male gametes to become the *endosperm*. Another one, referred to as the egg nucleus, fuses with the other male gamete to become the zygote. This is summarised in the diagram.

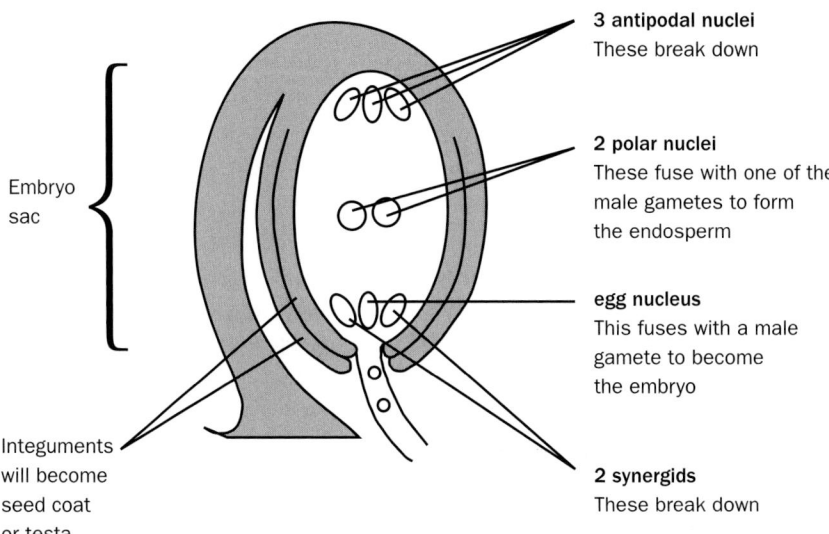

3 antipodal nuclei
These break down

2 polar nuclei
These fuse with one of the male gametes to form the endosperm

egg nucleus
This fuses with a male gamete to become the embryo

Embryo sac

Integuments will become seed coat or testa

2 synergids
These break down

m

The embryo sac and its development

emphysema: a lung disease in which the walls of the *alveoli* or air sacs break down. This causes the alveoli to enlarge. As a result, the total area of their walls is much less and gas exchange becomes less efficient. Severe emphysema results in breathlessness and the only way to treat a patient may be by providing oxygen. Although it is not understood how emphysema occurs, it is often associated with cigarette smoking and chronic *bronchitis*.

emulsion test: a *biochemical test* used to show the presence of a lipid.

endemic: a term used to describe a disease that occurs frequently in particular areas or among the members of a particular population. *Malaria*, for example, can be described as being endemic to many tropical countries. Although cases of malaria occur among people returning to the UK from the tropics, the disease is not endemic to the UK. People living in areas where a disease is endemic have often evolved adaptations that make them less susceptible to its effects. One such example is the occurrence of the sickle-cell *allele* (see *sickle-cell disease*) in parts of Africa where malaria is endemic. In contrast, diseases such as measles can have devastating effects if introduced to isolated populations, such as native Americans living in the Amazon rainforests. The term can also be used to refer to other organisms whose distribution is restricted to specific areas.

endocrine gland: a *gland* that secretes a *hormone*. Many glands have ducts through which they release the substances they secrete. Endocrine glands are different. They secrete hormones directly into the blood. They do not have ducts and are often called ductless glands. Endocrine glands may have several different functions. The *pancreas*, for example, acts as an endocrine gland because it secretes the hormones *insulin* and *glucagon* directly into the blood. It also secretes digestive juices through a duct into the *small intestine*.

endocrine system: the system in the body responsible for the production of *hormones*. *Endocrine glands* are distributed throughout the body. In addition, many organs which have other important functions are also part of the endocrine system. The *kidneys*, the *stomach*, the *ovaries* and the *testes* all produce hormones and are, therefore, part of the endocrine system.

endocytosis: the transport of large particles or fluids through the ***cell-surface membrane*** into the ***cytoplasm*** of a cell. The cell-surface membrane surrounds the particles concerned. A ***vesicle*** or ***vacuole*** is formed, which is pinched off and moves into the cytoplasm. The term phagocytosis is generally used where solid particles are involved. Pinocytosis refers to the cell taking in small droplets of fluid.

endoderm: the innermost layer of cells in a ***triploblastic*** animal. Cells in this layer form lining of the animal's gut and the digestive ***glands***.

endodermis: a ring of cells between the outer part of the root and the vascular tissue in the centre. A band of waterproof material called the ***Casparian strip*** runs round the walls of each of these endodermal cells. It prevents water and dissolved substances from going through the ***cell walls*** and intercellular spaces into the vascular tissue. As a result, these substances pass through the ***cytoplasm*** of the endodermal cells, which can control their movement into the ***xylem***.

endometriosis: a condition where tissue similar to the inner lining layer of the ***uterus*** is found in other organs in the pelvic area. This lining tissue may be found in the ***ovaries***, the ***oviducts*** and even in the wall of the ***large intestine***. It undergoes similar cyclic changes to the lining of the uterus and may cause considerable pain.

endometrium: the inner lining layer of the uterus. During the ***menstrual cycle***, the endometrium goes through a series of changes. In the first part of the cycle, the ***hormone oestrogen*** stimulates this lining layer to develop and thicken, preparing it for a possible pregnancy. Following ovulation, a second hormone, ***progesterone***, is secreted by the ***corpus luteum***. This maintains the endometrium. If pregnancy does not occur, the progesterone concentration falls and the lining is lost during menstruation.

endoparasite: an organism that lives inside the body of a host organism. The parasite gains a nutritional advantage from this relationship, but the host suffers a disadvantage. Tapeworms, ***roundworms*** and flukes are endoparasites. Endoparasites usually show more adaptations than ***ectoparasites*** to their parasitic way of life. Their life cycles often involve more than one host organism.

endopeptidase: a protein-digesting enzyme. It breaks peptide bonds in the middle of a polypeptide chain, rather than the peptide bonds at the ends of the chain. As a result, endopeptidases produce smaller polypeptides rather than individual ***amino acids***. This is shown in the diagram.

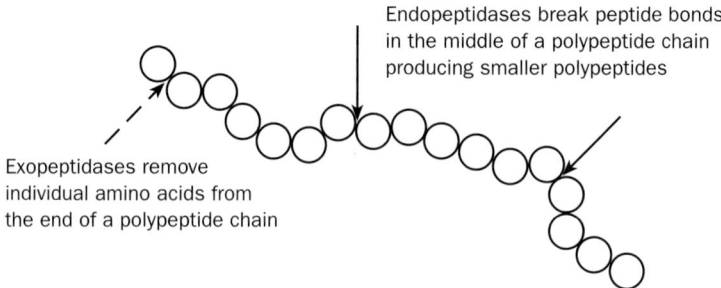

Endopeptidases break peptide bonds in the middle of a polypeptide chain producing smaller polypeptides

Exopeptidases remove individual amino acids from the end of a polypeptide chain

The action of endopeptidases

Pepsin and ***trypsin*** are endopeptidases found in the guts of mammals. It is more efficient to break a long polypeptide chain into shorter lengths before starting to remove individual amino acid molecules from its ends. Endopeptidases are therefore secreted and act early in the digestive process.

endoplasmic reticulum: a network of membranes found in the *cytoplasm* of a cell. It consists of a complex system of membranes enclosing flattened, fluid-filled spaces. In rough endoplasmic reticulum, these membranes may be covered with *ribosomes*. The ribosomes produce *proteins* that are transported through the spaces between the membranes. Smooth endoplasmic reticulum does not have ribosomes on its surface. Its main function is the production and transport of *lipids*.

endoscopy: a medical technique that involves the examination of the inside of body organs using an instrument called an endoscope. An endoscope is a long tube with a light at one end and a means of displaying an image at the other. Many modern endoscopes rely on fibre optics, so they are small in diameter and flexible. They enable detailed examination of the inside of an organ without the need for extensive surgery. Endoscopes are used for a variety of purposes, such as inspecting the external ear and eardrum for signs of damage or infection, or for looking at the inside of the gut.

endoskeleton: a skeleton that forms an internal supporting system for the body. It is the type of skeleton found in mammals. In contrast to an *exoskeleton*, such as that in arthropods, the muscles are attached to its outer surfaces. The skeleton of a mammal is composed of two materials: *cartilage* and *bone*. These are living tissues and are capable of growth as the animal gets older and increases in size. There is no need for the complex process of moulting, which is associated with the possession of an exoskeleton.

endosperm: a tissue that supplies nutrients for the developing embryo in a seed. During *fertilisation* in a *flowering* plant, one of the male nuclei from the pollen grain fuses with two of the nuclei in the ovule. The resulting nucleus contains three sets of *chromosomes*. It is triploid. This nucleus divides rapidly and gives rise to a mass of large cells that surround and nourish the developing embryo. In the seeds of *dicotyledons* such as peas and beans, the endosperm has been entirely absorbed by the time the seed has developed fully. In *monocotyledons* such as maize and onion, there is still a lot of endosperm present in the mature seed. In these species, it acts as a food store for the young plant during *germination*.

endotherm: an animal that maintains its body temperature by using physiological mechanisms. Mammals and birds are endotherms. The diagram summarises the way in which human body temperature is controlled.

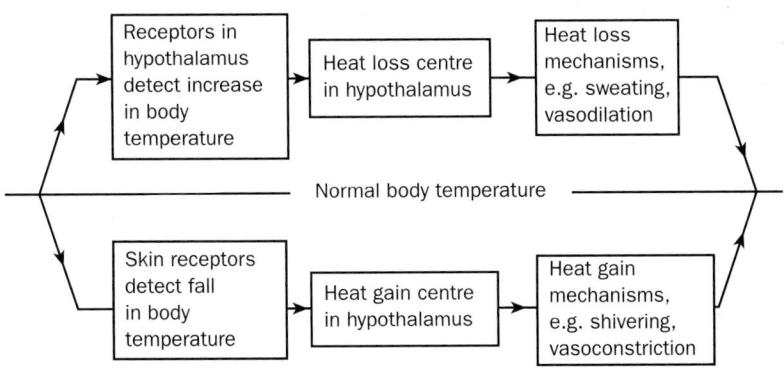

Control of body temperature in humans

A low temperature, usually the result of being in cold conditions, is detected by receptors in the skin. Information is sent by **nerve impulses** to the temperature control centre in the **hypothalamus** of the **brain**. The temperature control centre coordinates the responses that lead to the body producing more heat and limiting the amount of heat that it loses. If the body temperature rises, usually as a result of increased activity, there is an increase in blood temperature. This is detected by receptors in the hypothalamus itself and stimulates responses such as increased sweating and **vasodilation**, which lead to loss of heat. Temperature control in an endothermic animal provides a good example of **negative feedback**. A change in body temperature leads to the operation of various mechanisms which return the temperature to its resting value.

endotoxin: a poisonous substance or toxin contained inside the cells of certain bacteria. When the bacteria die and **cell walls** break down, the endotoxin is released and affects the tissues of the host. Endotoxins are produced by many bacteria including species of **Salmonella** and *Escherichia*. Whichever species is involved, the symptoms produced by the presence of endotoxins are similar and include fever, diarrhoea and damage to small **blood vessels**.

enterokinase: an enzyme secreted by cells in the upper part of the **small intestine**. *Trypsin* is an enzyme that digests **proteins** in the small intestine. It is secreted in an inactive form called **trypsinogen**. Enterokinase converts trypsinogen to trypsin.

enteron: the cavity enclosed by the body wall of an animal belonging to the phylum *Cnidaria*. It has a single opening that acts as both a mouth and an **anus**. The enteron has a number of functions other than acting as a digestive organ. Oxygen is exchanged between the water in the enteron and the cells that make up the inner layer of the body wall. In addition, as it is filled with water, it can act as a **hydrostatic skeleton**.

environment: the conditions surrounding an organism. These conditions can be divided into two main groups. **Abiotic** factors are concerned with the non-living part of the environment. Rainfall, soil pH, temperature and humidity are all examples of abiotic factors that affect the distribution of various organisms. **Competition** and **predation**, on the other hand, are concerned with the living part of the environment and are called **biotic** factors.

environmental impact assessment: an evaluation of the possible effect that a proposed development may have on the natural environment. It is meant to ensure that planning authorities take environmental issues into account before they give approval for the development to proceed.

environmental resistance: any condition that prevents a population increasing at its maximum rate.

enzyme: a protein that catalyses biochemical reactions. Without enzymes, many of the reactions that take place in living organisms would be very slow. Enzymes are vital in allowing these reactions to take place rapidly in the conditions that are found inside living cells. There are many different enzymes. A single mammalian cell, for example, may have as many as 2000 different chemical reactions taking place inside it. Each of these will be catalysed by a different enzyme. They all work in a similar way, however, by lowering the **activation energy** necessary to start the reaction. As less energy is necessary, biochemical reactions can take place at the temperatures and pressures found in living cells. The way in which an individual enzyme molecule works may be explained by referring to models, such as the **lock and key model** and **induced fit model**.

enzyme–product complex: formed when the products are bound with the *active site* of the *enzyme* controlling the reaction. In an enzyme-controlled reaction, a *substrate* molecule binds with the active site of the enzyme to form an *enzyme–substrate complex*. The reaction now takes place and the product molecules are formed. They remain bound in the active site as an enzyme-product complex. The product molecules are then released and the enzyme is able to accept another substrate molecule.

enzyme–substrate complex: formed when a *substrate* molecule binds with the *active site* of an *enzyme*. The reaction now takes place and the product molecules remain bound in the active site as an *enzyme–product complex*. The product molecules are then released and the enzyme is able to accept another substrate molecule.

epidemic: an outbreak of a disease that spreads rapidly through a population, affecting a large proportion of vulnerable members. *Influenza* is a disease that often results in epidemics. The influenza virus spreads easily from person to person and, in addition, it mutates frequently. As a result, the strain of influenza virus infecting people one year is likely to differ slightly from the strain that infected people in a previous year. People who have had influenza in the past will only be immune to one particular strain; they will be susceptible to the new strain. Since only the more vulnerable members of the community (such as the elderly and those with chest disease) are vaccinated against the current strain of influenza, a large proportion of the population is likely to be susceptible.

epidermis: the outer layer or layers of cells in a multicellular organism. In plants, the epidermis consists of a single layer of cells surrounding the tissues of the roots, stems and leaves. In the stems and leaves it is covered by a thin, waxy *cuticle* that helps to limit water loss.

In invertebrate animals, the epidermis is again made up of a single layer of cells that secrete a cuticle. This cuticle may consist largely of mucus as in earthworms or it may be more complex as in members of the phylum *Arthropoda*. Here, the *cuticle* forms an *exoskeleton*. In mammals and other vertebrates, there is no cuticle over the surface. The epidermis instead consists of several layers of cells, the outer ones of which are cornified and dead.

epigenetic imprinting: silencing of certain *alleles* that depends on the parent from which they were received. At *fertilisation* a *zygote* receives one set of *paternal chromosomes* from the father. It also receives one set of *maternal chromosomes* from the mother. Certain alleles in the zygote only become active if they are on a paternal chromosome. Others only become active if they are on a maternal chromosome.

epistasis: a genetic cross in which a single characteristic is affected by two or more different genes. These genes interact with each other. The diagram shows how two different genes may affect the colour of sweet pea *flowers*.

	Allele A		Allele B	
	↓		↓	
	Enzyme A		Enzyme B	
Colourless precursor molecule (White flowers)	→	Coloured intermediate molecule (Red flowers)	→	Coloured final molecule (Purple flowers)

In these flowers, a colourless precursor molecule is converted into a coloured final molecule by way of an intermediate molecule that has a different colour. This biochemical pathway is controlled by two genes. The **dominant** allele of the first gene, A, produces enzyme A. Enzyme A controls the first step in the reaction. The dominant allele of the second gene, B, produces enzyme B. Enzyme B controls the second step in the reaction. In each case the recessive **homozygote** produces a faulty enzyme that does not catalyse the relevant reaction.

For a flower to be purple, both alleles A and B must be present. A red flower is produced when allele A is present but allele B is absent. If allele A is absent, then the flower will be white. This is summarised in the table.

Allele A	Allele B	Precursor molecule	Intermediate molecule	Final molecule	Colour of flower
✗	✗	✓	✗	✗	White
✗	✓	✓	✗	✗	White
✓	✗	✓	✓	✗	Red
✓	✓	✓	✓	✓	Purple

Key

✓ = present

✗ = absent

epithelium: the **tissue** covering an organ, either on its inside or on its outside. Substances such as the products of **digestion**, oxygen, water and carbon dioxide must go through epithelial cells to get into or out of the body. Because of this, epithelial cells are often specialised for exchange. Those in the **small intestine**, for example, have their free **cell-surface membranes** folded to form **microvilli**. This provides a large surface area for the **absorption** of the products of digestion. As they act as lining cells, epithelial tissues are prone to damage and need constant replacement. Mitosis can often be seen in epithelial cells.

Epithelial tissues are classified according to shape. The three basic types are shown in the diagram.

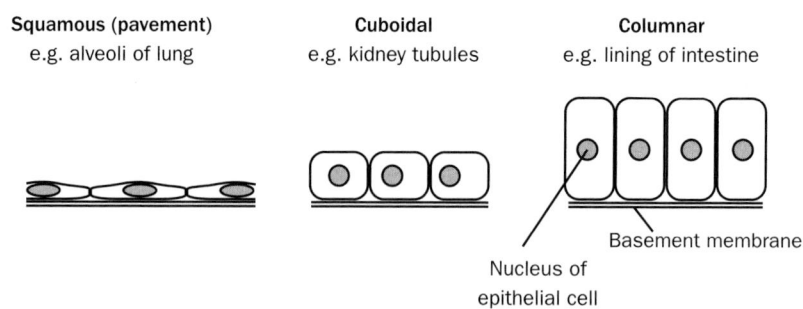

Squamous (pavement)	Cuboidal	Columnar
e.g. alveoli of lung	e.g. kidney tubules	e.g. lining of intestine

Basement membrane

Nucleus of epithelial cell

In addition to covering organs, epithelial tissue also forms **exocrine glands**.

EPOC: see **excess post-exercise oxygen consumption (EPOC)**.

erythrocyte: see *red blood cell*.

escape reflex: a reflex that involves the whole animal moving rapidly away from a source of danger. An escape reflex may be shown, for example, by an earthworm that is lying on the soil, partly out of its burrow. If its front end is touched, its longitudinal muscles contract and it rapidly shoots back into the ground. Similar types of response are shown by snails retreating into their shells and by squids escaping from predators. All these animals have large *neurones* that take *nerve impulses* rapidly to all the muscles involved so that they contract at the same time.

essential amino acid: an *amino acid* that must be supplied in food as it cannot be made in the body. Ten *amino acids* are essential for children but only eight for adults. *Proteins* that contain essential amino acids in appropriate proportions are called first-class proteins. Most foods derived from animals contain first-class proteins. Plant proteins, on the other hand, are second-class as they are often deficient in one or more of these essential amino acids. Because of this, people who are on strict vegetarian diets need to take care that they eat proteins from a variety of sources. In this way, they will obtain all the essential amino acids they require.

ester bond: a chemical bond formed as a result of *condensation* between glycerol and a *fatty acid*. There are three of these bonds in a *triglyceride* molecule.

ethene: a *plant growth substance* produced by most plant organs. Its best-known effect is its involvement in the ripening of *fruits* such as bananas and tomatoes.

euchromatin: *chromatin* that appears lightly stained when looked at with a microscope. It consists of *DNA* that is loosely packed. Much of the DNA that makes up euchromatin is undergoing *transcription*.

eukaryote: an organism with *eukaryotic* cells.

eukaryotic: a term used to describe cells that contain a nucleus. Animals and plants are *eukaryotes* and therefore have cells containing nuclei. Bacteria are *Prokaryotae*. They have cells that do not contain nuclei. Eukaryotic cells are different in other ways from *prokaryotic* cells. These differences are summarised in the table.

Eukaryotic cells	Prokaryotic cells
Large cells up to 50 μm in diameter	Cells small with a mean diameter under 5 μm
DNA linear and associated with *proteins* to form *chromosomes*. The chromosomes are found in a nucleus	Circular strands of DNA not associated with proteins and found in the *cytoplasm*. No nucleus present
Many membrane-surrounded organelles such as *mitochondria* present	Few organelles present. None are surrounded by a plasma membrane

eutrophication: an increase in the quantity of plant nutrients. The term is used when freshwater lakes or rivers are enriched with nitrates and phosphates, either as a result of the *leaching* of *fertiliser* from agricultural land or from sewage effluent. The flow chart overleaf shows how eutrophication can affect an *ecosystem* by causing the death of many of its organisms.

Increase in nitrate and phosphate concentrations

↓

Increase in the growth of algae and other photosynthetic organisms. These die faster than they can be eaten

↓

Increase in activity of **decomposers**. Their respiratory activity depletes the water of its oxygen

↓

Other organisms such as fish and the invertebrate animals on which they feed die due to a lack of oxygen

ex situ conservation: the protection of endangered species of animals and plants by moving some of the organisms concerned to another site. This may be to a zoo or botanic garden, or it may be to another suitable wild area. Examples of *ex situ* conservation include:

- the removal of populations of great crested newts to other areas when the ponds in which they breed have been drained
- the captive breeding and release of black stilts in New Zealand. The black stilt is an extremely rare bird with a wild population that is between 20 and 30 pairs. Without this breeding and release programme, it is likely that the species would soon be extinct in the wild
- **micropropagation** of lady's slipper orchids and their re-introduction to the wild.

excess post-exercise oxygen consumption (EPOC): the increased rate of oxygen consumption that follows a period of exercise. The excess oxygen is used to restore the body to its resting condition. It is used for processes such as:

- metabolising the **lactate** produced during **anaerobic respiration**
- replenishing muscle **glycogen** and **ATP**
- restoring the **resting potential** in nerve **axons**
- returning core body temperature to its resting level.

excretion: the removal of waste products that have been produced as the result of **metabolic** processes in an organism. Excretion involves the removal of nitrogen in waste products such as urea and **uric acid**, as well as the removal of carbon dioxide produced as a waste product of **respiration**. The table summarises some of the more important excretory substances in a mammal as well as the organs mainly responsible for their excretion.

Substance	Formed from	Organ mainly responsible for excretion
Urea	Breakdown of excess **amino acids**	*Kidney*
Carbon dioxide	Respiration	*Lungs*
Bile pigments	Breakdown of **haemoglobin**	**Liver**. Removed with the **faeces**

exocrine gland: a *gland* that produces a secretion which leaves through a duct. Glands, such as the *salivary glands*, the *pancreas* and the *liver*, produce secretions involved in *digestion*. They all have ducts so they are exocrine glands. The liver, for example, secretes *bile* which goes into the intestine through the bile duct.

exocytosis: a process that involves the transport of materials out of cells. Exocytosis takes place in many cells in the body. One example is provided by the cells in the mammary *glands*, which are responsible for secreting milk. *Proteins*, *lactose* and mineral ions such as *calcium* are transported to the *Golgi apparatus*. This is an *organelle* responsible for processing and packaging substances produced by a cell. *Vesicles* are formed. They pinch off and move towards the surface of the cell. Here, the membrane that surrounds them fuses with the *cell-surface membrane* and the contents of the vesicle are released outside the cell.

exogenous: coming from outside. Many organisms have a natural 24-hour *circadian rhythm* that governs their behaviour. This has a considerable effect on animals. In hot, dry climates, for example, it may be an advantage to be nocturnal, hunting for food during the cooler hours of the night and resting in the shade during the day. Organisms have a natural biological clock that enables them to adapt to conditions like this. Experimental evidence suggests that in some organisms this mechanism can be influenced by exogenous factors, that is, changes in environmental conditions.

Plant growth substances are chemicals produced by plants. They are involved in the control of various aspects of growth and development. Much early experimental work was done by treating plants with these substances and observing the consequences. However, there was one major difficulty with this approach. It involved making the assumption that the response of a plant to *exogenous* plant growth substances would be the same as that to the smaller concentration produced by the plant itself.

exopeptidase: a protein-digesting *enzyme* that breaks the peptide bonds at the end of a polypeptide chain releasing either single *amino acids* or dipeptides. In mammals, these enzymes are associated with the final stages in protein *digestion* and are located on the *cell-surface membranes* of the cells that line the *small intestine*.

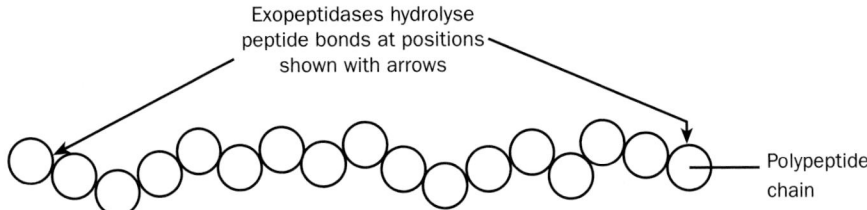

The action of exopeptidases

exon: a section of the *DNA* in a *gene* that codes for an *amino acid* sequence. A gene in a *eukaryotic* cell, such as one from an animal or a plant, consists of exons and pieces of DNA which do not give rise to amino acid sequences. These pieces of DNA are called *introns*. During transcription, the base sequence in the gene forms a template for producing an *mRNA* molecule. Before the mRNA leaves the nucleus however, it is edited, and the base sequences forming the introns are cut out. The mRNA base sequence therefore corresponds only to exons. The genes in many *prokaryotic* cells do not contain introns.

exoskeleton: a hard covering on the outside of an animal which acts as a *skeleton*. The body surface of an insect, for example, is covered in a *cuticle* that is of *chitin*, a nitrogen-containing polysaccharide. This tough layer is further strengthened by the addition of a hard type of protein. This means that the exoskeleton of an insect is like a system of hollow tubes. Where they bend, the skeleton is thinner and forms a joint which acts as a hinge. The insect's muscles are attached to processes on the inside of this skeleton. One of the disadvantages of having a rigid covering on the outside is that it creates problems with growth. When insects grow, therefore, they must shed this outer skeleton. Only while the new skeleton is soft can an increase in size take place.

exotoxin: a poisonous substance or toxin produced and secreted by certain bacteria as they grow. Exotoxins are produced by a number of bacteria including those responsible for botulism, tetanus and diphtheria. Diphtheria has a short incubation period before a sore throat and fever develop. The multiplying bacteria produce an exotoxin that is secreted into their surroundings. This exotoxin inhibits protein synthesis and may result in damage to the heart and nerves. Successful treatment of diphtheria involves the use of an *antibiotic* to kill the bacteria, together with an antitoxin.

expiration: when a person breathes out, the *diaphragm* muscle relaxes and returns to its original dome shape. This decreases the volume of the thoracic cavity and increases the pressure inside it. Air leaves the *lungs* down the resulting pressure gradient. During exercise, the internal *intercostal muscles* contract and help to push air out forcefully.

expiratory reserve volume: the additional amount of air that can be breathed out after a person has breathed out normally. For an average adult male, this is about 1500 cm^3. (See also *lung capacities*.)

exponential phase: see *log phase*.

extensive food production: methods of food production that require relatively large areas of land. Nomadic herding and ranching of cattle and sheep are examples of extensive methods of production. This method contrasts with intensive farming where the animals are confined in a relatively small area. Extensive farming has a number of advantages over intensive farming:

- It is considered to be more acceptable ethically as animals are kept under conditions that are thought of as being closer to those in their natural *environment*.
- Relatively low stocking densities minimise the spread of disease. Extensive food production also allows land that is considered unsuitable for other forms of agriculture to be used.
- It does not produce large concentrations of waste material, such as *urine* and *faeces*, which must be disposed of. Waste material is naturally recycled on the land itself.
- There is less need to have a system that can transport large quantities of such materials as food concentrates, *silage* and straw to the site of production. Much more food can be supplied naturally.

extensor: a muscle that straightens a limb. *Flexors* are muscles that bend limbs. An extensor and a flexor act as an *antagonistic muscle* pair. Muscles can only do work by contracting. Therefore, two muscles are required to move a limb. The flexor bends the limb when it contracts; the extensor straightens it.

external digestion: see *extracellular digestion*.

external fertilisation: *fertilisation* that takes place outside the body. In many aquatic animals, both male and female *gametes* are released into the surrounding water. Because the probability of fertilisation is not high, animals that have external fertilisation generally produce far more egg cells than those that use *internal fertilisation*.

extracellular digestion: the chemical breakdown of food molecules which takes place outside cells. Many bacteria and fungi are *saprobionts* and feed by extracellular digestion. They secrete digestive *enzymes* on to the surface of their food. These enzymes break down complex molecules, such as those of *starch* and *protein*, into smaller, soluble ones that are then absorbed. In addition to this, much of the digestion that takes place in the gut of a mammal is also extracellular. Enzymes are secreted by *glands* such as the *salivary glands* and the *pancreas*. They break down the individual food molecules. This process occurs in the gut *lumen*, so is extracellular.

Aiming for a grade A*?

Don't forget to log on to **www.philipallan.co.uk/a-zonline** for advice.

F₁: the first generation of offspring from parents that are homozygous (see **homozygote**) for a particular characteristic.

F₁ hybrid: offspring produced by crossing parents from strains that are inbred. They are characterised by having a much greater productivity than either of their parents. Much of the early work on producing F₁ hybrids was carried out on maize. It had been known for some time that repeated **self-fertilisation** of maize plants resulted in lower and lower productivity. Part of the reason for this was that a greater proportion of homozygous (see **homozygote**) plants resulted and there was an accumulation of unfavourable recessive **alleles**. If two such lines of repeatedly self-fertilised plants are developed, it is likely that the unfavourable alleles that accumulate will be different in each case. Neither strain will give a high yield. If they are now crossed, the offspring will be heterozygous for many of the genes concerned. The recessive alleles concerned will no longer exert their effect.

There is one major disadvantage with the use of F₁ hybrid varieties. The seeds produced by the F₁ plants do not breed true. They cannot be kept and used the following year, since a large number of different **genotypes** will be produced. Many of these will result in plants with low yields.

The technique of producing F₁ hybrids is widespread. Apart from maize, it has been used extensively to produce larger and more brightly coloured garden **flowers** and to increase the numbers of young reared by female pigs.

facilitated diffusion: **diffusion** that is helped by protein carrier molecules. Molecules such as **glucose** do not diffuse through the **phospholipid** layers in **cell-surface membranes**. The membranes, however, contain **carrier proteins**. These proteins have two important properties. They have binding sites for specific molecules. A glucose carrier protein, for example, only binds to glucose molecules. Other molecules do not fit into the binding site. In addition, the carrier molecules change shape many times each second. In one form, the binding site is exposed to the outside of the cell; in the other, it is exposed to the inside. Glucose molecules outside the cell are continually moving about. One eventually collides with the binding site of an exposed carrier protein. When the protein changes shape, the glucose molecule will be on the inside of the membrane where it will be released.

factor VIII: one of the factors or plasma **enzymes** involved in **blood clotting**. **Haemophilia** is an inherited condition where factor VIII is absent. The blood of a person with haemophilia therefore takes a much longer time to clot. Haemophiliacs are usually male because the condition is **sex-linked**. The recessive **allele** concerned is carried on the **X chromosome**.

FAD: a coenzyme that is important in respiration. FAD is reduced during the **Krebs cycle**. The reduced FAD transfers **electrons** and protons to the **electron transfer chain**. As the electrons are passed from one molecule to the next along this chain, energy is released. This energy is used to form **ATP**.

faeces: the waste products removed from the body during **egestion**. The main function of the last part of the gut or **colon** of a mammal is to remove water from the food that remains in the gut. This undigested material, together with dead cells that have been scraped off the gut wall and bacteria, form the bulk of the faeces. The distinctive colour is due to the presence of **bile** pigments.

fallopian tube: see **oviduct**.

family: a level of **classification**.

fast muscle fibre: see **skeletal muscle fibre**.

fast twitch muscle: see **skeletal muscle fibre**.

fasting blood glucose test: a test that measures blood **glucose** concentration after a person has not eaten for at least 8 hours. It is often the first test carried out on a person who is suspected of having **diabetes**. In a healthy person, the concentration of glucose is below 100 mg per 100cm^3. In someone with diabetes, the concentration will be higher than this because the excess glucose is not removed from the blood by **insulin**.

fat: a **triglyceride** that is solid at temperatures below 20°C. This is because it contains a high proportion of saturated **fatty acids**. Generally speaking, the triglycerides found in animals are fats. Those found in plants are oils and are liquid at temperatures of 20°C.

fatty acid: molecules with a COOH group at one end and a hydrocarbon tail at the other. Fatty acids with small numbers of carbon **atoms** in this tail are produced when **cellulose** is digested by **microorganisms** living in the guts of **ruminants** and other herbivorous mammals. They can be absorbed into the blood and are important **respiratory substrates** in these animals. Fatty acids with long hydrocarbon tails are important constituents of **fats** and oils (**triglycerides**). The diagram shows two fatty acids, stearic acid and oleic acid, that are commonly found in triglycerides.

Stearic acid

$$\begin{array}{c} O \\ \diagdown \\ HO \diagup \end{array} C-CH_2-CH_2-CH_2-CH_2-CH_2-CH_2-CH_2-CH_2-CH_2-CH_2-CH_2-CH_2-CH_2-CH_2-CH_2-CH_2-CH_3$$

Oleic acid

$$\begin{array}{c} O \\ \diagdown \\ HO \diagup \end{array} C-CH_2-CH_2-CH_2-CH_2-CH_2-CH_2-CH_2-CH=CH-CH_2-CH_2-CH_2-CH_2-CH_2-CH_2-CH_2-CH_3$$

Fatty acids

Both have hydrocarbon chains with 17 carbon atoms in them. Oleic acid, however, has a double bond present between two of the carbon atoms. Fatty acids with double bonds are described as unsaturated. They have much lower melting points than fatty acids, in which there are no double bonds. Triglycerides that contain **unsaturated fatty acids** therefore tend to be liquid at temperatures of around 20°C. They are called oils and are found mainly in plants. Those which contain **saturated fatty acids** are solid at this temperature. They are called fats and are mainly found in animal cells.

female: the organism, or part of the organism, that produces female sex cells or **gametes**. Female gametes have three characteristic features:

- they are larger in size than the male gametes
- they are produced in much smaller numbers than the male gametes
- they do not move. The male gametes either swim or are moved in some way to the female gametes.

fermentation: a process that occurs in the absence of oxygen in which small organic molecules are produced from larger ones. In animals, in the absence of oxygen, **glucose** is converted to **lactate**. This is sometimes called **anaerobic respiration**, but it is an example of a fermentation. Fermentation by **microorganisms** has been exploited by humans for many thousands of years. The ancient Egyptians, for example, used yeast to make beer. With modern techniques of **biotechnology**, **fermentations** by microorganisms have been used to produce a variety of different products. These range from foodstuffs such as yoghurt and cheese to drugs such as the **antibiotic** penicillin.

fertilisation: the process that involves the fusion of male and female **gametes** to produce a **zygote**. In **flowering** plants, this process must be distinguished from **pollination** (the transfer of pollen from the **stamen** to the **stigma**). When fertilisation takes place outside the body, it is called **external fertilisation**. When it takes place inside the body of the female, it is called **internal fertilisation**.

fertilisation membrane: a membrane formed round the egg after it has been fertilised by a sperm cell. This membrane prevents other sperm cells from entering the egg.

fertiliser: a substance that adds mineral ions to the soil. There are two types of fertiliser. Chemical or **artificial fertilisers** are made either from naturally occurring rocks or by industrial processes. The main soil nutrients found in these fertilisers are nitrogen, phosphorus and potassium (NPK). These are usually present in a fairly concentrated form. Organic or **natural fertilisers** include farmyard manure and **compost**. They are usually slower acting as the nutrients they contain must be released by the action of soil fungi and bacteria. In both cases, however, the ions taken up by the plants are exactly the same.

When a crop is harvested, it is taken from the soil and any nutrients that the plants contain are removed as well. Because of this, fertilisers are essential to restore the chemical balance of the soil. Too much can be added, however. A **law of diminishing returns** operates and excess fertiliser does not always lead to extra productivity. In addition, rainfall may cause **leaching** of excess nutrients into nearby streams and lakes. This may result in **eutrophication**.

fetal haemoglobin: the type of **haemoglobin** found in the blood of a **fetus**. There is an important difference between the properties of fetal haemoglobin and those of adult haemoglobin. The **oxygen dissociation** curve for fetal haemoglobin is further towards the left. This means that it can load up with oxygen at lower partial pressures. This is important as the mother's blood arriving at the placenta is not completely saturated with oxygen. The blood of the fetus, because of its special sort of haemoglobin, is still able to pick up oxygen and become almost saturated. Shortly after birth, fetal haemoglobin is replaced by the slightly different sort of haemoglobin that is normally found in the blood of adults.

fetus: a young mammal in the later stages of development inside the *uterus* of its mother. In humans, the term *embryo* is usually used to refer to the two months following conception. After two months, once it is becoming recognisable as a human, it is called a fetus.

FEV$_1$: see *forced expiratory volume in one second (FEV$_1$)*.

fibre: see *dietary fibre*.

fibrin: a protein involved in *blood clotting*. A protein called *fibrinogen* is present in blood *plasma*. When a wound occurs, this soluble protein is converted to insoluble fibrin. Fibrinogen can only be converted into fibrin by the *enzyme*, *thrombin*. Active thrombin is produced in the presence of a number of clotting factors, which are released at the site of tissue damage.

fibrinogen: a protein, found in the blood, that is converted to *fibrin* when blood clots. Fibrinogen is a soluble protein that is made in the *liver*. When a wound occurs, the *enzyme* *thrombin* converts fibrinogen to fibrin. It does this by breaking off part of the fibrin molecule. The remainder of the molecule polymerises to form long insoluble threads of fibrin. These form a mesh over the surface of the wound which traps *red blood cells* and forms a clot.

fibrosis: thickening of tissue, often as a result of an injury. Pulmonary fibrosis is thickening of the walls of the *alveoli*. This reduces the rate of *diffusion* of oxygen from the alveoli into the blood. The walls of the alveoli in people with pulmonary fibrosis are also stiffer, so people with this condition often have difficulty with breathing.

fibrous protein: a protein with molecules that consist of long chains of *amino acids*. The secondary structure of these *proteins* is important in determining their shape. They have little or no tertiary structure. Unlike *globular proteins*, fibrous proteins are insoluble. Fibrous proteins such as *collagen*, *elastin* and *myosin* have important structural functions in living organisms.

Fick's law: this states that the rate of *diffusion* of a substance is directly proportional to the surface area of the exchange surface and to the difference in concentration on either side. It is inversely proportional to the thickness of the exchange surface. This can be written in simple mathematical terms as:

$$\text{rate of diffusion} \propto \frac{\text{surface area} \times \text{difference in concentration}}{\text{thickness of exchange surface}}$$

Fick's law provides a useful framework with which to look at adaptations for diffusion. With gas exchange surfaces such as the *gills* of a fish, the rate of diffusion must be as great as possible. In order to achieve this, the values on the top line of the expression should be large, while that on the bottom line should be small. A careful look at the structure of a gill shows adaptations that produce a large gas exchange surface and maintain as big a difference in concentration as possible. The gas exchange surface, on the other hand, is extremely thin. Considerations like these can be applied to all exchange surfaces. However, the converse also applies in some cases. A low rate of diffusion is required, for example, in the case of water loss from plants living in dry areas. Here, various adaptations result in the figures on the top line of the expression being low and that on the bottom line being high.

filament: see *gill filament*.

filter feeding: a method of feeding on tiny organisms or particles of food suspended in the water. Filter feeders are all aquatic animals. They have a method of producing a current of water that flows over the feeding surface. In molluscs and marine worms, the food is trapped in mucus. **Cilia**, tiny hair-like structures on the **cell-surface membranes**, then waft this mucus towards the mouth. Larger organisms such as some sharks and whales also have filter-feeding mechanisms, although the mechanism by which they trap food particles is somewhat different.

first convoluted tubule: the part of the **nephron** between the **renal capsule** and the **loop of Henle**. The fluid that has been forced by **ultrafiltration** into the renal capsule contains substances which are useful to the body as well as toxic waste. In the first convoluted tubule, **active transport** results in useful substances such as **glucose**, **amino acids** and mineral ions being reabsorbed into the blood. Most of the urea remains in the tubule and is eventually excreted in the **urine**. This is, therefore, a process of selective reabsorption.

first messenger: a substance produced in the body that binds to a receptor on the **cell-surface membrane** of a target cell. This triggers activity inside the cell. **Hormones** and **neurotransmitters** are first messengers.

flaccid: a flaccid cell is a cell from which water has been removed. When water leaves a cell by **osmosis**, the protoplast shrinks. If the cell is an animal cell, the whole cell shrinks and shrivels. This is called **crenation**. If it is a plant cell, the protoplast pulls away from the **cell wall**. This is called **plasmolysis**.

flagellum (bacterial): a long, thin hair-like process (plural flagella) on the surface of a bacterium that enables it to move. It is made of molecules of a **protein** called flagellin that are spirally arranged to form a tube. The entire flagellum has a wave-like appearance but it is rigid and does not beat. It is its rotating action that moves the cell along.

flagellum (eukaryotic cell): a long, thin hair-like process on the surface of the cell. If a cross-section is cut through one of these structures it is seen to consist of a ring of nine **microtubules** surrounding two more microtubules located in the centre. The microtubules are made of **proteins** and enable the flagellum to beat and bring about movement in some single-celled organisms and **sperm** cells. **Eukaryotic** flagella are sometimes called undulipodia.

flatworm: a member of the phylum **Platyhelminthes**. This phylum includes some free-living species and some species such as **tapeworms** and the blood fluke, **Schistosoma**, that are important **parasites** of humans.

flavoprotein: a substance that acts as an **electron** acceptor in **electron transport chains**.

flexor: a muscle that bends a limb. **Extensors** are muscles that straighten limbs. A flexor and an extensor act as an **antagonistic muscle** pair. Muscles can only do work by contracting. Therefore, two muscles are required to move a limb. The flexor bends the limb when it contracts; the extensor straightens it.

flower: the structure on a **flowering** plant that contains the parts associated with sexual reproduction. For reproduction to be successful, pollen must be transferred from the male parts of the same or a different flower to the female parts, a process called **pollination**. There are a number of ways of doing this and although flowers are often divided into those that are wind-pollinated and those that are insect-pollinated, birds, bats and even water currents may be involved in the process. Flowers are adapted to the way in which they are pollinated. Because of this, there is a lot of variation in structure, but the diagram shows the main parts of a typical flower.

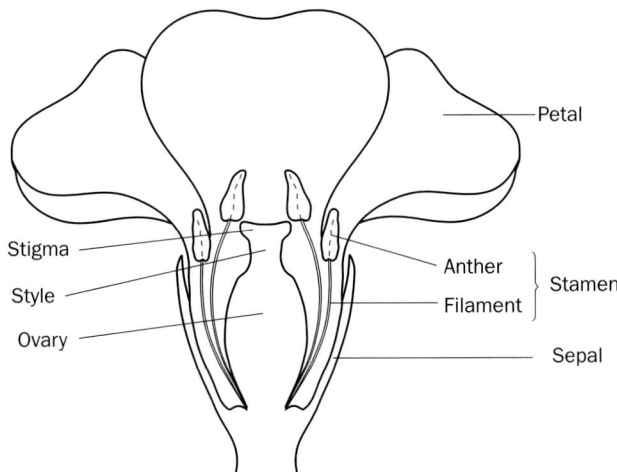

flowering: the process by which a plant forms flowers. The stimulus that usually triggers this is the relative proportion of daylight and darkness occurring over the course of a day. Three types of plants are recognised:

- *Long-day plants* develop *flowers* when the day length is longer than a certain critical value. They include species that flower in summer such as poppies and grasses.
- *Short-day plants* flower when day length is less than a certain amount. They include plants that normally flower in the spring, such as primroses, as well as those like chrysanthemums whose normal flowering time is during the autumn. Tropical plants such as poinsettias are also short-day plants.
- Day-neutral plants such as tomatoes are not affected by the length of daylight.

By changing the hours of daylight that a plant receives, it can be made to flower at a different time of the year. In this way, chrysanthemums, which normally flower in the autumn, can be made to flower at any time of the year.

fluid mosaic model: a model showing the way that scientists think that molecules are arranged in a *plasma membrane*. Even with an *electron microscope* it is not possible to see the molecular structure of a plasma membrane, so it was necessary to construct a model to explain its various properties. Today, the most widely accepted model is the fluid mosaic model. This was originally suggested by Singer and Nicolson in 1972. The diagram shows the main features of this model.

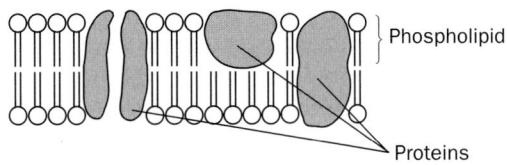

The fluid mosaic model of membrane structure. This is a simplified diagram. There are other molecules present such as cholesterol and glycoproteins, which have been omitted.

The diagram shows a double layer of *phospholipid* molecules within which protein molecules are scattered. Some of these *proteins* move freely within the phospholipid layers, while others are anchored rather more firmly to structures within the cell. The term 'fluid mosaic' refers to the membrane's structure as a patchwork of different molecules that are free to change in position. The protein molecules in the membranes have many important properties, among which are the following.

● They may act as *enzymes*. For example, there are important enzymes involved in *carbohydrate* and protein *digestion* on the *cell-surface membranes* of the epithelial cells in the *small intestine*.

● They act as *carrier proteins*. These are involved in the processes of *active transport* and *facilitated diffusion*.

● They may act as specific receptors for *hormones*.

fMRI: see *functional magnetic resonance imaging (fMRI)*.

folic acid: a *vitamin* that has many functions, including the synthesis of *nucleotides*. Because of this, it is particularly important during periods of rapid growth such as during pregnancy. A deficiency of folic acid during pregnancy may lead to *anaemia* in the mother and defects of the nervous system of the fetus. These defects include spina bifida.

follicle-stimulating hormone: see *FSH*.

food chain: a sequence that represents the way in which energy is transferred from one organism to another in a *community*. An example of a food chain is:

nettle plant → large nettle aphid → two-spot ladybird

The nettle plant is the *producer*. It converts the energy in sunlight into chemical potential energy in the organic molecules in the plant. The large nettle *aphid* is the *primary consumer* that feeds on the nettle and the two-spot ladybird is the secondary *consumer*. The arrows in a food chain represent the direction of energy transfer. Because energy is lost at each stage by processes such as *respiration*, it is rare for any food chain to have more than five links.

Food chains are simplifications of what really occurs in a community. There are many different species of animal, for example, that feed on nettle plants, while two-spot ladybirds feed on a number of different species of aphid. Food chains therefore interlink with each other to form *food webs*.

food irradiation: a method of preventing the spoilage of food by exposing it to *ionising radiation*. This destroys the bacteria and other *microorganisms* responsible for food spoilage by damaging their *DNA*. As a result they cannot multiply.

food vacuole: a small vesicle that contains food particles, found in the *cytoplasm* of the cells of some organisms. Food vacuoles are found in the cells of organisms, such as amoeba, which feed by *phagocytosis*. The *cell-surface membrane* engulfs particles of food and these are taken into the cytoplasm. *Lysosomes*, cell organelles that contain digestive *enzymes*, then fuse with the food vacuole and digest its contents. The products of *digestion* diffuse from the food vacuole into the cytoplasm.

food web: a diagram that shows the way in which all the different species of organism in a *community* depend on each other for food. A food web is made up of many interconnected *food chains*.

foramen magnum: the opening in the skull through which the *spinal cord* leaves the *brain*. It is where the vertebral column meets the skull. The position of the foramen magnum is important in interpreting evidence concerned with human evolution. It can be used to determine whether or not a fossil skull comes from an animal that was *bipedal* and walked on two legs or quadrupedal and walked on all fours. In bipedal animals, the foramen magnum is underneath the skull; in quadrupeds, it is farther towards the back.

foramen ovale: a hole situated in the wall between the right *atrium* and the left atrium in the heart of a *fetus*. It is covered by a membranous valve. While a fetus is developing inside the mother's uterus, its oxygen supply comes from its mother via the placenta. Its *lungs* do not function. Blood entering the right atrium of the heart opens the valve and some passes through the foramen ovale into the left atrium and on into the *aorta*. The remainder goes into the *pulmonary artery* but is not transported to the lungs. Instead, it goes through the *ductus arteriosus* into the aorta. At birth, the foramen ovale and the ductus arteriosus both close so all the blood then flows from the right side of the heart into the pulmonary artery and on to the lungs.

forced expiratory volume in one second (FEV$_1$): the maximum amount of air that can be breathed out of the *lungs* in one second. It is measured with a recording *spirometer* and can be compared with the predicted value for a person of the same age, sex and height. People with *chronic obstructive pulmonary disease (COPD)* have a low FEV$_1$ value.

forensic entomology: the study of the insects present on a decomposing body to estimate the length of time that the person has been dead.

founder effect: one of the causes of reduced genetic diversity in a population. Sometimes a new area is colonised by a few organisms from another population. These colonisers will not carry all the *alleles* present in the original population. They will therefore give rise to a new population with a lower genetic diversity. The founder effect is common where islands are colonised from the nearby mainland. It also occurs when species are deliberately introduced by humans into new areas.

fovea: part of the *retina* of the eye that has no *rod cells* but large numbers of *cone cells*. When attention is attracted to a particular object, the eye normally moves so that the image falls on the fovea. Because it has a large number of cone cells, it is the region of highest visual acuity. In other words, it is the part of the retina that enables the greatest degree of detail to be seen.

frame quadrat: see *quadrat*.

freezing: a method of preventing food spoilage by lowering the temperature at which food is stored. This slows the growth of *microorganisms*. Low temperature reduces the rate of the *enzyme*-controlled reactions in the microbial cells. It also makes less water available for their growth.

frequency: a way of measuring the abundance of organisms in a sample. Frequency is the number of *sampling* units in which a particular species is found. If ten *quadrats* are placed in the area being sampled and dandelion plants are found in three of them, the frequency of the dandelion plants will be three out of ten or 30%.

fructose: a *hexose* sugar – that is, a *monosaccharide* with six carbon *atoms* in each of its molecules. Fructose has the same molecular formula as *glucose*, $C_6H_{12}O_6$, but the atoms that make up the molecule are arranged in a different way. It is found naturally in many *fruits* and, together with glucose, forms the *disaccharide sucrose*. Fructose is an important constituent of diabetic diets as it tastes sweet but its metabolism does not depend on *insulin* (see *diabetes*).

fruit: the structure that develops from an ovary after *fertilisation* in a *flowering* plant. It encloses the *seeds*. Fruits are mainly concerned with the dispersal of seeds and many have adaptations which ensure that the seeds are dispersed at a considerable distance from the parent plant. Two important methods of dispersal in plants are:

- animal dispersal. Many ripe fruits have succulent flesh and are brightly coloured. Animals eat them and the soft tissue of the fruit is digested. The harder seeds resist the action of the *enzymes* in the gut. Consequently, they are deposited back on the soil in the animal's *faeces*. Other fruits have hooks or spines that result in them becoming temporarily attached to the fur or feathers of a passing mammal or bird.
- wind dispersal. Fruits often have tufts of hair or wings that increase their surface area and enable them to be blown by the wind. The fruits of other plants are capsules which move in the wind, shaking out large numbers of tiny seeds.

The development of fruits is controlled by *plant growth substances*.

FSH: follicle-stimulating hormone, a *hormone* produced by the *pituitary gland*. In females, it is produced during the first part of the reproductive cycle. It travels in the blood to its target organ, the *ovary*, where it stimulates one or more of the immature follicles to develop. FSH is also produced in males, where it is necessary for the development of sperms.

functional magnetic resonance imaging (fMRI): a specialised type of *MRI* scan. When nerve cells in the *brain* and *spinal cord* are active, they respire faster and use oxygen at a greater rate. Functional magnetic resonance imaging measures the changes in blood flow that result from supplying this extra oxygen.

Fungi: the *kingdom* containing fungi (see *classification*). As well as mushrooms and toadstools, this kingdom contains a variety of smaller moulds and yeasts. Fungi are *eukaryotic* organisms that share the following features:

- As they do not possess *chlorophyll*, they are all heterotrophic (see *heterotrophic nutrition*). Many are *saprophytes* but a considerable number are *parasites*.
- A mature fungus is made up of a number of thread-like structures called *hyphae*. These hyphae form a web called a *mycelium*.
- They have *cell walls* but these are made of *chitin* not *cellulose* as in plants. The main storage *carbohydrate* is the *polysaccharide*, *glycogen*.
- They reproduce by means of *spores*.

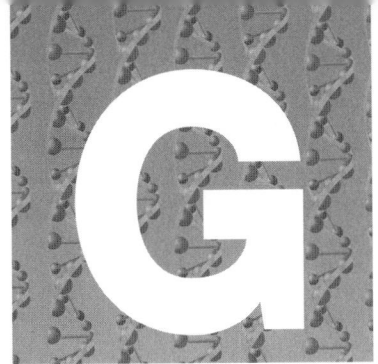

G₁ phase: a stage in the *cell cycle* at the start of interphase. During this phase, cells start to synthesise *proteins* and their *cytoplasm* increases rapidly in volume. This is the phase when the number of cell organelles also increases.

G₂ phase: a stage in the *cell cycle* at the end of interphase. This is the phase when the *proteins* involved in cell division are synthesised.

galactose: a *hexose* sugar – that is, a *monosaccharide* with six carbon *atoms* in each of its molecules. Galactose has the same molecular formula as *glucose*, $C_6H_{12}O_6$, but the atoms that make up the molecule are arranged in a different way. *Lactose*, the sugar found in milk, is a *disaccharide* formed by *condensation* of a molecule of glucose and a molecule of galactose.

gall bladder: a small muscular sac or bag found just under the *liver*. It stores *bile* between meals. Bile is produced in the liver and passes into the gall bladder. It is released through the bile duct into the first part of the *small intestine*.

gamete: a general name for a sex cell. Gametes are *haploid* cells that are formed by *meiosis*. In plants and some other organisms, however, mitosis takes place between meiosis and the formation of a mature gamete. In *sexual reproduction*, gametes fuse to produce a diploid *zygote*. In most organisms, gametes are either male or female. Male gametes move to the female ones and are smaller and produced in larger numbers.

gamete intra-fallopian transfer (GIFT): a method used to treat infertility. Eggs are removed from a woman's ovaries. They are put into one of her *oviducts* (fallopian tubes) together with sperms from the male partner. *Fertilisation* takes place, therefore, inside the woman's body. This technique is particularly useful when the male partner produces sperms that are not motile. It is also sometimes used when the cause of infertility is not known. An advantage of the technique is that the fertilised egg develops in its normal environment and not in *in vitro* surroundings.

gametogenesis: the process by which *gametes* are formed. In a mammal, gametogenesis takes place in the ovaries or the *testes*. Gamete-forming cells first divide by *mitosis*. Some of the cells that are formed replace these gamete-forming cells. Others develop into gametes. In a process that involves growth, *meiosis* and maturation, they form sperms or egg cells. In a *flowering* plant, gamete formation is more complicated but still involves meiosis. Although the detailed processes are different in animals and plants, there is an important similarity. In males, each gamete-forming cell develops to produce four individual daughter cells. In females, however, not all the products of meiosis go on to give rise to gametes. Usually only a single mature gamete will develop from each meiotic division.

This leads to the production of large numbers of small gametes in males and small numbers of large gametes in females.

gametophyte: the *haploid*, *gamete*-producing stage in the life cycle of a plant. Plant life cycles show an *alternation of generations* in which a gamete-producing stage or gametophyte alternates with a spore-producing stage or *sporophyte*. The gametophyte produces the gametes that will fuse and produce a sporophyte. In primitive plants, such as mosses and liverworts, the gametophyte is the *dominant* stage. The familiar green moss plant is therefore a gametophyte and will eventually produce either male or female gametes. In *flowering* plants, however, the sporophyte is the dominant stage in the life cycle and the gametophyte is very much reduced. Both male and female gametophytes are represented by a few nuclei found within the *flower*.

ganglion: nervous tissue consisting of a small group of cell bodies of neurones and *synapses*. The function of a ganglion (plural ganglia) is mainly one of integration. The nervous system in many invertebrate animals is made up of a nerve cord and a number of ganglia. In an earthworm, for example, there is a ganglion in each body segment. The nervous system of a mammal has a totally different structure, but there are ganglia in the *autonomic nervous system*. Additionally, the cell bodies of the *neurones* which convey sensory information from receptors to the *spinal cord* are found in the *dorsal root* ganglia.

ganglion cell: a type of nerve cell found in the *retina* of the eye. Each ganglion cell receives sensory information from a large number of *rod cells* or *cone cells*. Ganglion cells are involved in sorting out and processing this information before it is sent via the *optic nerve* to the *brain*.

gastric: an adjective meaning 'to do with the *stomach*'. *Gastric juice*, for example, is the digestive secretion produced by the stomach. The gastric *glands* are the glands in the lining of the stomach that are responsible for the secretion of gastric juice.

gastric juice: the secretion produced by glands in the lining of the *stomach*. Its main function is *digestion* and it contains digestive *enzymes*. The most important of these is *pepsin* which acts on *proteins* and breaks them down into smaller polypeptides. It also contains *mucus* and hydrochloric acid. The hydrochloric acid has several functions:

- it converts *pepsinogen*, the inactive form in which the enzyme pepsin is secreted, to pepsin
- pepsin works best in acid conditions and has an optimum pH of between 1.6 and 3.2. Hydrochloric acid ensures that this pH is maintained in the stomach
- it kills some of the bacteria that are ingested with food.

gel electrophoresis: see *electrophoresis*.

gene: a length of *DNA* that occupies a specific position on a *chromosome* called a *locus*. The sequence of nucleotide bases that makes up a gene gives it a particular function. This usually means that each gene will code for a specific protein or polypeptide. In humans, for example, there are genes that code for a variety of different *proteins*. These include all the various *enzymes* which control the biochemical processes in cells, *hormones* such as *insulin* and *growth hormone*, and other proteins such as channel proteins in *cell-surface membranes*, and *haemoglobin*. An individual gene may have more than one form that differs slightly from the others in its sequence of nucleotide bases (see *base*). Different forms of a gene are called *alleles*.

gene amplification: a process by which many copies may be made of a single gene. Biologists use the *polymerase chain reaction (PCR)* to make large amounts of identical *DNA* from very small samples.

gene library: a collection of living bacteria that contain pieces of *DNA* from another organism. DNA is extracted from an organism and cut into pieces using a *restriction endonuclease*. This DNA is then inserted into a *vector* such as a *plasmid*. The plasmids are taken up by bacteria. A complete gene library contains all the DNA in the organism's *genome*.

gene machine: a piece of equipment that can make artificial *DNA* with the required base sequence. It is an important tool in genetic research. It is often more economical to make a piece of DNA artificially than it is to use techniques based on cloning DNA fragments in living organisms.

gene mutation: a change in one or more bases in the *DNA* that forms an organism's genetic material. This changes the *genotype* of the organism and may be inherited. Mutations occur randomly and any part of the DNA present in an organism may mutate. The rate at which mutation takes place can be increased by exposure to various *mutagens*, such as ultraviolet radiation, *X-rays* and a wide range of organic substances.

There are various forms of gene mutation. The table shows three of these.

Normal base sequence on the coding strand of DNA		G C A T T C C A G
Type of mutation	**Description**	**Sequence**
Substitution	Replacement of one base by another	A C A T T C C A G G C C T T C C A G
Deletion	Removal of a base	C A T T C C A G
Addition	Addition of base	G G C A T T C C A G

The normal base sequence shown here consists of nine *bases* that code for three *amino acids*: GCA codes for arginine, TTC for leucine and CAG for valine. Look at the first example of a mutation where there has been a substitution. The first base in the original sequence, *guanine*, has been replaced by *adenine*. The first three bases now read ACA not GCA. This codes for a different amino acid, cysteine. This mutation will therefore result in cysteine replacing arginine in the protein produced. The second example of a substitution is a little different. Although the third base in the original sequence, adenine, has been replaced by *cytosine*, the altered sequence, GCC, also codes for arginine. A mutation involving the substitution of a single base may affect a single amino acid in the protein for which it codes. It may have no effect at all.

Where a deletion is involved, however, there may be a much greater effect. In the example above, the first base, guanine, has been lost. This results in a change to all the following codes. The first six bases on this length of DNA will now code for two different amino acids. CAT codes for valine and TCC for tryptophan. A mutation involving deletion, therefore, can affect the rest of the base sequence and, therefore, all of the following amino acids. A mutation involving addition has a similar effect. It changes all the following codes.

gene pool: all the *alleles* present in a particular population at a given time. If no selection takes place, the proportion of the alleles of a particular gene in a population remains the same from one generation to the next. In such a population, you can use the **Hardy–Weinberg principle** to predict the proportions of individuals with particular genotypes. In most populations, however, selection does take place. Some of the organisms in the population will have alleles that mean they are better adapted to particular conditions. These organisms will survive and breed, passing these alleles on to the next generation. This will produce a change in the proportion of alleles in the gene pool.

gene probe: see *DNA probe*.

gene sequencing: finding the order of nucleotide bases (see *base*) in the piece of *DNA* that makes up a *gene*.

gene therapy: treating inherited conditions by altering a person's genes. There are three basic ways in which this might be attempted. The faulty gene could be repaired, it could be replaced by a normal gene, or it could be left in place and a normal gene added as well. At present the most likely approach is the third of these options, leaving the faulty gene in place and adding a normal one. Some experimental work has been carried out on treating **cystic fibrosis** in this way. Cystic fibrosis patients lack the **plasma membrane** protein that transports chloride ions. As a result, they produce thick mucus in their *lungs*. This has to be dispelled by physiotherapy. By introducing the gene for normal chloride ion transport proteins into lung cells of the patient, it is hoped to be able to treat the condition successfully. Treatment like this involves **somatic cells**. These are normal body cells so changes cannot be passed on to the offspring. Gene therapy aimed at egg or sperm cells is not regarded as being ethically acceptable, as unknown and possibly harmful effects might be passed on in this way.

generator potential: the change that occurs in the electrical charge across the **cell-surface membrane** of a receptor as the result of a stimulus. There are many different sorts of receptor but most of them work in a similar way:

- At rest, there is a difference in electrical charge across the cell-surface membrane of a receptor. A sodium pump helps to maintain this difference.
- When the receptor is stimulated, the sodium pump stops working. This allows ions to pass through the membrane, producing a change in the electrical charge. This change is the generator potential.
- The larger the stimulus, the larger the generator potential. If the generator potential is large enough, it produces an **action potential** in the sensory nerve cell that leads from it.

genetic bottleneck: one of the causes of reduced genetic diversity in a population. Sometimes, almost all members of a population are killed. Only a few survivors remain. These survivors will not carry all the **alleles** present in the original population. If the population recovers, it will have a lower genetic diversity. Scientists believe that this is the explanation for the low genetic diversity of elephant seals. These animals were hunted for their blubber and only a few animals remained. All the elephant seals that are alive now have descended from these few survivors.

genetic code: the way in which information about the sequence of **amino acids** in a protein is coded by the bases on a molecule of *mRNA*. The genetic code has a number of characteristics:

- It is a triplet code. Each amino acid is coded by a sequence of three bases. The reason for this is not difficult to appreciate. If each of the four bases found in an mRNA molecule coded for an individual amino acid, it would only be possible to code for a total of four amino acids. If a group of two bases coded for an amino acid, there would be 4 × 4 or 16 possible combinations. Since 20 amino acids are found in **proteins**, this would not be enough. A group of three bases allows 4 × 4 × 4 or 64 combinations. This is enough to code for all 20 amino acids.

- The code is universal. The base sequence CGU, for example, codes for the amino acid, arginine, in all organisms. **Genetic engineering** would be impossible without this property. If we take a gene from one organism and insert it into an organism of a completely different species, it will code for exactly the same protein.
- It is a **degenerate code**. Because there are 64 different combinations of 3 bases and only 20 different amino acids, some combinations will code for more than one amino acid. For example, the combinations, CGA, CGC, CGG and CGU all code for the amino acid, arginine.
- It is a non-overlapping code. Each sequence of three bases codes for a separate amino acid. There is no overlap in the coding sequence.

genetic conservation: maintaining a range of genetic variation. Modern agriculture requires varieties that are productive and profitable. Consequently there has been a tendency towards fewer, more specialised breeds that have little genetic variation. If circumstances change, it is possible that **alleles** that would normally allow the evolution of a better adapted plant or animal are no longer present. Genetic conservation is the use of suitable techniques to maintain stocks of wild organisms or older breeds, so that their genetic potential is not lost. In plants, this is relatively easily done by establishing **seed banks**. With animals, the task is rather more difficult. It is possible, however, to maintain small stocks of older breeds and some success has also been achieved with freezing **gametes** and **embryos**. The concept of maintaining genetic variation is also important in conserving wild animals. In the long term, the breeding success of small isolated populations is unlikely to be high enough to save vulnerable species such as tigers from extinction.

genetic counselling: informing parents who are at risk about the likelihood and consequences of a child inheriting genetic disease, and discussing the various options that are open to them. In this way, they can arrive at an informed decision about a course of action. This will take into account their background and personal beliefs.

genetic drift: a change in the frequency of **alleles** in a population that occurs at random. Chance plays a big part in determining whether particular individuals will survive, mate with each other and produce offspring. Suppose we take an example of a particular allele that is very rare in a population. By chance, in one particular generation, none of the organisms with that allele mate and produce offspring. Obviously none of the offspring will contain that allele. By chance, its frequency will have changed. This is genetic drift. The effect of genetic drift is more important in small populations than in large ones.

genetic engineering: an aspect of **biotechnology** that involves altering the genetic make-up of an organism. Genetic engineering is a rapidly expanding branch of biology and is now used to make a considerable range of useful substances. A wide variety of different techniques is used in producing a particular protein from genetically engineered **microorganisms**, but these techniques generally involve some or all of the stages shown in the flow chart overleaf.

> Use of enzymes, such as *restriction endonucleases* and *reverse transcriptase*, and techniques, such as *electrophoresis*, to isolate the required gene. Alternatively, an artificial gene may be made

↓

> Insertion of the gene into a vector such as a *plasmid* or a *virus*. This process also makes use of enzymes including restriction endonucleases and *ligases*

↓

> Vector used to insert gene into the DNA of the microorganism forming *recombinant* DNA

↓

> Selecting organisms containing this gene and letting them reproduce to form *clones* all genetically identical and all containing the required gene

↓

> Separation of pure product by *downstream processing*

Scientists use similar techniques to insert genes into the DNA of plant and animal cells as well as into microorganisms.

genetic fingerprinting: a technique used to distinguish between individuals by looking at similarities and differences in part of their *DNA*. Not all of the DNA present in the nucleus of a *cell* codes for *proteins*. Some of this 'non-coding' DNA consists of short sequences of bases that may be repeated. The actual number of times these sequences are repeated varies from one person to another. Genetic fingerprinting compares these sequences. The flow chart summarises the main steps involved in the procedure.

> DNA is extracted from a suitable sample. In forensic work, this may be from blood, semen or even from the cells left on a cigarette end

↓

> Enzymes called *restriction endonucleases* are used to cut the DNA in the sample into smaller bits. Some of these will contain the repeated base sequences. The actual length of these pieces will be determined by the number of times that the particular base sequence is repeated in it. Few repeats will give rise to a small piece of DNA. A large number of repeats will produce a longer piece

↓

> The pieces of DNA are separated by *electrophoresis*. This produces a sheet of gel with bands of DNA arranged on it rather like the rungs on a ladder. The distance travelled by each piece of DNA will depend on its length. The smaller pieces will travel farther than the larger ones

↓

> Particular bands of DNA are now located by using a **DNA probe**. The position of these bands on the gel sheet can be used to compare individuals

genetic marker: see *marker gene*.

genetic screening: testing for the *allele* that causes a particular inherited condition. The analysis of family pedigrees only provides information about the probability of inheriting a particular condition. It cannot tell whether a particular person carries the allele for the condition. To do this, genetic screening is necessary. Scientists obtain samples of *cells*. They may get these from an adult by getting the person to wash out his or her mouth. They collect cheek cells from the washings. They can get fetal cells either by *amniocentesis* or by *chorionic villus sampling (CVS)*. The scientists then examine the DNA in these cells. They often use a *DNA probe* to identify the presence of the mutant allele responsible for the condition.

genetically modified (GM): organisms whose *DNA* has been altered by the process of *genetic engineering*. There is considerable public concern over the use of organisms that have been genetically modified, particularly when used in human food. Some of the biological concerns include:

- Genes coding for *resistance* to *antibiotics* are frequently used as *marker genes* to show that another gene has been successfully incorporated into the DNA of the host organism. These resistance genes could spread to pathogenic bacteria, making them difficult to control.
- Genes for *herbicide* resistance have been introduced into crop plants. Powerful herbicides can then be used that kill all species of *weed* but leave the crop unharmed. These genes might spread to closely related species of weed, making them difficult to control. The killing of all other species of plants also has serious ecological consequences.
- Foreign *proteins* resulting from the use of genetic engineering techniques may act as *antigens* and increase the likelihood of *allergy*.

genome: all the genetic material in a single *cell* from an organism. The word is usually used to describe the *genes* contained in its *chromosomes*, but it can also be applied to the genes on the *nucleic acid* of a *virus* or to the genes in the *DNA* of *mitochondria* and *chloroplasts*.

genotype: the genetic make-up of an organism. The genotype describes the organism in terms of the *alleles* that it contains. There are some simple rules that should be followed in choosing the letters that will represent these alleles:

- A single letter should be chosen to indicate the gene, with the capital representing the *dominant* allele and the small letter the recessive allele.

- If possible, the letter chosen should relate to one of the phenotypes. This makes it easier when it comes to translating genotypes into phenotypes.
- The letter should be chosen carefully so that there will not be any confusion between capital and small. It is best to avoid letters, such as c, o and s, where the only difference between the capital and the small is its size.
- Some genes have more than two alleles (see **multiple alleles**). Where this is the case, the gene should be given a capital letter and each allele should be represented by an appropriate letter in superscript. Therefore, with human ABO blood groups, I represents the gene and the alleles are I^O, I^A and I^B.
- Where there is **codominance**, the gene is represented by a single letter and the alleles as capitals written as superscripts. For example, some **flowers** are either white or red if they are homozygous (see **homozygote**) for a particular allele. If they are heterozygous, they are pink. In this case, the two alleles concerned are written as C^R and C^W.

genus: a level of **classification**. One or more different **species** belong to a genus. The scientific name given to a species contains two words. The first word gives the name of the genus and the second word relates to the name of the species. *Rattus norvegicus* and *Rattus rattus* are both rats. Although they have been classified as different species, they are similar in many ways. They are both members of the genus *Rattus*.

geotropism: a growth movement of a whole plant organ in response to gravity. Roots are usually positively geotropic. As a result, roots grow down into the soil away from the light. Stems are usually negatively geotrophic and grow upwards. (See also **tropism**.)

germ-line cell: a **cell** from the body of a multicellular organism that is able to undergo **meiosis** and form **gametes**.

germination: the process during which food reserves present in a seed are broken down and the embryo starts to grow. Germination begins when a **seed** starts to absorb water. The **carbohydrates**, **proteins** and **lipids** that form the food store are broken down by **enzymes**. The root of the embryo or radicle emerges and starts to grow down into the soil. The shoot or plumule grows upwards towards the light. Once the seedling starts to photosynthesise, germination is over.

Germination is controlled by two **plant growth substances**: **abscisic acid (ABA)** and **gibberellin**. As a seed ripens, it loses water. This process is controlled by abscisic acid. Many seeds do not germinate immediately they are produced. They undergo a period of dormancy during which they are prevented from germinating by the abscisic acid still present in the seed. When they are exposed to particular conditions, such as a period of heavy rain in desert plants or the cold weather of winter in temperate species such as apples, the concentration of abscisic acid falls and germination can begin. At the same time, there is an increase in gibberellin. This switches on the genes that are involved in the production of the enzymes which bring about the breakdown of the food reserves of the seed.

GI: see **glycaemic index (GI)**.

gibberellin: a **plant growth substance** that has a number of different functions. One of the most important of these is the promotion of stem elongation. Gibberellins are also involved in **germination**. The flow chart explains the part they play in this process.

A germinating seed such as a barley grain absorbs water

↓

Gibberellin diffuses from the embryo into the aleurone cells. These are the cells that surround the food reserves in the grain. The gibberellin switches on genes and the aleurone cells make large amounts of the enzyme amylase

↓

The amylase hydrolyses the starch stored in the grain to produce sugars. These are used by the growing embryo

Gibberellins are important commercially:

- They may be used to produce seedless *fruit* such as grapes.
- They can be used to alter the amount of space between individual grapes in a bunch. This results in the fruit being farther apart, so they are not as likely to be damaged by fungal infections.
- Some substances are known to prevent plants producing gibberellins. One of these substances is called CCC. This is widely used to produce dwarf plants.

GIFT: see *gamete intra-fallopian transfer (GIFT)*.

gill: an organ concerned with the exchange of respiratory gases. Gills are found in a variety of animals such as fish and some insects that live in water. In fish, water is taken in through the mouth and pumped over the gill surface. Oxygen diffuses from the water into the blood and carbon dioxide diffuses in the opposite direction. The gills have adaptations that make them efficient as gas-exchange surfaces:

- There is a large surface area over which *diffusion* can take place. This is provided by the huge number of thin *gill lamellae*. Each gill lamella contains a large number of *capillaries* that supply it with blood. These *blood vessels* provide a large combined surface area.
- There is a large difference in the concentration of respiratory gases. A *ventilation* mechanism ensures that a continual flow of oxygen-containing water is brought to the respiratory surface. The circulating blood is constantly bringing blood that has a low concentration of oxygen to the gills and removing blood with a high concentration of blood from the gills. In addition, a *counter-current mechanism* helps to maintain an oxygen gradient between the water and the blood.
- The total distance from the water surrounding the gills to the blood in the capillaries is small. Consequently, there is a short diffusion pathway.

gill filament: structures found on the *gills* of fish. Each gill is made up of thin gill filaments, stacked on top of each other rather like pages in a book. On top of each gill filament is a row of thin *gill lamellae*. Oxygen diffuses from the water into blood in *capillaries* in these lamellae.

gill lamellae: structures found in the *gills* of fish. Each gill is made up of two stacks of thin gill filaments. On top of each gill filament is a row of thin gill lamellae. These stand up vertically. Within each lamella is a network of blood *capillaries*. These capillaries are close to the surface, so oxygen in the water flowing over the gill lamellae only has a short distance to diffuse.

GL: see *glycaemic load (GL)*.

gland: a group of *cells* that secretes and releases a particular substance. *Endocrine glands* secrete *hormones* that always go directly from the cells into the blood. Other glands, however, pass their secretions into a duct. These glands are called *exocrine glands*. *Salivary glands* are exocrine glands. The *pancreas* is both an endocrine gland and an exocrine gland. It secretes the *hormones insulin* and *glucagon,* which it releases directly into the blood. It also secretes pancreatic juice. This is released into the pancreatic duct from where it is carried to the *small intestine*. Glands are *effectors*.

global warming: the rise in global temperature thought to have been brought about by an increase in the concentration of *greenhouse gases*, such as carbon dioxide and methane, in the atmosphere. Processes such as the burning of fossil fuels, the clearing of forests and the weathering of limestone have led to an increase in carbon dioxide from about 270 parts per million 200 years ago to over 350 parts per million today. Methane concentrations have also risen considerably. Radiation reflected from the Earth's surface is trapped by these and other gases present in the atmosphere preventing the escape of heat energy. It is difficult to predict exactly how global warming will affect biological processes such as crop production, but there is evidence to support the following suggestions:

- Increased carbon dioxide and temperatures could lead to a higher rate of *photosynthesis* in some plants. It may have no effect or even lead to a lower rate in others. Rate of photosynthesis affects crop yield.
- The life cycle of many insect pests is shorter at higher temperatures. This could lead to significantly greater damage to crops.
- Transpiration rate increases with temperature. This combined with major shifts in world climate could have a significant effect on crop distribution.

globular protein: a *protein* with a molecule that consists of a long chain of *amino acids*, forming a polypeptide. This polypeptide is tightly folded into a compact three-dimensional shape. It is the tertiary structure of these *proteins* that is most important in determining their shape. Unlike *fibrous proteins*, globular proteins are usually soluble in water. *Enzymes* and *hormones* such as *insulin* and *glucagon* are globular proteins.

globulin: a particular type of protein in blood *plasma*. Some globulins are made in the *liver*. They transport iron, *lipids* and some *hormones* around the body. Other globulins are made by *lymphocytes* and are involved in the body's immune response. They are *antibodies*.

glomerular filtrate: the liquid formed from filtering the blood into the *kidney* tubule. The pressure of the blood in the *capillaries* that make up the *glomerulus* forces fluid through the *basement membrane* into the space in the *renal capsule*. This fluid has the same composition as blood *plasma* except that it does not contain protein. Protein molecules are too large to go through the basement membrane.

glomerulus: a small ball of *capillaries* inside one of the *renal capsules* in a *kidney*. The glomerular capillaries differ slightly in structure from those that are found in other organs in the body. The endothelial cells that line them are perforated by small pores. This makes them more permeable than capillaries found elsewhere. The pressure of the blood in the glomerular capillaries is also quite high. There are two reasons for this. First, the *afferent arteriole* that brings the blood from a branch of the renal *artery* is short and straight, so there is only a small drop in pressure. As well as this, the *efferent* arteriole through which the blood leaves the glomerulus is slightly smaller in diameter than the afferent arteriole which brings blood in. The pressure of the blood in the capillaries that make up the **glomerulus** forces fluid through the pores in the walls and the **basement membrane** into the space in the renal capsule. The pores help to determine what can be passed out of the blood to form the **glomerular filtrate** during this **ultrafiltration** process. Blood cells and protein molecules are too large to pass through.

glucagon: a **hormone** produced by the **islets of Langerhans**, a group of **endocrine** cells in the **pancreas**. The secretion of glucagon is stimulated by a fall in blood **glucose** concentration. The main effect of the hormone on the body is to activate **enzymes** in the **liver** that catalyse the conversion of glycogen to glucose. Glucagon also stimulates the formation of glucose from other substances such as **amino acids**.

gluconeogenesis: a biochemical process in which **glucose** is made from substances that are not **carbohydrates**. The concentration of glucose in the blood is normally about 90 mg 100 cm^{-3}. Some of the organs in the body, such as the **brain**, can only use glucose as a **respiratory substrate**, so it is important that the blood glucose concentration is maintained at a constant level. However, glucose concentration falls following a period of exercise or when a person has gone without food for some time. Under these conditions, the **pancreas** secretes **glucagon**. This **hormone** stimulates the **liver** to convert stored **glycogen** to glucose, and it increases gluconeogenesis. Glucose is formed from molecules such as the **lactic acid** produced during **anaerobic respiration** and from **amino acids**.

glucose: a **hexose** sugar – that is, a **monosaccharide** with six carbon **atoms** in each of its molecules. **Carbohydrates** like this have the molecular formula $C_6H_{12}O_6$. The carbon atoms in a glucose molecule join up to form a ring-shaped structure. There are different ways of arranging the other atoms. This results in different **isomers** being formed. Two of these isomers are *α-glucose* and *β-glucose*. They are shown in full and in simplified form in the diagram.

α-glucose

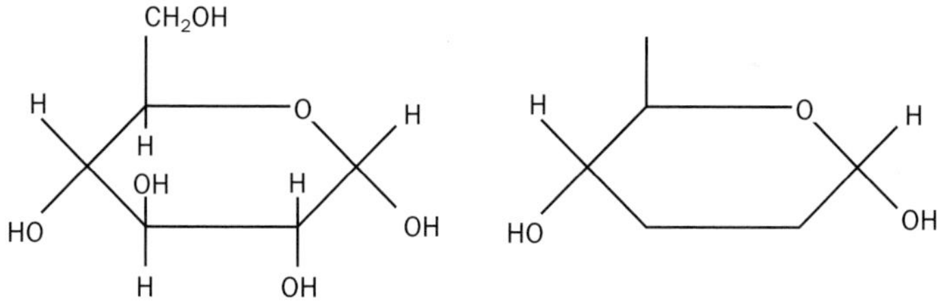

β-glucose

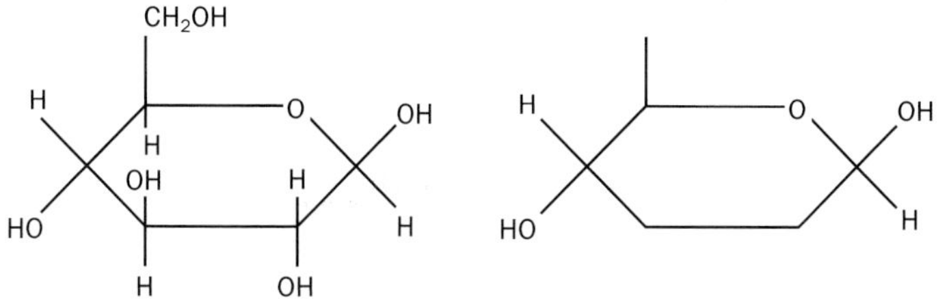

α-glucose is the substance commonly referred to as glucose. It is broken down to release energy in **respiration** and is the building block from which **starch** and **glycogen** are made. β-glucose is the sugar unit found in **cellulose**.

glucose oxidase: an *enzyme* that catalyses a reaction involving the oxidation of *glucose*. The reaction can be represented by the equation:

$$\text{glucose} + \text{oxygen} \rightarrow \text{gluconic acid} + \text{hydrogen peroxide}$$

This reaction is used in biosensors used to detect and measure the concentration of glucose in a solution.

glucose tolerance test: a test used in helping to confirm whether a person has *diabetes*. The person undergoing the test drinks a solution containing 75 g of *glucose* dissolved in 300 cm^3 of water. Samples of blood are then taken at half-hourly intervals for the next 2 hours. The concentration of glucose in these blood samples is measured.

glycaemic index (GI): a measure of the effect that 50 g of *carbohydrate* in a particular food has on the concentration of *glucose* in the blood. This index gives information about how quickly the glucose in a particular food is released in carbohydrate *digestion* and absorbed into the blood. Pure glucose has a glycaemic index of 100. The dietary carbohydrate in green bananas is *starch*. It is only digested very slowly. Green bananas therefore have a low glycaemic index.

glycaemic load (GL): a measure of the effect that the *carbohydrate* in a serving of a particular item of food has on the concentration of *glucose* in the blood. It is calculated from the *glycaemic index (GI)* using the following equation:

$$\text{glycaemic load} = \text{mass of carbohydrate in serving of food} \times \frac{\text{glycaemic index for the food}}{100}$$

glycerate 3-phosphate (GP): a substance containing three carbon *atoms*, which is formed when ribulose bisphosphate (RuBP) combines with carbon dioxide in the *light-independent reaction* of *photosynthesis*. Ribulose bisphosphate is a five-carbon molecule. It is able to combine with carbon dioxide to produce two molecules of glycerate 3-phosphate. Glycerate 3-phosphate is then reduced to triose phosphate. This requires *ATP* and reduced *NADP*, two substances that are produced in the *light-dependent reaction*.

Glycerate 3-phosphate is also formed during *respiration*. It is an intermediate substance in the biochemical pathway of *glycolysis*.

glycerol: a substance that is an important component of *triglycerides* and *phospholipids*. Triglycerides consist of three *fatty acid* molecules that are linked by *condensation* to a molecule of glycerol. In phospholipids, one of the *fatty acids* is replaced by a phosphate group.

glycocalyx: a *carbohydrate*-containing region on the surface of many cells. It consists of the sugars that form part of the *glycolipids* and *glycoproteins* which are embedded in the *cell-surface membrane*.

glycogen: a storage *carbohydrate* found in animals. Glycogen is a *polysaccharide* formed by linking large numbers of α-*glucose* molecules into branched chains. Branched chains are a characteristic feature of glycogen molecules. In a mammal, large amounts of glycogen may be found in the *liver*. Some is also stored in muscles.

glycogen loading: increasing the amount of glycogen present in muscles before an athletic event by eating *carbohydrate*-rich meals. Exercise involves muscle contraction and this requires large amounts of energy in the form of *ATP*. *Respiration* produces ATP and, to maintain a high rate of respiration in muscles for a considerable length of time, a good store of a suitable *respiratory substrate* is necessary. By eating carbohydrate-rich foods such as pasta before the event, more glycogen is stored in the muscles. This reserve can then be respired during the event, which may lead to an improved performance.

glycogenesis: the conversion of *glucose* to glycogen. When the blood glucose concentration rises following a meal, *insulin* is produced by cells in the *pancreas*. This *hormone* has a number of effects on the body, all of which reduce the blood glucose concentration. One of these effects is the activation of *enzymes* that catalyse the conversion of *glucose* to *glycogen*. Most tissues in the body contain some glycogen, but the main stores are in the muscles and the *liver*.

glycogenolysis: the conversion of glycogen to *glucose*. The blood glucose concentration falls following a period of exercise or when a person has gone without food for some time. Under these conditions, the *pancreas* secretes *glucagon*. This *hormone* activates *enzymes*, which catalyse the conversion of *glycogen* to *glucose*. This process is also stimulated by *adrenaline*.

glycolipid: a substance consisting of a *lipid* and a *carbohydrate*. Glycolipids form part of the *cell-surface membrane* surrounding a cell. Glycolipids are located on the outer surface of this membrane with the lipid parts of the molecules forming part of the lipid bilayer and the branched carbohydrate portions sticking out like antennae. Glycolipids have similar functions in membranes to *glycoproteins*. They help cells bind to each other to form tissues and they also allow cells to recognise one another by acting as *antigens*.

glycolysis: part of the biochemical pathway of *respiration* in which a molecule of *glucose* is broken into two three-carbon pyruvate groups. Glycolysis is the first part of the respiratory pathway. It takes place in the *cytoplasm* of the cell. It is common to both *aerobic respiration* and *anaerobic respiration*.

In glycolysis, phosphate groups are first added to the glucose molecule. In other words, the glucose is phosphorylated. This phosphate is supplied by *ATP*. The phosphorylated glucose is then broken down to give two pyruvate groups. The breakdown process produces a total of four ATP molecules, so there is a net gain of two ATP molecules. The complete reaction is an oxidation reaction and it is accompanied by the reduction of a *coenzyme* called *NAD*. The complete process of glycolysis is summarised in the diagram overleaf.

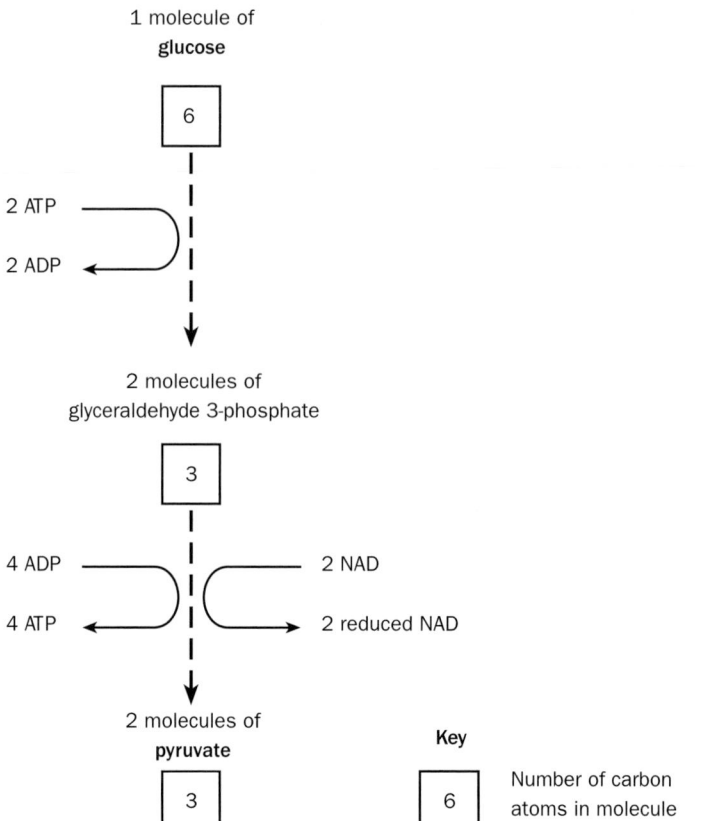

1 molecule of
glucose

2 ATP
2 ADP

2 molecules of
glyceraldehyde 3-phosphate

4 ADP — 2 NAD
4 ATP — 2 reduced NAD

2 molecules of
pyruvate

Key

| 6 | Number of carbon atoms in molecule |

glycoprotein: a substance consisting of a *protein* and a *carbohydrate*. Glycoproteins form part of the *cell-surface membrane* surrounding a cell. The protein parts of the molecules penetrate part or all the way through the membrane while the branched carbohydrate portions stick out like antennae. Glycoproteins have similar functions in membranes to *glycolipids*. They help cells bind to each other to form tissues and they also allow cells to recognise one another by acting as *antigens*. Mucin is an important glycoprotein. It is the main component of mucus.

glycosidic bond: a chemical bond formed as a result of *condensation* between two *monosaccharides*. When two *glucose* molecules are joined in this way, a glycosidic bond is formed between carbon atom 1 of one molecule and carbon atom 4 of the other. This produces a 1,4 glycosidic bond. *Polysaccharides* such as *amylose* are formed in this way. Their molecules consist of long, straight chains. Glycosidic bonds can also form between carbon *atom* 1 and carbon atom 6. This happens in *amylopectin* and *glycogen* molecules, and produces branched chains.

GM: see *genetically modified (GM)*.

goblet cell: a type of cell that produces *mucus*. Goblet cells are pear-shaped cells that are found in the lining *epithelium* of, for example, the *alimentary canal*, the trachea and bronchi and parts of the reproductive system.

Golden Rice™: a variety of rice produced by *genetic engineering*. By inserting genes from a daffodil and a species of soil bacterium, scientists produced *genetically modified (GM)* rice plants. These plants produced grain that contained a high concentration of beta-carotene. This is a substance that humans are able to convert into vitamin A. Golden Rice™ was developed to add to food in areas where there was a shortage of vitamin A in the diet.

Golgi apparatus: an *organelle* responsible for the processing and packaging of substances produced by a cell. When seen with an *electron microscope*, the Golgi apparatus consists of a series of flattened sacs, each one enclosed by a *plasma membrane*. These sacs are continually being formed on one side and pinched off to produce small *vesicles* at the other. Cells that are involved in the secretion of substances contain particularly large amounts of Golgi apparatus. Different processes are associated with this organelle:

- formation of *glycoproteins*. These are substances containing protein and carbohydrate. An example of a glycoprotein is mucin. This is an important component of the mucus produced by the cells that line the gut, reproductive and gas exchange systems
- packaging and secretion of particular *proteins*, such as the digestive *enzymes* produced by cells in the *pancreas*
- secretion of *carbohydrates* which make *cell walls* in plant cells
- formation of *lysosomes*.

gonad: a general name given to a *gamete*-forming organ in an animal. The gonads are the *ovaries* and the *testes*.

GP: see *glycerate 3-phosphate (GP)*.

Graafian follicle: see *ovarian follicle*.

Gram staining: a method of staining bacterial cells that enables two distinct types to be distinguished. The flow chart summarises the main steps in the process.

Bacteria smeared on a clean microscope slide and stained purple with crystal violet solution
↓
Slide washed with Gram's iodine solution then with ethanol
↓
Slide stained with the red stain, safranin

Gram-positive bacteria retain the purple dye when they are washed with ethanol. This hides the effect of safranin, so they appear purple in colour. Gram-negative bacteria lose the purple dye but take up the red one. They appear red. Whether or not the bacteria retain the purple dye depends on the structure of their *cell walls*. Gram-positive bacteria have thick cell walls; the cell walls of Gram-negative bacteria are much thinner.

grana: the stacks of *plasma membranes* found inside *chloroplasts*. They look rather like piles of coins. These membranes are associated with the *light-dependent reaction* of *photosynthesis* and contain the *chlorophyll*, *enzymes* and *carrier molecules* associated with this process.

granulocyte: a type of white blood cell or *leucocyte* that has a lobed *nucleus* and granular *cytoplasm*. The table overleaf summarises the key features of the three main types of granulocyte.

Type of cell	Main features	Function
Neutrophil	The commonest type of granulocyte with fine granules in its cytoplasm	Engulf bacteria by phagocytosis
Eosinophil	Large granules stain pink with a stain called eosin	Involved in allergic responses
Basophil	Large granules stain with basic stains	Produce substances which cause *inflammation*, such as *histamine*

greenhouse effect: the way in which heat energy is trapped in the atmosphere leading to a warming of the environment. Some of the radiation reaching the Earth from the sun has short wavelengths. It is reflected by the surface of the Earth as radiation with longer wavelengths. This is absorbed by various gases in the atmosphere and warms them up. Some of the gases that are normally present in the atmosphere absorb radiation with longer wavelengths. These include water vapour, ozone, carbon dioxide and methane, but some of them are much more effective than others. The rise in carbon dioxide and methane concentrations brought about by human activities causes particular concern, because these gases readily absorb the reflected radiation. It is thought that this is leading to *global warming*.

greenhouse gas: a gas that absorbs radiation that would otherwise escape from the Earth's atmosphere. Common greenhouse gases include water, carbon dioxide and methane. Human activities contribute to the concentration of greenhouse gases in the atmosphere. The table shows some information about some important greenhouse gases.

Gas	Percentage contribution to greenhouse effect	Human activities contributing to its increase in concentration
Water vapour	36–70	Not directly affected by human activity
Carbon dioxide	9–26	Burning of fossil fuels, deforestation and cement production
Methane	4–9	Cattle farming. Carbon dioxide is produced as a waste product by bacteria in the *rumen*
Ozone	3–7	Industrial processes such as electrical discharges

gross primary production: the total amount of *energy* in the *biomass* that the *producers* in a *community* form as a result of *photosynthesis*. Gross primary production is usually described as the amount of energy in the plant biomass in a given area in a given time and is expressed in units such as $kJ\ m^{-2}\ year^{-1}$. This allows a comparison of the production in different *ecosystems*. The gross primary production is the maximum amount of biomass produced. Some of this biomass goes into growth and goes to form new tisssues. This is the net primary production. Some is used in processes such as *respiration* that are needed to keep the producers alive.

gross primary production = net primary production + energy lost in respiration

growth: the addition of more body substance. This usually means that the organism gets larger. Growth excludes, however, temporary changes such as those that result from *osmosis*. The pattern of growth differs between organisms. In animals it is described as being diffuse, with the process taking part in almost all parts of the body. A human child, for example, grows because most of its organs grow. Not only is a ten-year-old taller than a three-year-old, but it has a larger *liver*, heart, *kidneys* and so on. Plants, however, tend to have particular regions of growth. These are called *meristems*. In mammals, *hormones* are involved in controlling the process of growth. These include *growth hormone*, *thyroxine (thyroid hormone)* and the sex hormones, oestrogen and testosterone. In plants growth is controlled by *plant growth substances*.

growth hormone: a *hormone* produced by the *pituitary gland*. Growth hormone controls the *growth* of children. It acts on the *cartilage* cells at the ends of *bones* and on muscle cells where it stimulates growth.

growth rate: the amount of growth made by an organism in a given period of time. The root of a germinating seed may, for example, grow 9 mm in a 24-hour period. Its growth rate would therefore be 9 mm per day. The graph shows the total amount of growth and the growth rate of a human male over the first 18 years of his life.

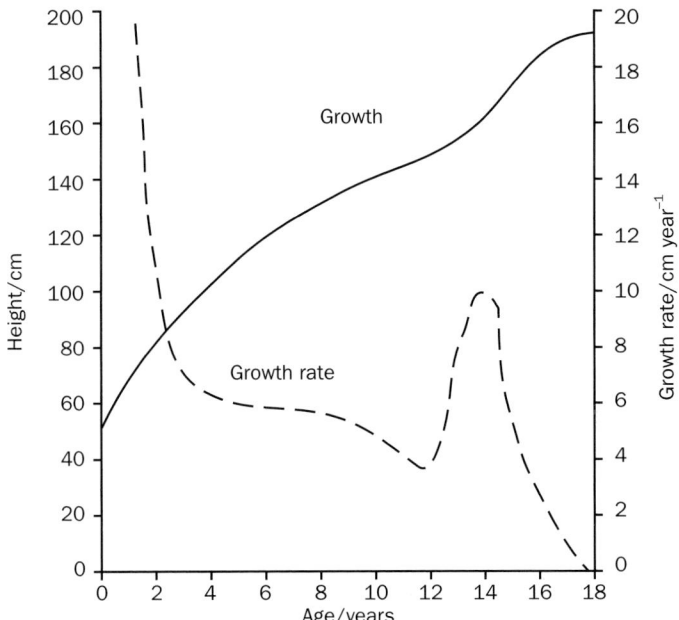

The curve showing the total amount of growth increases steadily before reaching a maximum where it levels out. This fits in with our common appreciation of the pattern of human growth. We get bigger over the first 18 years of our lives; we do not decrease in size. Growth rate, however, is the amount of growth in a particular time. It can decrease as well as increase.

Growth rate gives a much better idea of what is happening at a particular time. A useful analogy might be a car journey. Growth is like the total distance travelled. It only gives information about how far you have got at a particular time. Growth rate, on the other hand, can be compared with speed. It provides much more information about progress at particular times on the journey.

guanine: one of the nucleotide bases (see *base*) found in *nucleic acid* molecules. Guanine is a *purine*, which means that it has two rings of *atoms* in each of its molecules. When two polynucleotide chains come together, guanine always bonds with *cytosine*. The atoms of these two bases are arranged in such a way that three *hydrogen bonds* are able to form between them.

guard cell: one of a pair of cells on either side of a stomatal pore. Guard cells have unevenly thickened *cell walls*. Changes in the *water potential* inside guard cells cause them to change their shape. This change in shape controls the opening and closing of the *stomata* and affects the rates of water loss and gas exchange.

A–Z Online

Log on to A–Z Online to search the database of terms, print revision lists and much more. Go to **www.philipallan.co.uk/a-zonline** to get started.

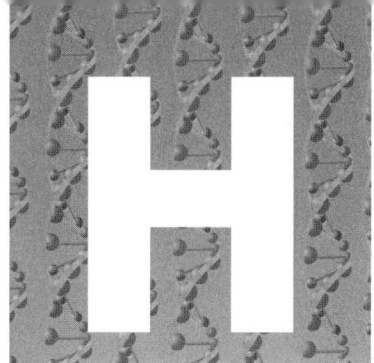

H band: see *H zone*.

H. pylori: see *Helicobacter pylori*.

H zone: a light-coloured area in the centre of the *A band* of a *myofibril* in a skeletal muscle. It is the area where the thick *myosin* filaments that make up the A band do not overlap with the *actin* filaments.

habitat: the particular place where a *community* of organisms is found. Rocky sea-shores, the surface layer of the soil and human skin are all habitats for particular communities of living organisms.

habituation: a form of behaviour in which an animal responds less and less to a repeated harmless stimulus. When a snail crawls across a sheet of glass and the glass is tapped, the snail withdraws into its shell. Eventually it will emerge again and carry on crawling. Tapping the glass again will cause it to withdraw once more into its shell. If this *stimulus* is repeated every time the snail emerges from its shell, eventually it will fail to respond. Other characteristics of this form of behaviour include:

- habituation does not last long if the stimulus is not repeated
- an animal will become habituated more quickly to a second stimulus if it is similar to the first.

Habituation has survival value for the animal concerned. It allows it to adapt to neutral stimuli by not responding. Suppose the snail in the example above always reacted to the tapping stimulus by withdrawing into its shell. Despite the fact that the stimulus was harmless to the animal, the snail would not have time for vital activities such as feeding.

haem: an *iron*-containing chemical group. A molecule of human *haemoglobin* is made up of four units. Each of these units contains a haem group and a polypeptide called globin. Haemoglobin combines with oxygen to form oxyhaemoglobin. In this reaction, oxygen becomes attached to the *iron* in the haem group. Haem is also part of another *respiratory pigment* called *myoglobin*.

After about 120 days in the circulation, red blood cells die and the haemoglobin that they contain breaks down. The haem group is split off from the globin part of the molecule. The iron from the haem group is used to make more haemoglobin and the rest of the group is converted into *bile pigments*. These are excreted in bile.

haemocytometer: a specially marked microscope slide, originally designed for counting the number of cells present in a sample of blood. The centre part of the slide is exactly 0.1 mm deep and it has, within this part, a number of fine lines engraved on the surface. This allows the number of cells in a particular volume of blood to be counted. Multiplication by the appropriate factor gives the number of cells per cubic millimetre. Although haemocytometers are being replaced by electronic methods in hospitals, they are useful in biology laboratories for counting *microorganisms* such as yeast cells.

haemodialysis: a method of removing *urea* and other substances from the blood of someone with *kidney* failure. It relies on the principle of *dialysis*. Blood from an *artery* is passed through a dialyser. The blood is kept apart from a suitable solution of salts by a *partially permeable membrane*. The small soluble molecules of urea pass out of the blood, through the membrane. Larger *protein* molecules and blood cells are unable to do so and remain in the blood. The treated blood is then returned to a *vein* in the patient's body.

haemoglobin: an oxygen-transporting *protein* found in the blood of many animals. In mammals, each haemoglobin molecule consists of four polypeptide chains. Each of these chains is associated with an *iron*-containing *haem* group. Oxygen molecules bind to these haem groups and form oxyhaemoglobin. A single molecule of haemoglobin can transport four molecules of oxygen because there are four haem groups. A graph can be plotted of the amount of oxygen carried by the haemoglobin in the blood against the concentration of oxygen present and a curve drawn through the points. This curve is called an *oxygen dissociation curve*.

There are many different types of haemoglobin. These result from differences in the sequence of *amino acids* which make up their polypeptide chains. They have slightly different oxygen-carrying properties and these properties can be related to the *environment* in which the animal lives. In addition to its oxygen-carrying function, haemoglobin is an important *buffer* and helps to keep the blood at a constant pH.

haemophilia: an inherited condition in which blood clots very slowly. This is because of a deficiency of one of two clotting factors in the blood. Clotting factors are substances produced in the body that are essential for normal *blood clotting*. The *alleles* for both of these clotting factors are *recessive* and are found on the *X chromosome*. Haemophilia is therefore sex linked and almost completely confined to males. Women, however, may act as carriers. Although they are unaffected, they may pass the condition on to their sons. (See also *sex linkage*.)

haploid: a cell, an organism or a stage in the life cycle in which the nucleus contains a single copy of each *chromosome*. The letter n is often used to represent the haploid number of chromosomes. In humans, for example, $n = 23$. In animals, the sex cells or *gametes* are haploid. They are formed by *meiosis*. When *fertilisation* takes place, the nuclei of the male and female gametes fuse and the *diploid* number of chromosomes found in the body cells is restored.

Hardy–Weinberg principle: a principle that can be expressed mathematically and used to predict the proportions of individuals with particular *genotypes* in a population. It relies on the basic idea that, in the absence of *selection*, the proportions of the *alleles* of a particular gene in a population remains constant from one generation to the next. This involves making a number of assumptions:
- It is a large population.
- Mating is random. Individuals with a particular genotype do not mate only with other individuals of the same genotype.
- There is no selection or *mutation*. This would affect the frequency of particular alleles.
- Immigration or emigration does not affect the frequency of particular alleles.

We will look at a specific example. In humans, the shape of the ear lobe is determined by a particular gene. This gene has two alleles. E is the allele for free ear lobes and e is the allele for attached ear lobes. If we consider all the ear lobe alleles in a particular population, a certain proportion will be the **dominant** allele E, and a certain proportion will be the recessive allele e. We will represent the frequency or proportion of the dominant allele by the letter p, and the frequency of the recessive allele by the letter q. Since the ear lobe alleles must be either E or e, we can say that:

$p + q = 1$

The Hardy–Weinberg principle gives the frequency of the genotypes of the individuals in the population. It is represented by the equation:

$$p^2 + 2pq + q^2 = 1$$

where

p^2 is the frequency of the dominant **homozygote** (in the case of the example given here, people with the genotype EE)

2pq is the frequency of the **heterozygote** Ee

q^2 is the frequency of the recessive homozygote ee.

This equation can be used to calculate the expected frequency of individuals with a particular genotype in a population.

In humans, ear lobes may be attached to the side of the head or they may hang free. In a population, 51% had free ear lobes. What percentage of this population would you expect to be heterozygous for the ear lobe gene?

Since the allele for free ear lobes is dominant to the allele for attached ear lobes:

- people with free ear lobes will have either the genotype EE or the genotype Ee. The frequency of these people in the population is 0.51
- people with attached ear lobes can only have the genotype ee. The *frequency* of these individuals in the population will be 1 − 0.51, or 0.49.

The Hardy-Weinberg equation is:

$p^2 + 2pq + q^2 = 1$

In this expression, the frequency of people with the genotype ee is given by q^2, so, combining this information:

q^2	= 0.49
and q	= 0.7
Since p + q = 1	
p + 0.7	= 1
p	= 1 − 0.7
p	= 0.3

Now that we have calculated p and q, it is simple to find 2pq, the expected frequency of the *heterozygotes* in the population:

$2pq = 2 \times 0.3 \times 0.7$
$2pq = 0.42$ or 42% of the population

hay fever: a type of *allergy* resulting from exposure to *pollen*. The pollen usually comes from wind-pollinated plants such as grasses and trees. As many of the plants that produce this pollen *flower* in late spring and summer, hay fever is most frequent at this time of year. The symptoms include sneezing and a running nose produced by the release of *histamine*.

hCG: see *human chorionic gonadotrophin (hCG)*.

HDL: see *high-density lipoprotein (HDL)*.

head louse: a small, wingless, parasitic insect that lives among the hairs on the human head and feeds on blood. Lice have flattened bodies and hooked claws. These modifications enable them to move through the hair and grip tightly. Each of their tiny white eggs is each cemented to a hair and is called a nit. Head lice are found throughout the world and are common among school children in the UK. Treatment involves applying an insecticidal shampoo.

Heaf test: a test for *tuberculosis (TB)*. An instrument that makes a ring of tiny punctures is used to inject a substance called tuberculin into the skin. Tuberculin is obtained from the bacteria that cause tuberculosis. It contains the tuberculosis *antigens* but will not cause the disease. After 3 days the results of the test are noted and interpreted. *Inflammation* indicates a positive response.

health: defined by the World Health Organization (WHO) as 'a state of complete physical, mental and social well being'. Like many everyday words, health is a difficult term to define in precise biological language. Even with the World Health Organization's definition, it is difficult to know exactly what is meant by a 'state of complete well being'. The best approach is probably to list its characteristics. The term health:
- means more than being free of disease
- describes a condition where all of the body's organs and systems are working properly
- takes into account mental as well as physical health.

heart block: a condition in which the heart beat is poorly coordinated and the heart fails to pump effectively. It is caused by abnormal conduction of the electrical impulses produced by the *sinoatrial node (SAN)*. There are three types of heart block:
- first degree heart block. The conduction of the *nerve impulse* from the *atria* to the *ventricles* is slowed
- second degree heart block. Some, but not all, of the impulses pass from the atria to the ventricles
- third degree heart block. No impulses pass from the atria to the ventricles, so the atria and the ventricles beat at different rates. The atria may, for example, beat at 70 beats per minute while the ventricles beat at only 40 beats per minute.

Heart block may be investigated from an *electrocardiogram (ECG)*.

heart cycle: see *cardiac cycle*.

heart sounds: the sounds that can be heard from the heart when a *stethoscope* is placed against the chest wall. The first of these heart sounds is the *atrioventricular valves* closing. The second heart sound is the closing of the *semilunar valves*.

Helicobacter pylori: a species of bacterium (see *bacteria*) that is important in causing ulcers in the *stomach*. In the stomach, *H. pylori* may cause *inflammation* and a condition called gastritis. This reduces the resistance of the stomach wall to damage by acid and *pepsin* and an ulcer may form. *H. pylori* is also associated with ulcers in the *duodenum*.

hepatic artery: the *blood vessel* that supplies oxygenated blood to the *liver*.

hepatic portal vein: the *blood vessel* that goes from the *small intestine* to the *liver*. It takes the products of *digestion* to the liver for further processing. Most *veins* in the body start from a *capillary* network. Blood drains from this capillary network into larger vessels known as venules and then into the veins. These return the blood to the heart. However, portal veins are different. They start with a capillary network and they also end in one. The concentration of *glucose* in the hepatic portal vein varies more than it does in other blood vessels. This is because of the glucose that has been absorbed from the small intestine after a meal.

hepatic vein: the *blood vessel* that returns blood from the *liver* directly to the heart.

hepatitis: a disease involving *inflammation* of the *liver*. Infectious hepatitis is caused by one or other of the hepatitis *viruses*. These include:

- hepatitis A. This is usually spread by contaminated food or water and is associated with poor hygiene and sanitation. A *vaccine* can be used to protect travellers to areas where hepatitis A is common
- hepatitis B. This is contracted from infected blood or blood products or from sexual contact. It may develop into long-term chronic hepatitis
- hepatitis C. The commonest source of infection is contaminated blood or blood products. The hepatitis C virus is an *RNA*-containing virus that has only recently been distinguished from other viruses that cause hepatitis.

herbicide: a substance used to kill *weeds*. Weeds are economically important as they reduce crop yields by competing for light, water and inorganic ions. From an agricultural point of view, it is therefore important that they are controlled. There are many different types of herbicide available, but they can be classified by the way in which they act:

- contact herbicides kill only those parts of the plant with which they come into contact. They are not effective in dealing with the underground roots and stems of weeds such as thistles and nettles
- *systemic herbicides* are absorbed by leaves and roots and transported to all parts of the plant where they cause death by interfering with the natural growth pattern. They have little effect on *monocotyledons* such as cereals and other grasses, but they rapidly bring about the death of dicotyledonous weeds (see *dicotyledons*). They are suitable, therefore, for controlling weeds on lawns or in cereal crops.
- residual herbicides can be added to bare soil before the crop emerges or can be used in providing total weed removal from small areas of ground. They remain active in the soil and kill the seedlings as they emerge.

herbivore: an animal that feeds on plants. In a *food web*, a herbivore is a *primary consumer*.

herd effect: indirect protection from a disease when a proportion of the population has been vaccinated against it. *Infectious diseases* are spread from one person to another. When a high proportion of people have been vaccinated, there will be a lot of immune people in the population. This makes the probability low that a person who is susceptible will come into contact with an infected person. The percentage of people that need to be vaccinated for herd immunity to be effective differs from one disease to another, but it is estimated to be between 75% and 95%.

heterochromatin: *chromatin* that appears densely stained when looked at with a micro-scope. It consists of *DNA* that is tightly packed. Little of the DNA that makes up heterochro-matin is undergoing *transcription*.

heterotroph: see *heterotrophic nutrition*.

heterotrophic nutrition: a method of nutrition in which an organism gains its nutrients from complex organic substances. These are broken down into simple soluble molecules by digestive *enzymes* and then built up again to form the organic molecules that the organism requires. As well as this being the method of nutrition in free-living animals, *saprobionts* and *parasites* also gain their nutrients this way. Heterotrophs are the *consumers* in *food chains*. They ultimately depend on *autotrophs*.

heterozygote: an organism in which the *alleles* of a particular *gene* are different from each other. In peas, pod colour is determined by a gene with two alleles. The allele for green pods, G, is *dominant* to the allele for yellow pods, g. The heterozygote will have the *geno-type* Gg. It will have green pods as the allele for green pods is the dominant one.

hexose: a sugar that has six carbon *atoms* in each of its molecules. Hexoses are *mono-saccharides*. Each molecule consists of a single sugar unit. Biologically important hexose sugars include *glucose* and *fructose*.

hexose bisphosphate: a sugar that has six carbon *atoms* and two phosphate groups in each of its molecules. The biochemical reactions of *glycolysis* start with converting *glucose* to hexose bisphosphate by adding phosphate groups. These phosphate groups come from *ATP*.

high-density lipoprotein (HDL): a *lipoprotein* containing a high concentration of *cholesterol*. High-density lipoproteins are synthesised in the *liver* and the wall of the intestine. They remove cholesterol from tissues and transport it back to the liver cells from where it is excreted from the body in *bile*. High concentrations of HDL are associated with a lower risk of *coronary heart disease*.

histamine: a substance produced as part of the immune response. Histamine is a local chemical mediator and only affects cells in the immediate vicinity. It dilates *blood vessels* and increases the permeability of *capillaries* to white blood cells and to *proteins*. Histamine produces many of the features associated with *inflammation* round a cut or a scratch, or with an insect sting or bite. *Hay fever*, *urticaria* and *asthma* are conditions in which the body is sensitive to particular *allergens*. In a person suffering from an *allergy*, *mast cells* in the tissues are stimulated by these allergens and secrete histamine. Histamine is mainly responsible for the characteristic symptoms of the particular allergy.

histocompatibility: tissues that can be transplanted from one person to another without being rejected by the immune system are histocompatible. In the same way that red blood cells have *antigens* on their surfaces (see *blood group*), other tissues have similar mol-ecules called histocompatibility antigens on the *cell-surface membranes*. These antigens are determined genetically. Even if two people are closely related to each other, they may possess different histocompatibility *alleles*. However, they are much more likely to be geneti-cally similar and have compatible antigens if they are closely related. The closer the match between donor and recipient, the more likely it is that a transplant will be successful. If the histocompatability antigens are not compatible, there is a strong possibility that the transplanted organ or tissue will be rejected.

histone: a type of protein found with *DNA* in *chromosomes*. In *eukaryotic* cells, such as those of animals and plants, half of the total mass of each chromosome is DNA and half is protein. Histones are the commonest type of protein and are important in the packaging of DNA to form chromosomes. *Prokaryotic* cells such as *bacteria* do not contain histones.

HIV: human immunodeficiency virus is the virus that is responsible for AIDS (acquired immune deficiency syndrome). Each *virus* is made up of an *enzyme*, called *reverse transcriptase*, and an *RNA* molecule. The RNA molecule carries the genes of the virus. The enzyme and the RNA are surrounded by a protein coat. Finally, the whole virus particle is enclosed in a piece of *plasma membrane*, which comes from its human host. The virus is transmitted in blood *plasma* and sexual fluids. When a person is infected by HIV, *proteins* in the membrane surrounding the virus bind to specific *proteins* on the surface of a type of white blood cell known as a *T helper cell*. The membrane surrounding the virus fuses with that of the white blood cell, and the RNA molecule and the enzyme enter the host cell. Reverse transcriptase is an enzyme that makes a *DNA* copy from the RNA of the virus. This DNA copy is now inserted into the human DNA in the white blood cell where it can remain for a long time. Sooner or later, however, new viruses are made using the genetic information on this inserted piece of DNA. These viruses infect other white blood cells.

hives: see *urticaria*.

HLA system: four groups of genes that encode *proteins* present on the surface of most human cells. These proteins are *histocompatibility antigens*. They determine whether tissues can be transplanted from one person to another without being rejected by the immune system.

homeobox sequence: a sequence of 180 *DNA* bases that encodes a sequence of *amino acids* found in DNA-binding *proteins*. These DNA-binding proteins are involved in switching on and off *genes* that control the development of organisms. Homeobox sequences are highly conserved. This means that they are similar in all organisms.

homeostasis: the maintenance of constant internal conditions. All living organisms regulate their internal *environment* and show homeostasis to some extent. Even a simple organism like an amoeba has a *contractile vacuole* that regulates the amount of water in the *cytoplasm*. Mammals, however, have a very extensive range of homeostatic mechanisms. These maintain the levels of features such as temperature, pH, *water potential* and blood *glucose* concentration. Many of these mechanisms rely on *negative feedback*. This is the process in which departure from a set level is detected by receptors. These convey information to *effectors* which bring about a return to the original value.

Homeostasis is important to living organisms because:
- biochemical reactions in living organisms are controlled by *enzymes*. Fluctuations in pH and temperature affect the rate of enzyme-controlled reactions. In extreme cases, they lead to denaturing of enzymes and other *proteins*
- most biochemical reactions are reversible and an equilibrium is maintained between the various substances in a cell. If there were major fluctuations in the concentrations of these substances, it would not be possible to maintain equilibrium
- the external environment in which many organisms live shows wide fluctuations in *abiotic* factors such as temperature. A constant internal environment allows independence from

these fluctuations and lets organisms such as mammals live in areas ranging from the arctic to the tropics

- water moves into or out of cells by *osmosis*. By maintaining a constant water potential in the surrounding tissue fluids, osmotic problems are avoided.

home base: the central core area of the *home range*. In human *hunter-gatherer societies* the home base is the place to which the group returns at the end of the day.

home range: the area in which a group of animals normally lives and carries out its day-to-day activities, such as getting food and water. A part of the home range that is defended against other members of the same species is known as a *territory*.

hominid: modern humans and their closely related fossil relatives. Some biologists think that, as humans are closely related to the gorilla and chimpanzee, these two animals ought to be classified as hominids as well. There is some uncertainty over which definition we should adopt. It is probably better, however, to regard hominids as the group containing modern humans and their direct relatives only. If we adopt this approach, hominids can be charac-terised by two particular features. They are able to walk upright and they have relatively large *brains*. Associated with these features is another important characteristic: the ability to use tools. The first hominids appeared in the fossil record around 4 million years ago. From these ancestral species, two distinct genera appeared: *Australopithecus* and *Homo*.

Homo: the *genus* to which modern humans and their immediate ancestors belong. Their most obvious characteristic is that they have a large *brain* when compared to the size of the body. The large brain is associated with changes in the shape of the skull. The cranium or brain case is large in relation to the face which is consequently much flatter than in early *hominids*. Many of the characteristics that help to define the genus, such as the use of language and the development of complex societies, are behavioural rather than structural. They create difficulties because they do not leave any direct fossil evidence.

homologous chromosomes: *chromosomes* that pair with each other during the first division of *meiosis*. Each member of a pair of homologous chromosomes has the same sequence of gene loci (see *locus*), but the individual *alleles* are not always identical. For example, in humans there is a protein that is responsible for the transport of chloride ions. The gene responsible for producing this protein is on chromosome number 7. Both copies of chromosome 7 present in a cell will have the locus for the gene for this protein. However, if a person is *heterozygous* for the gene, one chromosome will have a copy of the allele for making the protein while the other chromosome will have the allele for the defective protein. If a person has two copies of the allele for the defective protein, he or she will have *cystic fibrosis*.

homologous structures: structures that have different functions but a common evolu-tionary origin. An example is provided by the wing of a bird and the front leg of a dog. They look very different and they have different locomotory functions. However, they are both varia-tions of the basic vertebrate limb. Careful examination of the skeleton of both of these limbs reveals this similarity.

homozygote: an organism in which the *alleles* of a particular *gene* are identical to each other. In peas, pod colour is determined by a gene with two alleles. The allele for green pods, G, is *dominant* to the allele for yellow pods, g. There are two different homozygotes. The dominant homozygote has the *genotype* GG and green pods. The *recessive* homozygote has the genotype gg and yellow pods. Homozygotes may be described as pure breeding as, when crossed, they produce offspring identical to their parents.

horizontal gene transmission: the passing of *DNA* from one mature cell to another. Horizontal gene transmission is common in bacteria. *Antibiotic resistance* may spread from one species of bacterium to another when DNA is transferred during *conjugation*.

hormone: a substance that is transported in the blood and acts as a chemical messenger. Hormones are produced by the cells and *glands* that make up the *endocrine system*. The blood transports them throughout the body, but they only produce responses in cells in target tissues. Many of the hormones found in a mammal are either *proteins* such as *insulin*, or *steroids* such as *oestrogen* and *progesterone*. The type of substance determines the way in which the hormone works. Protein hormones attach to receptor molecules on the *cell-surface membrane*. This triggers a series of reactions leading to the activation of particular *enzymes* within the cell. Because there is a *cascade effect*, a small number of hormone molecules leads to the activation of relatively large numbers of enzyme molecules and the production of even larger amounts of product inside these cells. Steroid hormones, on the other hand, work in a different way. They penetrate the cell-surface membrane. Once inside the cell, they activate particular genes.

hormone replacement therapy (HRT): treatment with female sex *hormones* to overcome the unpleasant symptoms that occur after a woman passes through the *menopause* or after her *ovaries* have been removed surgically. As a woman reaches the end of her reproductive life, her ovaries stop working. The lowered concentrations of *oestrogen* and *progesterone* result in unpleasant symptoms. These may include circulatory and psychological disorders as well as an increased tendency to develop *osteoporosis* and some types of heart disease. A combination of oestrogen and progesterone, or oestrogen only, is prescribed in the form of tablets, implants or skin patches.

hPL: see *human placental lactogen (hPL)*.

HRT: see *hormone replacement therapy (HRT)*.

human chorionic gonadotrophin (hCG): a *hormone* produced during the early stages of pregnancy by the cells that will become the *placenta*. Following ovulation, the *corpus luteum* in the ovary produces the hormone *progesterone*. This hormone is responsible for maintaining the lining of the uterus. If *fertilisation* does not take place, the corpus luteum starts to break down. Progesterone is no longer produced and menstruation takes place (see *menstrual cycle*). If fertilisation takes place, therefore, it is important that the corpus luteum continues to produce progesterone. The main function of human chorionic gonadotrophin is to maintain the corpus luteum. Pregnancy can be diagnosed early by testing for this hormone. It usually appears in measurable quantities in the *urine* about 2 weeks into the pregnancy.

human genome project: an international research project involving the determination of the base sequence of the genes on all 23 pairs of human *chromosomes*. As there are about 100 000 genes and 3000 million *base pairs* in the human *genome*, this is a massive undertaking. The knowledge derived so far from this project has resulted in the identification and sequencing of many of the genes associated with human genetic disorders. In theory, it could lead to the development of successful treatments. There are fears, however, that the information revealed could be abused.

human placental lactogen (hPL): a polypeptide *hormone* produced by the *placenta* during pregnancy. It alters the mother's metabolism and increases the supply of *glucose* to the fetus. Human placental lactogen promotes the breakdown of *fats* in the mother's body. She uses the *fatty acids* produced as a *respiratory substrate*. This makes more glucose available to diffuse across the placenta to the fetus (see *diffusion*). The hormone also decreases the sensitivity of the mother's cells to *insulin*. This increases her blood glucose concentration and also makes more glucose available for the developing fetus.

hunter-gatherer society: a human society in which the main means of getting food involves hunting animals and gathering wild plants. Until about 9000 years ago, many human societies were hunter-gatherer societies. There are a few groups of people living today who still get a substantial proportion of their food in this way. Evidence from these modern people and from archaeological remains suggest that early hunter-gatherer societies had a number of features in common:

- they involved relatively small groups of 40 or so people living in a closely integrated community
- individual women had relatively few children and often went from 4 to 6 years between births
- the incidence of *infectious disease*, particularly in earlier societies, was extremely low
- there was a marked *division of labour*. The men hunted and the women and older children gathered plant foods.

Members of a group exploited an area around a *home base* to which they returned in the evening. When this area was exhausted, they moved on. Largely because of the need to move from place to place, they had few personal belongings.

Huntington's disease: an inherited condition in which an affected person shows symptoms that include involuntary jerky movements and progressive mental degeneration. These symptoms do not develop until the person reaches middle age. Huntington's disease is a *dominant* condition and the *allele* responsible is on chromosome 4.

hydrogen bond: a type of chemical bond. Hydrogen bonds require relatively little energy to break. These bonds are important in the three-dimensional structure of biological molecules, such as *proteins*, *polysaccharides* and *DNA*. Hydrogen bonding also gives water a number of its properties such as the ability of its molecules to stick to each other by cohesion.

hydrolase: an *enzyme* that is responsible for catalysing reactions which involve the splitting of larger molecules into smaller ones by *hydrolysis*. Digestive enzymes are hydrolases. *Amylase*, *lipase* and *protease* are important hydrolases.

hydrolysis: the splitting of a large molecule into smaller molecules by the addition of water. The digestive *enzymes* found in the gut are *hydrolases* and hydrolyse their *substrates*. Chemically, hydrolysis is the opposite reaction to *condensation*.

hydrophyte: a plant that is adapted to living in water. General features shown by hydrophytes include:

- their stems contain *aerenchyma*, a tissue that contains a mixture of cells and large, air-filled spaces. Aerenchyma provides buoyancy and its air-filled spaces form a pathway allowing the *diffusion* of oxygen and carbon dioxide to parts of the plant that are under water
- the leaves of many hydrophytes float on the water surface. The *stomata* are generally on the upper surface of these leaves
- support is provided by the water. The stems of hydrophytes contain little if any of the woody tissue that provides support in larger land-living plants.

hydrostatic skeleton: the skeleton that is found in soft-bodied animals such as worms. A worm has a body cavity or *coelom* that contains coelomic fluid. Round the outside of this are two layers of muscle: longitudinal muscle that goes along the length of the animal and circular muscle that runs round the circumference. Squeezing a liquid does not change its volume. Therefore, when the muscles contract, the animal will not get smaller; it will only change its shape. Contraction of the longitudinal muscles results in it becoming shorter and fatter, while contraction of the circular muscles results in it becoming longer and thinner. These changes in shape allows the animal to move.

hyperglycaemia: a condition in which the concentration of *glucose* in the blood is higher than normal. In healthy people, the rise in blood glucose concentration that follows a meal stimulates the secretion of *insulin*. Insulin then reduces the blood glucose concentration. People with *diabetes* either secrete insufficient insulin or the insulin that they secrete is not effective. Hyperglycaemia is therefore characteristic of untreated diabetes.

hyperpolarisation: an increase in the potential difference across a *cell-surface membrane*. Because the inside of a cell is normally negatively charged, hyperpolarisation leads to the inside of the cell being even more negative. Hyperpolarisation of *rod cells* is important in producing *action potentials* in optic neurones.

hypersensitivity: an abnormal reaction to a particular *antigen*. Exposure to this antigen may result in a variety of symptoms. These may involve local effects such as those associated with *hay fever*. Occasionally symptoms may involve the whole body and anaphylactic shock may result (see *anaphylaxis*).

hypertension: occurs when arterial blood pressure remains high even when a person is at rest. Hypertension can have serious health effects, as shown in the flow diagram.

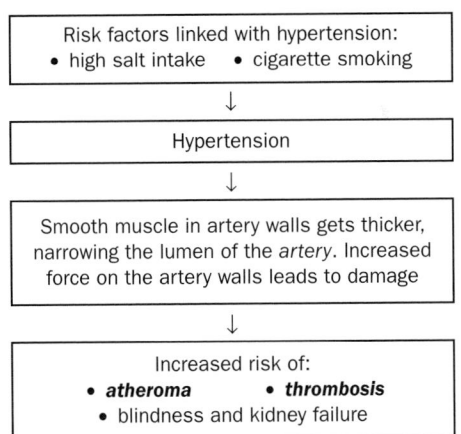

Risk factors linked with hypertension:
- high salt intake
- cigarette smoking

↓

Hypertension

↓

Smooth muscle in artery walls gets thicker, narrowing the lumen of the *artery*. Increased force on the artery walls leads to damage

↓

Increased risk of:
- *atheroma*
- *thrombosis*
- blindness and kidney failure

hyperthermia: a condition in which the core body temperature is higher than normal. This may occur when a person is exposed to a hot, humid environment and may result in heat stroke. Sweating stops and *arterioles* taking blood to the skin dilate. This deprives internal organs of oxygen and may result in *brain* and *liver* damage. Hyperthermia can also be produced by some so-called recreational drugs and as a result of fever produced by infections.

hypertonic: one solution is hypertonic to another if it has a higher solute concentration. When cells are put in a hypertonic solution, the concentration of dissolved substances in the *cytoplasm* of the cell will be lower than that in the surrounding solution. The **water potential** of the cytoplasm will, therefore, be higher than the water potential of the surrounding solution. As a result, water will move out of the cell by **osmosis**. In plant cells, this results in **plasmolysis**.

hyphae: the thread-like structures which make up a fungus (see **Fungi**). Each hypha is surrounded by a **cell wall**. This cell wall is made of the nitrogen-containing polysaccharide **chitin**. There are one or more nuclei and a number of vacuoles in the **cytoplasm**. The cell wall at the tip of a hypha is thin and this is where growth takes place. If you use a microscope to look at a living hypha, you will see that the cytoplasm is constantly moving, flowing towards the tip and away from it. This is called cytoplasmic streaming and is the process which transports food materials to the growing tip. Further back along the hypha, the walls are much more rigid and growth does not take place here.

hypoglycaemia: a condition in which the concentration of **glucose** in the blood is lower than normal. Hypoglycaemia occurs most often in people with **diabetes**. It may result from injecting too much **insulin** or from failing to eat sufficient **carbohydrate**.

hypothalamus: an area in the **brain** that has a number of important functions. These include:

- control of many aspects of **homeostasis**. For example, the hypothalamus contains receptors which monitor the temperature of the blood. **Nerve impulses** go from these receptors to the temperature control centres. The temperature control centres are also in the hypothalamus. They coordinate the body's response either by increasing the amount of heat that is lost or by conserving it within the body
- integration of the nervous and hormonal systems. The reproductive behaviour of some mammals is initiated by a change in day length. Information concerning day length reaches the hypothalamus from sensory receptors. This information stimulates the hypothalamus to produce substances known as release factors. These travel in the blood to the anterior lobe of the **pituitary gland**, where they trigger the production of the **hormones** which control the sexual development and activity concerned with breeding.
- coordination of the activity of the **autonomic nervous system**.

hypothermia: a condition in which the core body temperature is lower than normal. In humans, prolonged exposure to the cold can produce hypothermia, particularly in the very young and the elderly. There is a slowing down of metabolic and physiological processes. Breathing and heart rate fall, blood pressure drops and finally the person may lose consciousness. Reducing the body temperature or induced hypothermia is sometimes used when surgical operations are carried out on the heart or **brain**. This is an advantage because the heart can be stopped for relatively long periods, as the rate of **respiration** is much lower and the tissues require less oxygen. In addition, the resulting low blood pressure reduces the dangers of extensive bleeding.

hypotonic: one solution is hypotonic to another if it has a lower solute concentration. When cells are put in a hypotonic solution, the concentration of dissolved substances in the **cytoplasm** of the cell will be higher than that in the surrounding solution. The **water potential** of the cytoplasm will, therefore, be lower than the water potential of the surrounding solution. As a result, water will move into the cell by **osmosis**.

hypoxia: a low amount of oxygen reaching the tissues. There are a number of reasons why the amount of oxygen reaching the tissues may be inadequate to meet the needs of *respiration*. These include:

- a low partial pressure of oxygen in the blood. This may happen at high altitudes, where there is not enough oxygen present to saturate the *haemoglobin*. Some diseases which affect the *lungs* can also result in hypoxia. These include *emphysema* and pneumonia
- the amount of haemoglobin in the blood is insufficient to transport the oxygen. This occurs when a person is anaemic
- reduced blood flow to the tissues concerned.

Do you need revision help and advice?

Go to pages 306–314 for a range of revision appendices that include plenty of exam advice and tips.

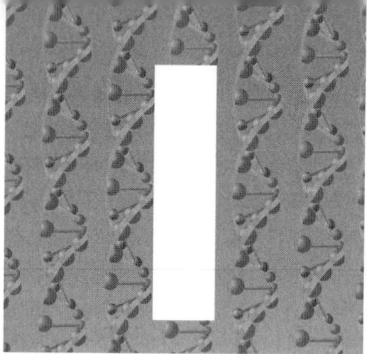

I band: one of the light bands that runs across a *myofibril* in a skeletal muscle. It corresponds to filaments of the protein *actin*.

IAA: see *indoleacetic acid (IAA)*.

IgE antibody: antibodies that are involved in allergic responses. They are bound to the surface of *mast cells*. These cells are particularly abundant in the walls of the airways in the *lungs* and in the gut wall. When a person with an *allergy* encounters the *allergen* that causes the condition, it binds to these antibodies. This causes the mast cell to release substances such as *histamine*. Histamine is mainly responsible for the characteristic symptoms of the particular allergy.

ileum: the last part of the *small intestine*. The human ileum is a tube about 5 m long. Its inside surface is covered in folds. Each of these folds has many tiny finger-like projections known as *villi* on its surface. In turn, the epithelial cells that line these villi are covered with *microvilli*. The ileum, therefore, has a very large surface area. This is an adaptation for absorbing the products of *digestion*. The villi contain *capillaries* and small *lymph* vessels called *lacteals*. The capillaries take the *amino acids* and *glucose* produced by digestion to the *hepatic portal vein* and the *liver*. The lacteals absorb the products of *fat* digestion. The cells that line the ileum also produce *enzymes* that digest *proteins* and *carbohydrates*. These enzymes are on the *cell-surface membranes* and in the *cytoplasm* of the epithelial cells. Layers of circular and longitudinal muscle push digested food along the ileum by waves of muscle contraction called *peristalsis*.

immobilisation: *enzymes* or *microorganisms* that are bound to larger particles which are then used in *biotechnological* processes. There are several advantages in using immobilised enzymes. If the enzyme is attached to a large insoluble particle instead of being free in solution, it is much easier to remove at the end of the reaction. This is cheaper as the enzyme can be used again and it helps to avoid contamination of the product. Immobilisation results in enzymes being able to work at higher temperatures and over a wider pH range than enzymes that are free in solution.

immune response: see *immunity*.

immunisation: an artificial method of producing immunity. Passive immunisation is the injection of *antibodies* made by another organism. It is useful where the body would not have enough time to produce antibodies of its own. Passive immunisation is used to treat snake bite and certain diseases such as tetanus. Antibodies produced by one organism become foreign *proteins* when they are introduced into the body of another. They act as *antigens* and are usually destroyed rapidly. Protection by passive immunisation is therefore only short-lived. Active immunisation (see *active immunity*) involves stimulating the body to produce its own antibodies. This is done by injecting live *microorganisms* that have been treated in such a way as to make them harmless, dead microorganisms or the toxins produced by microorganisms. Active immunisation is often referred to as vaccination.

immunity: the way in which white blood cells and the *antibodies* that some of them produce enable the body to fight infection.

immunosuppression: lowering or suppressing of the activity of the immune system. This results in a person being more susceptible to some diseases. AIDS results from the destruction of a particular group of white blood cells by the human immunodeficiency virus (*HIV*). As a result, AIDS patients are much more likely to become infected by pathogenic organisms that do not normally affect healthy people. Some conditions, such as athlete's foot (a fungal infection of the skin between the toes), are usually only mild infections in healthy people. They may be severe in AIDS patients.

When an organ such as a *kidney* or heart is transplanted from one person to another, it is likely that it will be rejected by the recipient's immune system. In order to reduce the risk of this happening, the recipient is treated with immunosuppressive drugs. Unfortunately, this treatment increases the susceptibility of the patient to infections and to some types of cancer. (See also *tumour*.)

Immunosuppressive drugs are also used in the treatment of *autoimmune diseases* such as rheumatoid arthritis.

immunotherapy: the prevention or treatment of a disease by therapy that affects the immune system. Applications of immunotherapy include:
* *hypersensitivity*. People who are hypersensitive can be made less sensitive by giving gradually increasing doses of the relevant *allergen*. This has been found to be effective in treating allergies to wasp and bee stings (see *allergy*)
* *immunosuppression*. Drugs that suppress the activity of the immune system are important in preventing the rejection of organ transplants
* *monoclonal antibodies*. *Antibodies* can be produced that bind to specific *antigens* found only on the surface of cancer cells. They can be used to target drugs in the treatment of cancer. (See also *tumour*.)

implantation: the attachment of a developing mammalian *embryo* to the wall of the uterus. *Fertilisation* of an egg cell results in a zygote. The zygote divides and develops into a hollow ball of cells called a *blastocyst*. During implantation the blastocyst burrows into the wall of the *uterus*. In humans, implantation leads to:
* hormonal changes in the mother that prevent the breakdown of the *corpus luteum* and stop menstruation from taking place (see *menstrual cycle*). A *hormone* called *human chorionic gonadotrophin* (hCG) is produced. Its main function is to maintain the corpus luteum. Pregnancy can be diagnosed early by testing for the presence of this hormone. It is usually present in the *urine* about 2 weeks after fertilisation
* growth of the embryo and the development of the placenta.

imprinting: a form of learnt behaviour shown by the young of some species. When orphaned lambs are hand-reared, they learn to follow the person who looks after them. When they are returned to the flock, they still try to find the person who reared them and stay close by. Similar behaviour can be seen in birds, such as ducks and geese, where the young leave the nest almost immediately after hatching and remain near the parent bird. This increases their chances of survival.

impulse: see *nerve impulse*.

in situ **conservation:** the protection of endangered species of animals and plants in their natural **environment**. *In situ* conservation involves maintaining the **habitat** of the organism so as to allow it to exist in sufficiently large numbers to maintain genetic **diversity**.

in vitro: experiments carried out in test-tube conditions rather than on living materials. Inside a cell, there may be many hundreds of biochemical reactions taking place at the same time. It is obviously much simpler to look at individual reactions and study these on a test-tube scale. Care must be taken, however, in interpreting the results of *in vitro* investigations. Something may happen in a test-tube, but this it does not mean that it will occur in exactly the same way in a living organism.

in vitro **cloning:** the use of the **polymerase chain reaction (PCR)** in cloning fragments of **DNA**.

in vitro **fertilisation (IVF):** a technique that may be used to treat women who are infertile because, for example, they have blocked **oviducts**. The woman is given **hormone** treatment. This causes several of her egg cells to mature at the same time. These egg cells are then removed from her ovary and mixed with sperm cells from her partner. The fertilised eggs are incubated outside her body until they have developed into hollow balls of cells called **blastocysts**. One or more of these blastocysts are put into her **uterus** and the pregnancy is allowed to continue normally.

in vivo: refers to experiments carried out on living materials rather than in test-tube conditions. (See also *in vitro*.)

in vivo **cloning:** the use of **restriction endonucleases** and **ligases** to insert **DNA** fragments into **vectors** which are then transferred to host cells. The host cells reproduce and **clone** the fragments of DNA.

incipient plasmolysis: when a plant cell is put into a solution that has a lower **water potential**, water leaves the cell by **osmosis**. As a result, the cell's **protoplast** shrinks away from the **cell wall**. Incipient plasmolysis is the point where the protoplast is just beginning to shrink away from the cell wall. In practice, this point is difficult to judge. We therefore say that it is the point where 50% of the cells in a plant tissue are plasmolysed and 50% are not plasmolysed.

incomplete metamorphosis: the process in which an organism gradually changes from a *larva* to an adult. Incomplete metamorphosis is a feature of the life cycles of insects such as cockroaches and locusts. It is also the type of metamorphosis found in frogs and toads as they change from tadpoles into adults.

independent segregation: the way in which sister **chromosomes** divide during the first stage of **meiosis**. During this stage, one member of each pair of chromosomes is pulled to each end of the cell, so there is one complete set of chromosomes at one pole of the cell and one complete set at the other pole. Some of the chromosomes that have gone to a particular pole will have come from one parent and some from the other. This is called independent segregation or random assortment. It is one reason why the cells that are produced as a result of meiosis differ from each other in the genetic information that they carry.

index of diversity: a mathematical way of expressing species diversity. There are many different ways of doing this. Simpson's index of diversity is one way. Another way is shown by the example below.

The table shows the number of birds of different species encountered on a walk across a Manchester park.

Species	Number of birds of this species encountered
Magpie	11
Black-headed gull	4
Carrion crow	4
Blackbird	1
Starling	37
House sparrow	7
All species	*64*

The index of diversity may be calculated from the formula:

$$d = \frac{N(N-1)}{\sum n(n-1)}$$

where
N = total number of organisms of all species
n = total number of organisms of a particular species
$\sum$ means 'the sum of'.

$$d = \frac{(64 \times 63)}{(11 \times 10) + (4 \times 3) + (4 \times 3) + (1 \times 0) + (37 \times 36) + (7 \times 6)}$$

$$d = \frac{4032}{110 + 12 + 12 + 0 + 1332 + 42}$$

$$d = \frac{4032}{1508} = 2.7$$

On its own this figure does not mean much, but it does allow the species diversity to be compared with other areas. In this particular case, the species diversity is fairly low, but it is higher than the value for the city centre.

indicator organism: an organism that is sensitive to a particular *abiotic* factor. The presence or absence of this organism can therefore be used as an indicator of this factor. Indicator organisms are used in monitoring pollution. For example, numerous species of lichens are found growing on tree trunks, rocks, stone walls and gravestones all over Britain. They are sensitive to the concentration of sulphur dioxide in the atmosphere. We can link sulphur dioxide concentration with the presence of particular lichens. By identifying the species of lichens present in a particular area, we can estimate the mean sulphur dioxide concentration.

indoleacetic acid (IAA): an important *auxin*. It is produced by cells in the tips of plant shoots and roots. It controls growth responses of plants such as those that occur towards or away from light.

induced fit model: a model that explains the way *enzymes* work. The *lock and key model* is a rather simple model. It suggests that the *active site* of an enzyme has a rigid shape into which the *substrate* fits. In the induced fit model, the shape of the active site is not completely complementary to the shape of the substrate molecule. When a substrate molecule collides with the active site of the enzyme, the active site moulds round the substrate. This is rather like putting a sock on your foot. The sock moulds round to fit your foot perfectly. In other words, the substrate induces the active site of the enzyme to fit.

infectious disease: a disease that is spread from one organism to another. In humans, this may occur by direct or sexual contact, through contaminated blood or blood products, or by droplet infection. Droplet infection is where infected droplets of *saliva* or *mucus* are coughed, sneezed or breathed out into the air. The most dangerous infectious diseases must be notified to the relevant health authorities so that control measures may be taken. These diseases are called *notifiable diseases*.

inflammation: the response of the body to injury or infection. The infected part swells, becomes red and feels hot. *Blood vessels* near the site of the injury dilate. White blood cells *(leucocytes)* enter the tissue. Some of these are *phagocytes* and engulf bacteria and dead tissue cells. Pus may be produced. This is a thick yellowish liquid that contains dead white cells, bacteria and fragments of dead cells from the site of the injury.

influenza: an *infectious disease* of the gas-exchange system caused by one of several *viruses*. The symptoms include a rise in temperature, headache and more general aches and pains. These are often accompanied by a cough. Most people do not develop complications and start to get better after 3 to 5 days. Different strains of influenza virus have different *antigens* on their surface. Because of this, *vaccines* produced against one strain are not effective against other strains. Elderly people, and those with heart and breathing problems, are usually vaccinated each year with a vaccine against the strain of virus thought most likely to cause an epidemic.

ingestion: taking food into the mouth.

innate behaviour: an aspect of behaviour that an animal shows from birth. It is genetically determined. In practice, it is difficult to say how much of an animal's behaviour is a direct result of its genes and how much is influenced by its *environment* or by learning from other members of its species. It is probably safe to say that simple responses, such as *taxes* and *kineses*, are innate, as they are always shown by the organism concerned in appropriate circumstances. Other aspects of behaviour may be partly innate but are modified by learning. Many birds, for example, can sing a recognisable song even when they have been isolated from other members of their species. This song, however, is modified once they hear other members of their species singing. Such behaviour is part innate and part learned.

innate releaser mechanism: the mechanism involved when a specific external stimulus triggers a particular pattern of behaviour. As an example, the red belly of a male stickleback will stimulate an aggressive behaviour in a rival male.

insight learning: a type of learning that relies on reasoning to solve problems. Kohler was a biologist who studied behaviour in chimpanzees. In one investigation he hung bananas from the top of the cage but they were too high for the chimpanzees to reach. The chimpanzees piled boxes on top of each other, then climbed up and knocked the bananas down with a stick.

inspiration: when a person breathes in, the *diaphragm* muscle contracts and flattens. This increases the volume of the thoracic cavity and decreases the pressure inside it. Air enters the *lungs* down the resulting pressure gradient. When we breathe in deeply, the external *intercostal muscles* also contract. They pull the rib cage upwards and outwards and increase the volume of the chest cavity.

inspiratory reserve volume: the additional amount of air that can be breathed in after a person has breathed in normally. For an average adult man, this is about 1500 cm^3. (See also *lung capacities*.)

insulin: a *hormone* produced by the *islets of Langerhans*, a group of *endocrine* cells in the *pancreas*. The secretion of insulin is stimulated by the rise in blood *glucose* concentration that follows a meal. The hormone has a number of effects on the body, all of which tend to reduce the blood glucose concentration. Two of these effects are:

- insulin speeds up the rate at which glucose is taken into cells from the blood. Glucose normally enters cells by *facilitated diffusion* through protein carrier molecules in the *cell-surface membrane*. Cells have extra carrier molecules present in their *cytoplasm*. Insulin causes these carrier molecules to be sent to the membrane where they increase the rate of glucose uptake by the cell
- it activates *enzymes* responsible for the conversion of glucose to glycogen.

Insulin used to treat diabetics is now produced by *genetic engineering*.

insulin-dependent diabetes: see *diabetes*.

insulin-independent diabetes: see *diabetes*.

integrated pest management: managing pests by making use of all available methods of control. Pests are managed in the most economical way and with the lowest risk to people and to the environment. Sugar beet is a crop widely grown in the UK. It is harvested at the end of one year of growth. Sugar beet seed, however, is only obtained from 2-year-old plants. *Aphids* are important pests of sugar beet because they spread the virus disease, beet yellows. These aphids come from sugar beet plants that have survived the winter. Growing beet for sugar a long way away from beet that is grown for seed (and from closely related plants) reduces the number of aphids reaching the crop. However, when the number of aphids exceeds a certain number, insect sprays are used. The advantage of this method of pest management is that much less insecticide is needed. This saves on cost and minimises environmental damage.

intensive food production: the production of large amounts of food from relatively small areas of land. Much of the farming of pigs, cattle and sheep in lowland Britain is intensive. This method contrasts with extensive farming, where the animals are more thinly spread over a wider area. Intensive farming has some advantages over extensive farming:

- it enables measures to be adopted that result in greater profitability
- it allows greater control of the environment. By providing shelter or heating, for example, less food goes towards maintaining body temperature and more into production
- reproduction can be better controlled because it is possible to monitor individual animals. It is much easier, for example, to identify *oestrus* in cattle and to make sure that artificial insemination takes place at the correct time.

inter-: a prefix meaning 'between different'. For example, interspecific variation is variation between different species. Intercellular communication is communication between different cells.

intercostal muscles: two layers of muscles that connect the ribs to each other. The external intercostal muscles form the outer layer and the internal intercostal muscles form the inner layer. When we breathe in deeply, the external intercostal muscles contract. They pull the rib cage upwards and outwards and increase the volume of the chest cavity. During exercise the internal intercostal muscles contract and help to push air out forcefully.

interferon: a substance produced by the cells of a mammal when they are attacked by *viruses*. There are a number of different interferons, each produced by a different type of cell. They do not have an effect on the viruses themselves, but they appear to protect other cells from infection. Interferons have other effects on the body and it is thought that they could have possible uses in medicine, including the treatment of certain cancers.

internal fertilisation: *fertilisation* that takes place within the body of the female. In animals, *gametes* are small. This means that they have a large surface area to volume ratio and they dry out rapidly when exposed to air. This, and the fact that male gametes have flagella that enable them to swim, means that they need an aquatic environment if fertilisation is to take place. For animals living on land, this is provided when the male of the organism concerned copulates with the female and introduces sperm cells into the female reproductive tract. Fertilisation will therefore take place internally.

interphase: see *cell cycle*.

interspecific competition: *competition* between organisms of different species. The main reason why *weeds* are removed from crops is that they compete with crop plants for resources such as light, water and soil nutrients. Crop yield therefore falls as a result of interspecific competition.

intra-: a prefix meaning 'within'. For example, intraspecific variation is variation within a species. Intracellular communication is communication within a cell.

intracytoplasmic sperm injection (ISCI): a technique used in *in vitro fertilisation (IVF)* in which a single sperm is injected into the *cytoplasm* of an egg cell. Intracytoplasmic sperm injection is often used when the male parent is not fully fertile.

intraspecific competition: *competition* between organisms of the same species. Farmers plant seeds of crop plants at particular population densities. If they plant them too far apart, they will not get the best yield. If, on the other hand, they are planted too close together, they will compete with each other for light and the substances they need in order to grow. Each will get less than the required amount and the yield from each individual plant will be less. The aim is to plant seeds at the optimum density, so that the best possible crop yield can be obtained. The same principle holds whether yields of crop plants are being discussed or the number of animals kept in a field.

intrinsic protein: a protein associated with the *plasma membrane*. Intrinsic *proteins* pass completely through the *phospholipid* bilayer, from one side of the membrane to the other. They are often involved with the passage of different substances into and out of a cell:

- *Voltage-gated ion channels* are intrinsic proteins in the *cell-surface membranes* of *axons*. *Action potentials* depend on the passage of sodium and potassium ions through these channels.
- One of the intrinsic proteins present on a cell-surface membrane is the CFTR protein. In healthy people, this protein transports chloride ions out of the cell. People with *cystic fibrosis* have a faulty CFTR protein.

- Antidiuretic *hormone* helps to control water balance when the *water potential* of the blood falls. It causes more water to be absorbed from the *collecting ducts* in the *kidney* through intrinsic proteins called aquaporins.

intron: a non-coding piece of the *DNA* in a *gene*. A gene in a *eukaryotic* cell consists of introns and *exons*. Exons are pieces of DNA that code for the *amino acids* which make up proteins. During transcription, the base sequence of the gene forms a template for producing a molecule of *mRNA*. Before the mRNA leaves the nucleus, however, it is edited, and the *base pairs* forming the introns are cut out. The genes in *prokaryotic* cells do not contain introns.

involuntary muscle: muscle that is not under conscious control. The movements of most of the internal organs of the body are brought about by this type of muscle. Involuntary muscle forms the outer wall of the uterus or *myometrium*. It produces *peristalsis* in various parts of the gut and its presence in the walls of *blood vessels* allows them to change diameter and divert blood flow according to the needs of the body. Involuntary muscle contracts more slowly than *skeletal muscle* and has two important properties. It is able to contract spontaneously if it is stretched. It also responds to substances released from nerve endings or to *hormones* brought to it in the circulation. Involuntary muscle is also called visceral muscle or smooth muscle.

iodine: a *micronutrient* required by mammals. It is an important part of *thyroxine (thyroid hormone)*, the *hormone* secreted by the *thyroid gland* that helps to control the *basal metabolic rate (BMR)* of the body. If there is not enough iodine in the diet, the thyroid gland enlarges producing a condition known as goitre.

iodine-potassium iodide test: a *biochemical test* used to show the presence of *starch*.

ion: a charged particle. *Atoms* or parts of *molecules* can lose one or more *electrons*. If this happens, they acquire a positive charge and are known as anions. Other atoms or parts of molecules can gain electrons. They have a negative charge and are known as cations. Many substances in the body exist as ions. One example is provided when carbonic acid is formed in the reaction between carbon dioxide and water.

$$\text{carbon dioxide} + \text{water} \rightarrow \text{carbonic acid} \rightarrow \text{hydrogen ions} + \text{hydrogencarbonate ions}$$
$$CO_2 \quad + \quad H_2O \quad \rightarrow \quad H_2CO_3 \quad \rightarrow \quad H^+ \quad + \quad HCO_3^-$$

The positively charged hydrogen ion and the negatively charged hydrogencarbonate ion behave in different ways in the body.

ion-channel protein: a type of *protein* found in the *cell-surface membrane* of a cell, which allows the passage of *ions* into or out of the cell. Ion-channel proteins are *intrinsic proteins*. Some of them are able to open and close and are therefore described as being gated. There are important *voltage-gated ion channel* proteins in the cell-surface membranes of nerve cells. Other gated channels are found in nerve *synapses*. They open in response to the presence of *neurotransmitters*.

ionising radiation: radiation produced when radioactive elements decay. Some of this radiation consists of electromagnetic waves and some involves parts of *atoms* moving at high speed. Radiation of this type is called ionising radiation because, when it strikes atoms in materials through which it passes, it causes the loss of *electrons* so that the atoms concerned are ionised. Cells are vulnerable to ionising radiation. High doses kill them, but even small amounts can cause damage to **DNA** molecules. This may result in increased *mutation* or, in some circumstances, may lead to cancer (see *tumour*).

iris: a ring of muscle surrounding the pupil of the eye. In bright light, circular muscle round the iris contracts. This makes the pupil smaller and reduces the amount of light passing through. In dim light, radial muscles contract. This makes the pupil larger and increases the amount of light entering the eye.

iron: a *micronutrient* required by animals and plants. Iron is:
- contained in **cytochromes**. These are substances that form part of the *electron transport chain*. A cytochrome is a protein that is combined with another chemical group called a *cofactor*. This chemical group may contain iron
- part of *haemoglobin*
- essential for the synthesis of *chlorophyll*.

irradiation: see *food irradiation*.

ischaemic heart disease: heart disease caused by an inadequate supply of blood to the heart muscle. (See *coronary heart disease*.)

ISCI: see *intracytoplasmic sperm injection (ISCI)*.

islets of Langerhans: small groups of cells found in the *pancreas*. They secrete *hormones* involved in the control of blood *glucose* concentration. The pancreas is an important organ found in the abdomen. The islets contain different types of cells. Small numbers of large α cells secrete the hormone *glucagon*. The more numerous but smaller β cells produce *insulin*.

isomers: molecules that contain the same types and numbers of *atoms*, but these are arranged in different ways. All *hexose* sugars have a molecular formula that may be written as $C_6H_{12}O_6$, so each molecule contains 6 carbon atoms, 12 hydrogen atoms and 6 oxygen atoms. These atoms, however, can be arranged in different ways to give different sugars. The diagram shows two of them: α-glucose, which is the basic building block for *starch*, and β-glucose, the basic building block for *cellulose*.

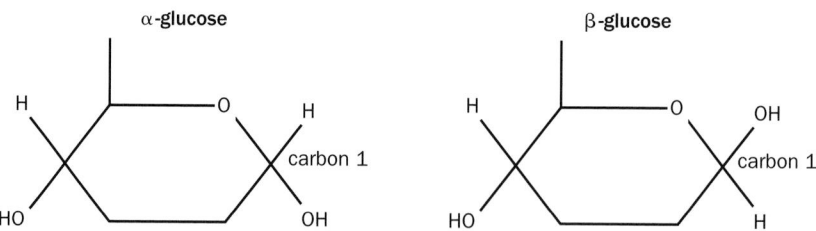

Isomers of glucose: both of these molecules have the formula $C_6H_{12}O_6$. Note the different positions of the –H and –OH groups on carbon 1

isotonic: solutions that have the same solute concentration as each other. If isotonic solutions are separated by a partially permeable membrane, the *water potential* will be the same on either side. There will be no net osmotic movement of water between the two solutions. The amount of water which moves in one direction will be exactly balanced by the amount which moves back in the other.

Mitochondria may be separated from the rest of the organelles present in a *liver* cell by *cell fractionation*. The first stage in the procedure is to grind up or homogenise the liver tissue in an isotonic *buffer* solution. The buffer solution maintains a constant pH. Being isotonic prevents damage to the mitochondria because of water entering or leaving by *osmosis*.

isotope: a form of an element that has a different number of neutrons in the nuclei of its *atoms*. Because of this, different isotopes differ in mass. This property can be useful in marking or labelling atoms to see what happens to them during particular biological processes. Meselson and Stahl used two isotopes of nitrogen, ^{14}N and ^{15}N, to provide evidence for the semi-conservative replication of *DNA*. Some isotopes emit particles from their nuclei as they break down. These are referred to as being radioactive and can be detected by using a Geiger counter or because they will blacken a photographic plate. Radioactive isotopes have proved to be useful in investigating physiological processes such as the transport of various substances in *phloem* and in investigating biochemical pathways such as the *light-independent reaction* of *photosynthesis*.

IVF: see *in vitro fertilisation (IVF)*.

Are you studying other subjects?

The *A–Z Handbooks (digital editions)* are available in 14 different subjects. Browse the range and order other handbooks at **www.philipallan.co.uk/a-zonline.**

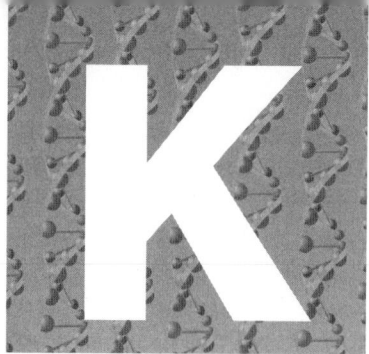

K selected species: species that live in *habitats* which are stable, with little random variation in environmental conditions. They often form large populations that are more or less constant in size. There is therefore intense *competition* between adults. K selected species:

- produce few young and provide a substantial amount of parental care. Each offspring therefore has a high probability of surviving to become an adult
- mature late and have a long period over which they are able to reproduce.

Examples of K selected species are elephants, humans and oak trees.

karyotype: a complete set of *chromosomes* from a single cell. It shows the number and the structure of these chromosomes. A karyotype from a person with *Down syndrome*, for example, would show 47 chromosomes instead of the usual 46. There would be three copies of chromosome 21.

keyhole surgery: surgery that involves as little physical damage to the patient's body as possible. A tiny incision is made and the operating instruments passed into this opening. Several types of operation are commonly performed in this way. They include removal of egg cells from the *ovary* for *IVF* treatment and repair of *ligament* damage. Keyhole surgery usually allows the patient to resume normal activities after a short recovery time.

kidney: an organ responsible for the formation of *urine* in a mammal. A human kidney is about 10 cm in length and divided into an outer, dark-coloured cortex and an inner, lighter *medulla*. Each kidney contains approximately a million kidney tubules or *nephrons* that produce the urine.

kidney tubule: see *nephron*.

kinesis: a form of behaviour in which the response of the animal is proportional to the intensity of the stimulus. In dry conditions with low humidity, a woodlouse moves fast and changes direction very little. When the humidity is higher, it moves more slowly and turns much more. The advantage of this pattern of behaviour to the woodlouse is that it helps the animal to stay in favourable, humid conditions.

kingdom: a level in the biological system of *classification*. All living organisms can be classified into one of five kingdoms:

- *Prokaryotae* – simple organisms that do not have nuclei in their cells. Bacteria are members of the Prokaryotae
- *Animalia* – this is the animal kingdom. You should take care to refer only to members of this kingdom as animals
- *Plantae* – the plant kingdom
- *Fungi* – the group of organisms that includes moulds and yeasts
- *Protoctista* – many different organisms are put in this kingdom, mainly because they do not fit anywhere else. This kingdom contains single-celled organisms such as

amoeba. Protoctista also contains the algae. They range in size from single cells to giant seaweeds.

Klinefelter's syndrome: a genetic condition in which each cell in the person affected has three *sex chromosomes*, XXY. A person with this condition appears male but has small testes and is unable to produce sperms. *Secondary sexual characteristics* such as facial and body hair are also poorly developed. Klinefelter's syndrome arises from the failure of sex chromosomes to separate properly during *meiosis*.

knockin mouse: a genetically engineered mouse that has genetic information inserted at a particular place in its *DNA*.

knockout mouse: a genetically engineered mouse which has one or more of its *genes* permanently switched off. Scientists use knockout mice as animal models to study the function of genes. They are particularly useful where scientists have found the base sequence of the gene but do not know its function.

Koch's postulates: a set of basic principles put forward by the nineteenth-century biologist, Robert Koch, and used for confirming that a particular *microorganism* is the pathogen that causes a disease. There are many different microorganisms found on and in the human body. Most of these do no harm, but some cause disease. Finding a bacterium or virus present in a person with a disease does not, therefore, prove that it causes the illness concerned. Koch suggested four principles that would identify the pathogen:

- The microorganism must always be present in a person with the disease. If a person does not have the disease, then the microorganism must be absent.
- It should be possible to isolate the microorganism from an infected person and culture it.
- The disease should develop when the cultured microorganism is introduced into a healthy person.
- The microorganism can be re-isolated from the new host.

Krebs cycle: a cycle of biochemical reactions that forms part of the pathway of *aerobic respiration*. Acetylcoenzyme A is produced in the *link reaction*. It contains two carbon *atoms* which are involved in the process of *respiration*. Acetylcoenzyme A is fed into the Krebs cycle. It combines with a four-carbon compound to form a six-carbon compound. The six-carbon compound is then broken down via intermediate compounds to produce the four-carbon compound once again. This process is shown in the diagram.

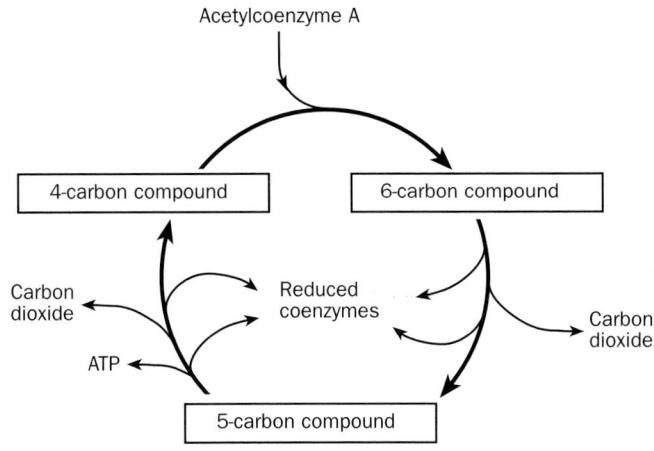

The Krebs cycle involves both *decarboxylation* and oxidation. Decarboxylation is the removal of carbon atoms as carbon dioxide. Oxidation involves the loss of *electrons* and the reduction of *coenzymes*. The importance of the Krebs cycle is in the production of reduced coenzymes, from which *ATP* may be generated in the *electron transport chain*.

kwashiorkor: *malnutrition* resulting from a diet that is deficient in both *protein* and energy-containing foods. Kwashiorkor is common in parts of Africa where children are weaned on to an inadequate diet after prolonged breast feeding. The child may be in reasonable health until its protein requirements are increased by an infection. Diarrhoeal diseases, measles and *malaria* often lead to the development of kwashiorkor. One of the characteristic symptoms of the condition is *oedema*. The protein deficiency leads to a low concentration of protein in the blood. Consequently, the *water potential* of the blood is relatively high and *tissue fluid* is not absorbed effectively back into the blood at the venous end of the *capillaries*.

Aiming for a grade A*?

Don't forget to log on to **www.philipallan.co.uk/a-zonline** for advice.

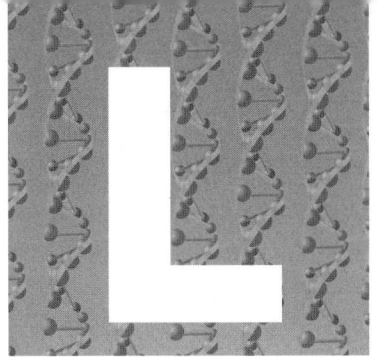

lac operon: a group of genes that control the uptake and metabolism of *lactose* in the bacterium, *Escherichia coli*. This bacterium can use either *glucose* or lactose as its *respiratory substrate*. When glucose is present, it uses this. It would clearly be wasteful to devote resources to producing substances necessary to take up lactose and metabolise it in these conditions. The set of genes that make up the lac operon act as a control mechanism and ensure that the *proteins* necessary for transporting lactose into the cell and the *enzymes* required for its metabolism are only produced when lactose is present.

lactase: an *enzyme* that breaks down the milk sugar, *lactose*. It hydrolyses lactose and forms two *monosaccharides*: *glucose* and *galactose*. In most mammals, it is produced in the *small intestine* only when the animal is young. Most adult mammals do not produce lactase and cannot, therefore, digest lactose. If they are given milk, it is likely to make them ill (see *lactose intolerance*).

lactate: a waste product of *anaerobic respiration* in animal cells. The first step in *anaerobic respiration* is *glycolysis*. In glycolysis, *glucose* is converted to pyruvate and some energy is transferred to produce *ATP*. Oxidation is involved and this is associated with the reduction of a coenzyme called *NAD*. When oxygen is available, this reduced NAD is reconverted to NAD in the reactions of aerobic respiration. In anaerobic respiration, however, the reduced NAD is reconverted to NAD during the formation of lactate from pyruvate. This is shown in the diagram.

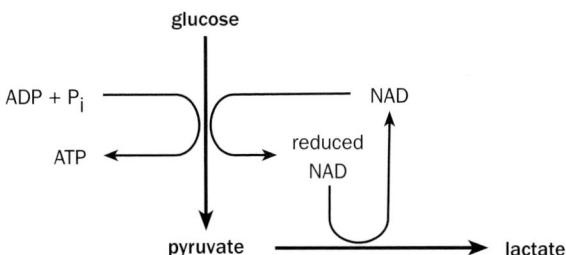

The formation of lactate in anaerobic respiration

The build-up of lactate in muscles is one of the factors that contributes to fatigue during vigorous exercise. Only so much lactate can be tolerated before we have to take a rest. Once exercise stops and sufficient oxygen becomes available again, this lactate is removed. There are several *metabolic pathways* involved in lactate removal. These include conversion to *carbohydrate* in the *liver* and oxidation in the respiratory pathway.

lactation: the production of milk by the mammary glands of a female mammal. Lactation involves three main steps:

- preparation of the tissues of the mammary gland. This takes place during pregnancy. *Hormones* produced by the *placenta* and the *ovary* cause the glands and ducts that make up the mammary glands to develop and begin their secretory activity
- secretion of milk. This is controlled by the hormone *prolactin*
- the release of milk when the young animal suckles. Another hormone, *oxytocin*, plays an important part in this process.

lacteal: a small *lymph* vessel inside one of the *villi* in the wall of the *small intestine*. It absorbs the products of *fat digestion* and transports them to one of the veins in the neck.

lactic acid: a substance formed by *anaerobic respiration* in animal cells. In the conditions in the body, lactic acid is present as *lactate* ions. Because of this, the term lactate is often used instead of lactic acid.

lactose: the main *carbohydrate* found in milk. Lactose is a *disaccharide* that is made in the mammary glands from two *monosaccharides*: *glucose* and *galactose*.

lactose intolerance: the inability to digest *lactose*, the sugar found in milk. The *enzyme lactase* is found in the *small intestine* of young mammals. It digests lactose, breaking it down to the *monosaccharides glucose* and *galactose*. In many adults, however, this enzyme is not present. These people are described as lactose intolerant. They suffer considerable discomfort and diarrhoea if they drink milk. The condition is inherited and is common in many parts of the world. Approximately 70% of the black American population, for example, is lactose intolerant.

lag phase: the early part of a curve showing the growth of a population of *microorganisms*. The microorganisms concerned may increase in size but they do not increase in number. If, for example, yeast cells are put into a sucrose solution, the lag phase is the time when the cells are adapting to their new medium. During this stage, various *genes* are switched on, *mRNA* is synthesised and the relevant *enzymes* are produced and secreted.

lamellae: see *gill lamellae*.

large intestine: the last part of the *alimentary canal*. It is the part between the *small intestine* and the *anus*. (See also *colon*.)

larva: an immature stage in the life cycle of an animal. Many larvae (plural larva) are different in appearance from the adults. This is particularly true of the larval stages of insects such as the caterpillars of moths and butterflies. Larvae may develop gradually into adults or there may be a more rapid transition known as *metamorphosis*. The possession of a larval stage enables an animal to be adapted in different stages in its life cycle:

- The larval stage in the life cycle of insects is the main growth stage. Many insect larvae are specialised for feeding on protein-rich foods. The caterpillars of most moths and butterflies, for example, eat the protein-rich young leaves of plants. Adult butterflies do not grow. They can survive on nectar which, although containing *carbohydrates*, has little protein.
- *Sessile* animals are animals that do not move. They are usually aquatic and spend their adult lives in one place. Many sessile aquatic animals have larvae that can swim and help to ensure effective dispersal.

latent learning: a form of learning in which the outcome is not seen immediately.
Scientists put three groups of rats into mazes. They recorded their behaviour every day:

- Rats in group 1 always received food when they reached the end of the maze. They
  quickly learnt the way through the maze.
- Rats in group 2 never received food when they reached the end of the maze. They did
  not learn the way through the maze.
- For the first 10 days, rats in group 3 did not receive food when they reached the end
  of the maze. After day 11 they received food. They then quickly learnt the way through
  the maze. They were soon as good at finding the way through the maze as the rats in
  group 1.

The rats in group 3 showed latent learning.

lateral geniculate nucleus: an area in the *brain* that processes sensory information
from the *retina*.

law of diminishing returns: a basic economic rule usually associated with the applica-
tion of *fertiliser* to crop plants. The growth of many crop plants is limited by soil nutrients
such as nitrates and phosphates. Therefore, if fertiliser is added, yield is increased. When
a certain point is reached, however, the addition of even more fertiliser results in no further
increase in crop yield. There are several reasons for this. These include:

- other factors may limit plant growth. The addition of fertiliser during the winter, for
  example, has little benefit. Low temperatures limit crop growth and most of the fertiliser
  added is leached (see *leaching*) from the soil when it rains
- if high concentrations of nitrate are added to cereal crops, the leafy parts of the plant
  grow as well as the developing grain. The plants are then much more likely to be blown
  over in summer storms. This results in difficulties with harvesting and a lower yield
- high concentrations of fertiliser reduce *germination* and result in reduced root growth.

LDL: see *low-density lipoprotein (LDL)*.

leaching: the loss of soluble substances such as nitrates and phosphates from the top
layer of the soil when water drains through. Leaching removes important plant nutrients from
the soil. If these nutrients enter lakes and rivers they may cause *eutrophication*.

learning: a process that involves changes in behaviour which result from experience of new
situations. At one stage it was thought that behaviour was either learned or it was *innate
behaviour* and shown by the animal from birth. In practice, it is difficult to say how much of
an animal's behaviour is innate and genetically determined and how much is influenced by
its *environment* or by learning from other members of its species. Several types of learned
behaviour are recognised. These include *habituation*, *imprinting*, *classical conditioning*
and *operant conditioning*.

leguminous plants: an agriculturally important group of plants that contain species
such as peas, beans and clover. They have seeds that are generally rich in protein and are
contained in pods. They also have nodules on their roots which are associated with *nitrogen
fixation*. These nodules contain bacteria that belong to the *genus Rhizobium*. The bacteria
convert nitrogen gas into ammonium compounds, some of which they use and some of
which they pass on to the plant itself. In return, the bacteria gain *carbohydrates* that are
made by *photosynthesis*. This relationship is an example of *mutualism*. Leguminous plant

crops are often ploughed back into the soil as green manure. This rots down and increases the soil nitrate concentration.

lens: a transparent structure at the front of the eye that helps to focus light rays on the *retina*. Because it is able to change shape, the lens is able to change the amount of refraction, allowing both close and distant objects to be brought into focus. The lens is composed of living cells and these require a supply of nutrients. The *aqueous humour* provides these nutrients.

lenticel: a small pore found on the surface of a plant stem. In older stems, a layer of cork develops round the outside. Mature cork cells are impermeable to water and respiratory gases. If this cork layer completely covered the stem, the cells inside would die as they would be unable to gain oxygen from or lose carbon dioxide to the atmosphere. Lenticels are small pores in the surface of the cork layer. They are very loosely packed with cells, so respiratory gases can diffuse through the large intercellular spaces from the atmosphere to the cells of the stem.

leucocyte: another name for a white blood cell or white cell. There are several different types of leucocyte but all of them are involved in some way in fighting infection. They are found in smaller numbers than red blood cells and, because their *cytoplasm* is transparent, they need to be stained before they can be examined in detail. The term white blood cell is a little misleading because some types of leucocyte leave the blood and are found in the *lymphatic system* and other parts of the body. The table shows the characteristics of some different types of leucocyte.

Type of leucocyte	Appearance	Function
Lymphocyte	Relatively large, round nucleus and small amount of cytoplasm	B cells secrete antibodies. T cells have a number of functions. These include killing infected cells and controlling different aspects of the immunological process
Monocyte	Large, kidney-shaped nucleus	These cells are phagocytes and engulf bacteria
Granulocytes	Possess granular cytoplasm and a lobed nucleus	These have a number of different functions. Some engulf bacteria by phagocytosis. Others are involved in allergies and in *inflammation*

LH: see *luteinising hormone (LH)*.

ligament: a strip of *connective tissue* that attaches *bones* to each other. Some ligaments contain a lot of closely packed fibres of a protein called *collagen*. The properties of collagen make it ideally suited to its functions in a ligament. It is flexible but very resistant to stretching. Other ligaments, such as those between the vertebrae in the spine, contain *elastin*. This is another protein. It can be stretched.

ligase: a type of *enzyme* that is used to join lengths of *DNA* together. Ligases are used in *genetic engineering* to produce *recombinant* DNA.

light-dependent reaction: the process in which light energy is absorbed by *chlorophyll* and is used to produce *ATP* and reduced *NADP* during *photosynthesis*. When light strikes a chlorophyll molecule, some of its *electrons* become excited. These are lost from the chlorophyll molecule and pass through a series of molecules that act as electron acceptors. As a result, energy is progressively lost and some of this is used to make ATP in the process of *photophosphorylation*. The electrons are eventually accepted by a coenzyme called NADP and convert this to reduced NADP.

Another reaction is also involved. This is *photolysis*. Water molecules break down to produce electrons, hydrogen ions and a molecule of oxygen. Photolysis can be summarised by the equation:

$$2H_2O \rightarrow 4H^+ + 4e^- + O_2$$

The electrons produced as a result of photolysis replace those which were lost from the chlorophyll molecule as a result of absorbing the light energy. The hydrogen ions or protons help in the formation of reduced NADP and the oxygen is given off as a waste product. The ATP and reduced NADP produced as a result of this light-dependent reaction play an important part in the conversion of carbon dioxide to carbohydrate in the *light-independent reaction*.

light-independent reaction: the process in which carbon dioxide is reduced to form carbohydrate during *photosynthesis*. The biochemical reactions that form the light-independent reaction of photosynthesis are summarised in the diagram.

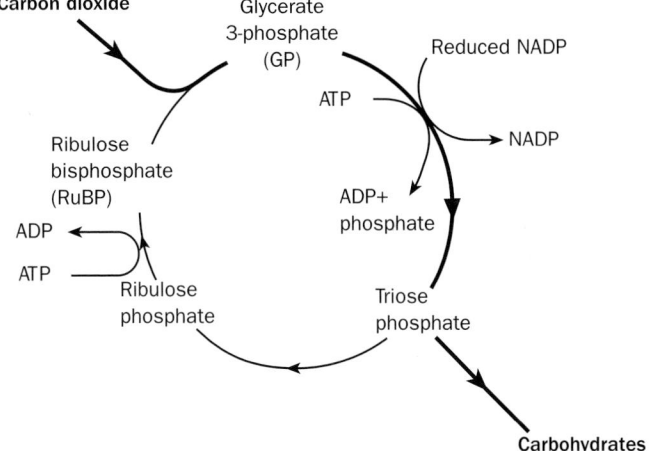

Ribulose bisphosphate (RuBP) is a five-carbon compound. It combines with a molecule of carbon dioxide to produce two molecules of the three-carbon compound, *glycerate 3-phosphate (GP)*. GP is converted into the sugar, triose phosphate. This requires the reduced *NADP* and *ATP* produced in the *light-dependent reaction*. Some of this triose phosphate is then converted into other *carbohydrates*, such as *glucose* and *starch*, while the rest is used to form more ribulose bisphosphate. This cycle of reactions involving ribulose bisphosphate is called the *Calvin cycle*.

lignin: a complex biological *polymer* found in some plant *cell walls*. A newly formed plant cell has a cell wall containing *cellulose* microfibrils cemented together with other substances. This is called the primary cell wall. In some cells, such as those in the *sclerenchyma* and *xylem*, lignin is then laid down. It further helps to cement the cellulose microfibrils together and makes the wall even stronger and more resistant to the forces on it. However, it also makes the wall impermeable, so important substances are unable to pass through. As a consequence, cells that are heavily lignified do not have living contents.

limiting factor: a variable that limits the rate of a particular process. If it is increased, then the process will take place at a faster rate. The graph shows how *substrate* concentration affects the rate of an *enzyme*-controlled reaction.

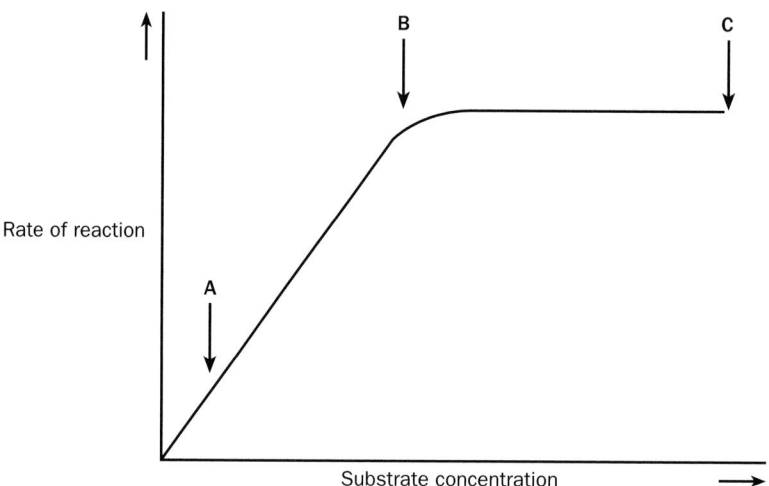

At low substrate concentrations, such as between points A and B on the graph, the rate of the reaction is limited by substrate concentration. An increase in substrate concentration produces an increase in the rate of the reaction. Between points B and C, however, increasing the substrate concentration has no effect on the rate of reaction, so something else must be limiting. In this case it is probably the concentration of the *enzyme*. There are many other biological processes that show a similar pattern. Examples are the effects of light intensity and carbon dioxide concentration on the rate of *photosynthesis,* and the rate of uptake of *glucose* by cells.

Lincoln index: an alternative name for the *mark–release–recapture* method of estimating the size of an animal population.

linkage: refers to two or more genes that are found on the same chromosome. This is important in genetics as it means that linked *alleles* are normally inherited together. During the first division of *meiosis*, the chromosomes come together in their pairs, with each chromosome split longitudinally into two daughter *chromatids*. If two alleles are linked, they will both be situated on a single chromatid, so they will be inherited together when these chromatids are eventually pulled apart. They will both go to the same daughter cell. The only way in which linked genes can be prevented from being inherited together is when *crossing over* occurs.

Here, the chromatids break and rejoin so that alleles from one chromatid become attached to alleles on another chromatid.

link reaction: the part of the respiratory pathway linking *glycolysis* with the *Krebs cycle*. In the link reaction, pyruvate is converted into acetylcoenzyme A. This conversion involves the loss of a molecule of carbon dioxide. It is also an oxidation reaction, and a molecule of reduced *NAD* is formed. The link reaction is summarised in the diagram.

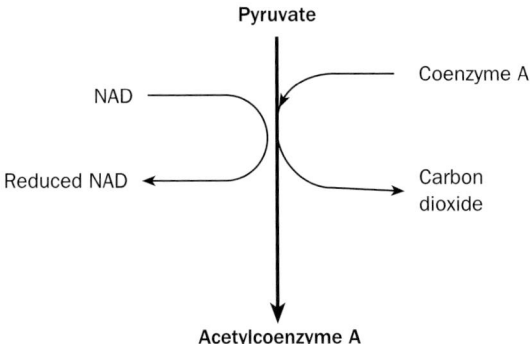

lipase: an *enzyme* that breaks *lipids* down into glycerol and *fatty acids*. Lipases are used in industrial processes to digest *fats* in, for example, some biological washing powders. They are also used in the food industry, where they are involved in chemical reactions that swap fatty acid chains between different lipids. Cocoa butter is an expensive, naturally occurring lipid, used in making chocolate. By using lipase and exchanging fatty acid chains, a cheaper substitute can be made from vegetable oils.

lipid: a large and varied group of organic substances that are insoluble in water but are soluble in organic solvents such as ethanol. Lipids include:
- *triglycerides* such as fats and oils
- *phospholipids*. These substances play an important role in *plasma membranes*
- *steroids* including *hormones* such as *progesterone* and *oestrogen*.

lipoprotein: a mixture of *protein* and *lipid* molecules. Lipoproteins are found in blood *plasma* and in *lymph* where they play an important part in the transport of lipids. There are different sorts of lipoproteins. *Low-density lipoprotein (LDL)* contains a relatively low proportion of protein. *High-density lipoprotein (HDL)* contains rather more protein. Different lipoproteins have different functions in the body. LDLs, for example, play an important role in the transport of *cholesterol*.

liver: a large, reddish-brown organ found in the body of a mammal. Like all other organs in the body, the liver is supplied with oxygenated blood. This reaches it through the *hepatic artery*. The liver, however, is also supplied with blood by another vessel. This is the *hepatic portal vein* which brings blood from the intestine. The *hepatic vein* returns blood to the heart. The liver has many different functions, but it is possible to group these together.
- It is concerned with *homeostasis*. It helps to regulate the blood *glucose* concentration. If the concentration of glucose rises, the excess is converted into *glycogen* in the liver. The *hormones insulin* and *glucagon* play an important part in this process.

- It synthesises substances such as *plasma proteins* and it produces *bile*.
- It stores substances such as *fat*-soluble *vitamins*.
- It is involved in *excretion*. It is in this organ that *deamination* takes place. In deamination, the amino groups are removed from surplus amino acid molecules. The ammonia formed by this process is then converted into *urea*. *Haemoglobin* from broken down red blood cells is processed in the liver and the unwanted part is excreted in the bile.

lock and key model: used to explain the way in which *enzymes* work. An enzyme molecule has an *active site* formed by a group of amino acid molecules. This active site has a particular shape into which the *substrate* molecule fits to form an *enzyme–substrate complex*. The *activation energy* of the reaction is lowered and the reaction takes place under normal cell conditions. This complex then breaks down to form the products and release the enzyme molecule.

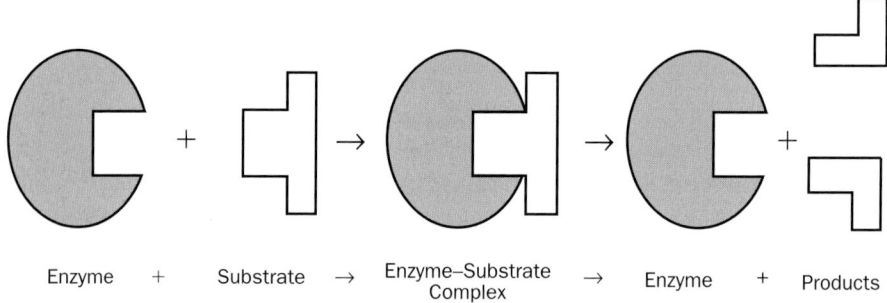

| Enzyme | + | Substrate | → | Enzyme–Substrate Complex | → | Enzyme | + | Products |

How an enzyme-controlled reaction takes place

It is important to realise that this is only a model. The actual situation is rather more complicated. However, a thorough understanding of the lock and key model lets us explain a number of factors that affect the rate of reaction of enzymes, such as substrate concentration and *competitive inhibition*. (See also *induced fit model*.)

locus: the position of a *gene* on a *DNA* molecule.

log phase: the part of a *population growth curve* where *microorganisms* increase rapidly in numbers. Suppose a bacterium reproduces by dividing in two and, in the conditions present in a particular culture, it does this once every hour. After an hour, there will be 2 bacteria present; after 2 hours there will be 4, then 8, 16, 32, 64, 128 and so on. Plotted on a graph, this gives a curve that gets steeper and steeper. During this log phase, nutrients will be in plentiful supply and there will be little build-up of toxic waste products.

long-day plant: a plant that flowers when day length is greater than a certain amount. Long-day plants include plants such as poppies that flower naturally in the summer.

longitudinal study: a method of collecting information about growth by measuring the same individual over a period of time. A *cross-sectional study* involves measuring different individuals, each on a single occasion. Longitudinal studies have an important advantage over cross-sectional studies because they show differences in individual growth patterns. Study of variation between individuals can produce evidence about the influence of genetic and environmental factors on growth. However, there are some disadvantages as well:

- longitudinal studies take rather a long time to carry out as it is necessary to wait for the individuals in the study to age
- it is necessary to keep careful track of individuals throughout the study. This can be very difficult.

loop of Henle: part of a *nephron* that forms a loop into the *medulla* or inner part of the *kidney*. The descending limb is permeable to water. The ascending limb is permeable to sodium and chloride ions. As a result of this, a concentration gradient is set up in the medulla. The deeper you go into the medulla, the higher the salt concentration becomes. Therefore, the deeper you go, the lower the *water potential* becomes. This means that the water potential of the fluid in the *collecting duct* is always higher than that in the medulla. Water can be removed by *osmosis* from the fluid in the collection duct all the way down the duct. The length of the loop of Henle is associated with the amount of water that can be reabsorbed in the kidney. Small mammals that live in deserts have long loops of Henle. They can reabsorb a lot of water and therefore produce small amounts of very concentrated *urine*.

low-density lipoprotein (LDL): a *lipoprotein* that transports *cholesterol* in the blood. These lipoproteins bind to receptors on *cell-surface membranes* and are taken into cells. Other mechanisms lead to the cholesterol in low-density lipoproteins being incorporated into *atheroma*. A high concentration of low-density lipoprotein is therefore a risk factor associated with *coronary heart disease*.

lumen: the cavity inside a hollow structure. The lumen of the gut, for example, is the space in the gut through which food passes.

lumpectomy: a surgical operation for breast cancer in which only the *tumour* and the surrounding tissue are removed. This treatment is usually followed by radiotherapy. It is used when the tumour is less than 2 cm in diameter and cancer cells have not spread to other organs.

lung: one of a pair of gas exchange organs found in the chest of mammals and some other chordates. When you look at lung tissue with a microscope, you will see three distinct types of structure:

- *alveoli*. These are the tiny thin-walled air sacs in the lungs where gas exchange takes place
- airways. Air is taken into and expelled from the lungs through the trachea and bronchi (see *bronchus*). The bronchi branch repeatedly, finally leading into a system of small tubes called *bronchioles*
- *blood vessels*. The lungs are well supplied with blood. You can see arteries, veins and *capillaries* in suitable prepared sections.

The total volume of air in the lungs of an adult man is about 5 litres, but the actual volume exchanged each time he breathes depends on a number of factors (see *lung capacities*).

lung capacities: the volumes of air that are either contained in the *lungs* or are taken in or breathed out under particular circumstances. The diagram overleaf shows these values in an average man.

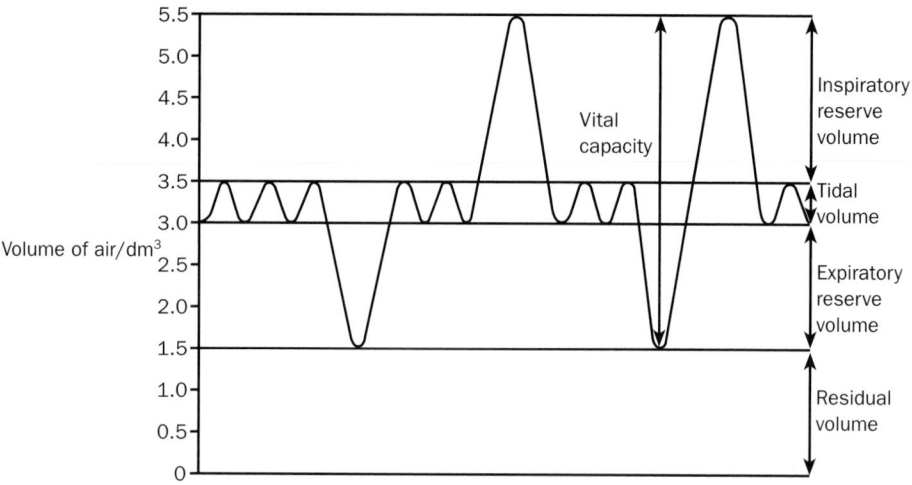

Human lung capacities

The total volume of air contained in the lungs is approximately five litres or 5000 cm³. Its precise value will obviously vary with a number of factors, including the age of the person, his physical fitness and state of health. When breathing quietly at rest, the volume of air taken in and given out at each breath is much less than this. It is only about 500 cm³. This is the tidal volume. If, after having breathed in normally, the person continues to breathe in as deeply as possible, another 1500 cm³ of air may be taken into the lungs. This is the *inspiratory reserve volume*. Similarly, if after breathing out normally, the person concerned continues to breathe out as much as possible, it is possible to expel another 1500 cm³. The additional volume of air breathed out is the *expiratory reserve volume*. Finally, the total of the amount of air that can be breathed out following the deepest intake of air possible is the vital capacity. This is the sum of the tidal volume, the inspiratory reserve volume and the expiratory reserve volume. It is not as much as the total volume of the *lungs*. There is always a small volume that cannot be expelled. This is the *residual volume*.

luteinising hormone (LH): a *hormone* that is produced by the *pituitary gland*. In females, it is produced during the first part of the reproductive cycle. It travels in the blood to its target organ, the *ovary*. In the ovary, it stimulates ovulation and causes the follicle cells that remain in the ovary to develop into the *corpus luteum*. The corpus luteum secretes *progesterone* during the second half of the reproductive cycle. Luteinising hormone is also produced in males, where it stimulates *interstitial cells* in the testes to produce *testosterone*.

lymph: *tissue fluid* after it has drained into the *lymphatic system*. Lymph is similar to *plasma* in chemical composition but it contains less protein. *Lipids* are absorbed from the *small intestine* into the lymphatic system and lymph from this part of the body may have a milky appearance due to the large amount of *fat* that it contains.

lymph node: one of a number of small swellings found in the *lymphatic system*. It contains channels in which there are white blood cells or *leucocytes*. Some of these engulf foreign material while others produce antibodies. Lymph nodes act as filters and help to prevent the spread of infection.

lymphatic system: the system of vessels that returns excess *tissue fluid* to the blood system. During the course of a day, more fluid leaves the blood *capillaries* as tissue fluid than drains back into them. If this tissue fluid accumulates it results in swelling, a condition called *oedema*. In a healthy person, the excess tissue fluid drains into a system of tiny, blind-ending tubes, the lymphatic capillaries. This fluid, now called *lymph*, is collected by a series of larger lymph vessels before finally returning to the blood in veins in the neck. Lymph vessels are surrounded by skeletal muscles. When these muscles contract they squeeze the lymph along the vessels. Valves in the larger lymph vessels, like those found in veins, ensure that the flow is in one direction. At intervals through the lymphatic system, there are swellings called *lymph nodes*.

lymphocyte: a white blood cell that has a relatively large nucleus and a small amount of *cytoplasm*. There are several different types of lymphocyte but they can only be identified by the way they respond to *antigens*. Broadly, they can be divided into *B cells* and *T cells*.

lysis: the destruction of cells caused by breakdown of their *cell-surface membranes*. Penicillin is an *antibiotic* that prevents the formation of bacterial *cell walls*. When bacteria enter a person's body, they are in an environment that has a higher *water potential* than that of their own *cytoplasm*. Water therefore enters the bacterial cell by osmosis. Without a strong cell wall, the bacterial cells swell and burst. This is called osmotic lysis.

lysosome: an *organelle* that contains digestive *enzymes*. These enzymes are separated from the rest of the cell contents by the *plasma membrane* that surrounds the lysosome. This is essential otherwise these enzymes would digest the *proteins* and *lipids* normally found in the cell and destroy them. Lysosomes are usually associated with animal cells where they have a number of different functions:

- *Phagocytes* are white blood cells that surround and ingest bacteria. Once inside the phagocyte, these bacteria are destroyed by enzymes produced by the lysosomes. Food taken in by *microorganisms* such as amoeba is also digested by lysosomal enzymes.
- During the life of an animal, most organs and structures increase in size. Sometimes, however, they may be broken down during the course of development. *Bone* is a tissue that has to be continuously remodelled to cope with the changing stresses which occur as an animal grows and increases in size. This can involve the breakdown of bone in some areas as well as its formation in others. Breakdown is carried out by lysosomes in special bone cells.
- *Metamorphosis* is a process that occurs in the life cycle of some animals and involves a major change in the form and shape of the animal concerned. When a frog changes from a tadpole into an adult, its tail disappears as it is gradually reabsorbed. This is another process carried out by lysosomes.

lysotroph: an organism that obtains its nutrients by secreting *enzymes* on to the surface of its food. The enzymes digest the large insoluble substances and the organism absorbs the smaller molecules that result. *Decomposers* are lysotrophs.

lysozyme: an *enzyme* that catalyses the destruction of bacterial *cell walls*. It is found in secretions such as tears and *saliva*. Babies who are breast fed receive lysozyme in their mother's milk. Those who are bottle fed have much lower concentrations of lysozyme in their *alimentary canals*. This is one reason why bottle-fed babies are more prone to diarrhoeal disease.

M line: a thin line in the middle of the *H zone* in a muscle *sarcomere*.

macromolecule: a very large molecule. Many of the organic substances found in living organisms are made up macromolecules. These substances include *proteins*, *polysaccharides*, *nucleic acids* and *triglycerides*. Proteins, polysaccharides and nucleic acids are *polymers* because they are made up of many smaller similar molecules called *monomers*. Triglycerides are macromolecules but they are not polymers.

macrophage: large cells that form part of the immune system. A few are found in the blood system, but most are found in other tissues. There are many macrophages in the walls of the gut, in the reproductive system and in the *lungs*. They destroy pathogens by *phagocytosis*. Macrophages are also important *antigen*-presenting cells (see *antigen presentation*).

magnetic resonance imaging (MRI): a technique used in the diagnosis of a range of diseases. It is based on the absorption and transmission of radio waves by water molecules in tissues that have been put in a strong magnetic field. This information is then analysed by a computer to produce an image. One advantage of magnetic resonance imaging is that it does not use *X-rays*. X-rays can be harmful.

magnification: the amount by which an image is enlarged. The magnification of an object in a photograph is its length in the photograph divided by its real length. We can write this as a simple equation:

$$\text{magnification} = \frac{\text{length of cell in photograph}}{\text{real length of cell}}$$

When using a microscope, magnification on its own is not enough. Magnification only makes things larger. *Resolution* involves distinguishing between objects that are close together.

malaria: an *infectious disease* caused by a *microorganism* that lives inside the red blood cells of its human host. The microorganism concerned is one of several species of *Plasmodium* and it is spread from person to person by anopheline mosquitoes. When a female mosquito bites an infected person, blood containing parasites is taken into her *stomach*. The parasites reproduce inside the mosquito and enter her *salivary glands*. When she bites another person, parasites are injected into the blood along with the *saliva*. Once inside a new human host, the parasites migrate to the *liver* where they multiply. After an incubation period which varies from approximately 2 weeks to several months, they return to the blood where they enter the red blood cells. A cycle of multiplication, destruction of blood cells and infection of further blood cells gives rise to the characteristic pattern of fever associated with malaria.

male: the organism, or part of the organism, that produces male sex cells or male *gametes*. From whatever organism they come, male gametes have three characteristic features:

- they are smaller in size than the female gametes
- they are produced in larger numbers than the female gametes
- male gametes either move themselves or are moved in some way to the female gametes.

malignant: a *tumour* that destroys the tissues around it and can spread to other areas in the body. Cells become detached from the primary or original tumour and are transported to other sites in the body by the *blood system* or *lymphatic system*. A malignant tumour in the breast, for example, can produce secondary tumours in the lymph *glands* or in the *bones*. If untreated, malignant tumours will result in death.

malnutrition: a condition resulting from eating a diet that is not *balanced*. It may result from eating too little, but it can also result from eating more than is required to maintain health. *Kwashiorkor* and *obesity* both result from malnutrition.

maltase: an *enzyme* that breaks down *maltose*. It is a *hydrolase* and splits each maltose molecule to produce two molecules of *glucose* by adding a molecule of water. In a mammal, the maltase enzymes found on the *cell-surface membranes* of epithelial cells in the *small intestine* break down the maltose produced by the *digestion* of *starch*. Maltase enzymes are also found in germinating seeds and are secreted by some fungi.

maltose: a *disaccharide* that is made up of two *glucose* units joined together by *condensation*. Maltose is a *reducing sugar* and therefore produces a positive result with *Benedict's test*. Maltose formed by the *hydrolysis* of *starch* during the *germination* of barley grain is fermented by yeast in the production of alcoholic drinks such as beer and whisky.

M

mammal: a member of the class *Mammalia*. You should be careful to distinguish between animals and mammals. The term mammal should only be used when referring to organisms such as humans, sheep, bats and whales that belong to this class. The term animal refers to any member of the animal *kingdom Animalia*.

Mammalia: the class containing mammals (see *classification*). Mammals along with fish, amphibians, reptiles and birds belong to the phylum Chordata. In addition to its more familiar members, the Mammalia include kangaroos and other marsupials and some primitive egg-laying species. Members of this class share the following features:

- they feed their young on milk secreted by mammary *glands*
- their body temperature fluctuates very little, although this is also a feature of birds
- they have hair and sweat glands. These are distinctive mammalian features that play an important role in temperature control.

Most mammals possess different types of teeth that have different functions. Other chordates have teeth that are all similar to each other.

mammography: the use of *X-rays* for the detection of *tumours* in breast tissue. Mammography allows breast cancer to be detected early, which leads to a greater probability of it being treated successfully.

MAOA: see *monoamine oxidase (MAOA)*.

mark–release–recapture: a method of estimating population size by marking a number of animals, releasing them and then counting the number of marked animals in a second sample. The method is particularly suitable for estimating the population of species that move about. It makes a number of assumptions:

- the number of organisms in the population does not change significantly between the time when they were captured and marked and the time when the second sample was taken. This applies to births, deaths and migration
- the marked animals mix randomly in the population
- marking does not affect the animal in any way such as, for example, making it more conspicuous to predators.

The following example shows how population size can be estimated using this method.

Pitfall traps were set and 60 beetles were trapped. These were marked and released. In a second sample of 66 beetles trapped 4 days later, 20 were marked. Estimate the population of beetles.

The calculation relies on the simple relationship:

Proportion of marked beetles in population = Proportion of marked beetles in the second sample

or

$$\frac{\text{Number of marked beetles in population}}{\text{Total number of beetles in population}} = \frac{\text{Number of marked beetles in the second sample}}{\text{Total number of beetles in the second sample}}$$

$$\frac{60}{\text{Total number of beetles in population}} = \frac{20}{66}$$

$$\text{Total number of beetles in population} = \frac{60 \times 66}{20}$$

$$\text{Total number of beetles in population} = 198$$

---

marker gene: a *gene* used to find out whether a piece of *DNA* has been taken up by another organism. When organisms are **genetically modified (GM)**, only a few cells take up this piece of DNA. Marker genes are used so these cells can be identified easily. Some marker genes cause the cells that take them up to appear different in some way. One that is commonly used codes for green fluorescent protein. A cell that contains this gene makes the protein which glows green when it is illuminated with UV light. Other marker genes code for resistance to **antibiotics**. The antibiotic can then be used to kill all the cells that have not taken up the gene. Only those with the marker gene for antibiotic resistance will survive. These cells will also contain the piece of DNA that genetic engineers wanted to insert.

mass transport: the transport of substances in bulk from one part of an organism to another. Large organisms cannot rely only on *diffusion* as this process is much too slow to meet their needs. They require a way of moving substances rapidly from one place to another. Mass transport systems are linked with exchange surfaces. The main function of these exchange surfaces is to establish and maintain differences in concentration and pressure. It is because of these differences that substances can be moved by a mass transport system from one place to another. Examples of mass transport systems are the *blood system* of a mammal, and *phloem* and *xylem* in a plant.

mast cell: a cell that is produced in the **bone** marrow and is part of the immune system. Mast cells release **histamine** and other substances that produce the responses characteristic of **inflammation** and **allergy**.

mastectomy: a surgical operation for breast cancer in which all the breast tissue is removed. Radical mastectomy involves removing the breast tissue and the **lymph nodes** in the nearby armpit. Mastectomy is used for large **tumours** and is usually followed by radiotherapy or **chemotherapy**.

maternal chromosome: the member of the pair of **homologous chromosomes** that originally came from the egg cell produced by the female parent.

mechanoreceptor: a sensory **receptor** that responds to mechanical stimuli, such as touch, pressure, sound and gravity. Many mechanoreceptors are cells which have hair-like structures called **cilia**. When one of these cilia is moved, a **nerve impulse** is produced. Mechanoreceptors provide information about many different kinds of stimulus. For example, mechanoreceptors in muscles detect stretching, while others in the ear are concerned with balance and hearing.

medulla (brain): part of the hindbrain. It contains the centres that are responsible for controlling heart rate, blood pressure and breathing.

medulla (kidney): the inner part of the **kidney**. As well as a large number of **blood vessels**, the medulla contains the **loops of Henle** and the **collecting ducts** of the kidney tubules. Many small mammals that live in desert regions have kidneys which produce concentrated **urine**. An animal such as jerboa, for example, can produce urine which is between two and three times as concentrated as that of humans. In order to do this, the loops of Henle are much longer than in closely related animals that live in areas where water is more readily available. This is reflected in the appearance of the kidney, which has a much thicker medulla.

meiosis: a type of nuclear division in which the number of **chromosomes** is halved. This results in the **haploid** number in each of the daughter cells. Meiosis involves two separate divisions which are referred to as meiosis I and meiosis II.

In animals and many plants, each body cell contains the diploid number of chromosomes. This number varies from organism to organism, but in humans there are 23 pairs. One member of each pair originally came from the female **gamete**. These are the **maternal chromosomes**. One member of each pair came from the male gamete. These are the paternal chromosomes.

In meiosis I, the chromosomes come together in their homologous pairs, with each chromosome split longitudinally into two sister **chromatids**. One member of each pair of chromosomes is now pulled to each end of the cell, so we have one complete set of chromosomes at one pole of the cell and one complete set at the other pole. Some of the chromosomes which have gone to a particular pole will have come from one parent and some from the other. This is called **independent segregation**. It is one reason why the cells that result from meiosis carry different genetic information from each other. Another process called **crossing over** means that there are further genetic differences in these cells. At the end of meiosis I, there will be two daughter cells. Each contains one complete set of chromosomes. In meiosis II, each of these daughter cells divides again. This time the chromatids separate and are pulled to the poles of the cell. The diagram

M

shows how this process results in four daughter cells, each containing half the number of chromosomes that was present in the original cell.

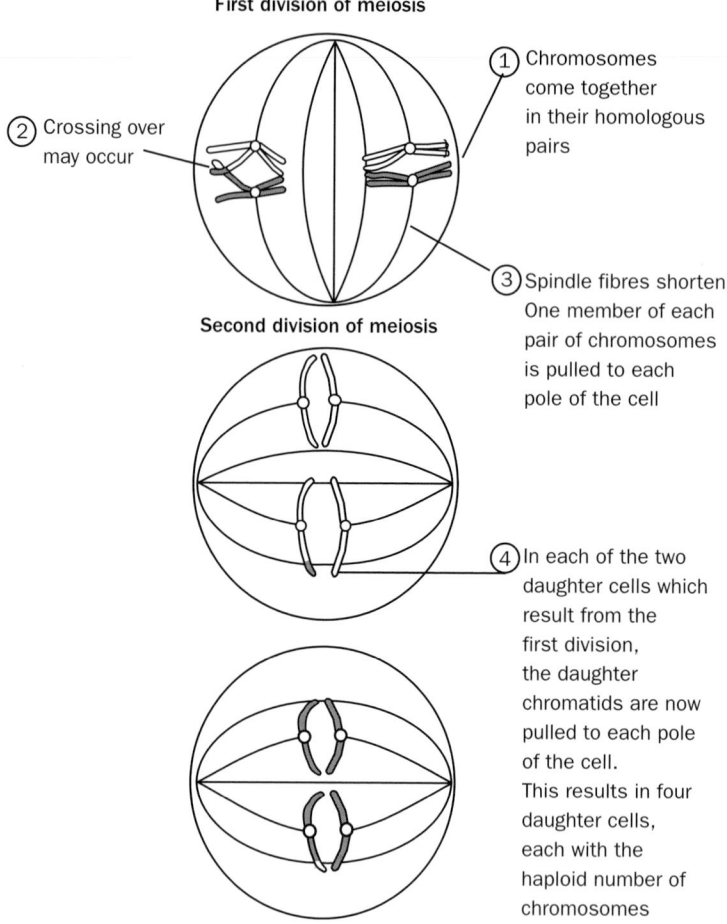

First division of meiosis

① Chromosomes come together in their homologous pairs

② Crossing over may occur

③ Spindle fibres shorten. One member of each pair of chromosomes is pulled to each pole of the cell

Second division of meiosis

④ In each of the two daughter cells which result from the first division, the daughter chromatids are now pulled to each pole of the cell. This results in four daughter cells, each with the haploid number of chromosomes

Because **sexual reproduction** involves the fusion of the nuclei of two different cells, the number of chromosomes is first halved. Otherwise, the number would double with each successive generation. In most animals, **gametes** are formed by meiosis. Plants, however, have very complex life cycles involving an **alternation of generations** and meiosis takes place some time before gamete formation. The haploid cells which result from meiosis in plants often go through several stages before the gametes are finally produced.

melanin: a dark-brown pigment that is found in many animals. The colour of human skin results from the amount and type of melanin that is present. Dark skin contains a lot of melanin. It is found mainly in people who originated in tropical regions. The melanin in the skin of these people protects them from ultraviolet light. It reduces the probability of damage to their **DNA** and of developing skin cancer as a result. In people in temperate regions, light skin containing less melanin is an advantage. People with light skin can absorb more sunlight and produce more vitamin D. In mammals, albinos do not produce melanin. Because of this, they are light skinned and have white hair and pink eyes.

memory cell: a *lymphocyte* that remains in the blood after the original infection has gone. When the *antigen* has been destroyed by the *plasma cells*, most of these plasma cells die. The memory cells, however, remain in the blood and respond rapidly if infection with the same antigen occurs again. Multiplication takes place and more plasma cells are produced. In this way the response to a second infection is much more rapid.

Mendelian ratio: the ratio of offspring phenotypes expected in a simple genetic cross such as that carried out by Mendel in his work on inheritance in peas.

Mendel's laws of inheritance: two laws produced by Gregor Mendel as a result of his early experiments on inheritance in peas:
- The law of segregation states that when organisms produce *gametes*, the *alleles* of a particular *gene* separate. Each gamete receives one allele from the two in the parent cell.
- The law of *independent segregation* states that the alleles of different genes segregate independently of each other during the formation of gametes. We now know, however, that this is only true if the gene loci (see *locus*) concerned are on different chromosomes. It is not true if they are on the same *chromosome* (see *linkage*).

menopause: the stage in a woman's life when her *ovaries* stop producing *gametes* and she is no longer fertile. There is a fall in the amount of *oestrogen* secreted by her ovaries and this fall produces the symptoms associated with menopause. These symptoms may be sufficiently disturbing to require treatment with *hormone replacement therapy (HRT)*. The fall in oestrogen is also associated with an increased risk of *osteoporosis*.

menstrual cycle: the name given to the type of *oestrous cycle* found in humans, apes and some monkeys. It is slightly different from the cycle found in other mammals. The lining of the uterus called the *endometrium* is shed together with some blood and lost from the body between cycles. In other mammals, this lining is usually reabsorbed back into the body if pregnancy does not take place.

M

meristem: a group of plant cells that are able to divide by *mitosis*. When a mammal grows, each organ grows. The *liver*, for example, is present at birth but it is smaller than in an adult. It contains liver cells which divide and grow. In this way, the mature liver is identical to that in the young animal; it just contains more cells. Plant growth is different. In the root tip, there is a group of unspecialised cells. These form the root meristem. These cells divide, and only then do they become specialised for particular functions. Once they have differentiated, they do not continue to divide. Another meristem near the tip of the shoot controls the growth of the stem. These two groups of cells are called primary meristems, as they are present when the plant begins its growth. They are concerned mainly with increase in length. Secondary meristems develop in woody plants and increase the diameter. An example of a secondary meristem is the *cambium*. The fact that there are some undifferentiated cells in a mature plant which still retain the ability to divide is important in allowing *micropropagation* of plants.

mesoderm: the middle layer of cells in *triploblastic* animals such as those that belong to the *Annelida* and the *Chordata*.

mesogloea: a jelly-like layer found in the body wall of animals that belong to the phylum *Cnidaria*. The mesogloea does not contain cells and is sandwiched between the outer *ectoderm* and the inner *endoderm*.

mesophyll: the types of cells found in the middle layers of a leaf. In leaves of plants which are not specialised for living in water or for particularly dry conditions, there are two sorts of cell in the mesophyll. Palisade cells are found near the upper surface of the leaf. They are tall and thin and packed tightly together. They contain many *chloroplasts* and have adaptations for efficient *photosynthesis*. Underneath the palisade cells is the spongy mesophyll. This consists of irregularly shaped cells in between which are large air spaces. These spaces play an important role in gas exchange between the leaf and the surrounding atmosphere. Plants found in full sunlight tend to have leaves with a thicker layer of palisade cells than those growing in the shade.

mesosome: an infolding of the *cell-surface membrane* of a bacterium. Various functions have been suggested for mesosomes. Some scientists thought that they were concerned with bacterial *respiration*. Others considered them to be involved in cell division. Scientists now think that mesosomes are *artefacts* that form as a result of the processes involved in preparing specimens for examination with an *electron microscope*.

messenger RNA: see *mRNA*.

metabolic pathway: a series of biochemical reactions taking place in a living cell. They may involve substances being broken down to smaller molecules (*catabolic reactions*) or they may involve substances being built up into more complex ones (*anabolic reactions*). The diagram shows a typical metabolic pathway in which substance A is converted to substance E via a number of intermediate steps. Each reaction in the pathway is controlled by a separate *enzyme*.

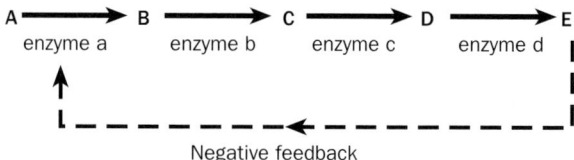

A metabolic pathway

Pathways may be regulated by *negative feedback*. When sufficient product, in this case substance E, has been produced, it acts as an inhibitor. It slows down or stops the action of an enzyme at the start of the pathway. As a result, less product is formed. If there is not enough substance E, there will be no negative feedback and no enzyme inhibition. The enzyme will continue to work at its normal rate so more substance E is produced.

metabolic rate: the rate at which energy is released in the body. Because this depends to a considerable extent on body size, it is usual to give figures for metabolic rate in energy units per m^2 of body surface per hour. Metabolic rate is influenced by a number of factors. These include:

- physical activity. Minimum energy release occurs when the body is in a complete state of rest. It rises with an increase in physical activity. The average amount of energy released in an adult man lying down is about 4.5 kilojoules per minute. This can rise to 70 kilojoules per minute with exceptionally demanding physical work
- reproductive state. Metabolic rate rises during pregnancy. This is partly due to an increase in the resting level of energy release, partly to the growth of the fetus and partly to an increase in the *fat* stores of the mother. *Lactation* involves an even higher level of energy expenditure
- *hormones*. Increased production of *thyroxine (thyroid hormone)* causes an increase in metabolic rate
- time after eating. Energy release is increased immediately following a meal.

As so many factors influence metabolic rate, it is usual to compare the energy released under standard conditions. The **basal metabolic rate (BMR)** of a person is the metabolic rate at complete rest. It is a measurement of the amount of energy required for vital activities, such as the action of the heart and the muscles associated with breathing.

metabolism: all the chemical reactions that take place in living organisms. In a single cell, there are many hundreds of such reactions. Processes involving the breakdown of large molecules into smaller ones, such as many of those which occur in **digestion** and respiration, are known as **catabolic reactions**. Others, such as those involved in building **proteins** from **amino acids** and in **photosynthesis**, are concerned with combining smaller molecules to form larger ones. These are called **anabolic reactions**. Catabolic reactions release energy while anabolic ones require energy.

metabolite: a substance produced or used in one of the chemical reactions that take place in the body of a living organism.

metamorphosis: changes that take place when the **larva** of an organism changes into an adult. In animals such as frogs, these changes occur progressively. The larval stage is a tadpole. Metamorphosis involves a series of gradual changes that are shown externally by the growth of legs and the absorption of the tail. There are internal changes as well. These involve, for example, the transition from **gills** to **lungs** as gas exchange surfaces and modifications to the gut that are associated with the change from a vegetarian to a carnivorous diet. Many insects go through a more drastic change. The larva first moults into a pupa. Inside the pupa most of the existing tissues are broken down. This involves **lysosomes**. The new tissues that make up the adult then develop. Metamorphosis is usually controlled by **hormones**.

metastasis: the process by which cancer cells spread from the original **tumour** to other parts of the body. A clump of cells breaks free from the original, or primary, tumour and is carried in the blood or by the **lymphatic system** to another part of the body where it may form a secondary tumour.

metaxylem: *xylem* that is formed after **protoxylem**. It has vessels that are larger in diameter than those of protoxylem. Its walls also contain more **lignin**.

methicillin-resistant *Staphylococcus aureus*: see *MRSA*.

methylation: the adding of methyl groups to **DNA**. Methylation may prevent **transcription** of particular genes and therefore helps to regulate gene action. Methylation of **tumour suppressor genes** may be important in the development of **tumours**.

Micrococcus: a *genus* of spherical bacteria. Micrococci are common in the soil and are often found on human skin. They may appear as contaminants in cultures of bacteria on *agar* plates. Micrococci do not cause disease.

microevolution: small-scale evolutionary changes. They usually involve changes in the frequency of **alleles** in a population, but do not, on their own, result in the formation of new species. An example is the evolution of copper-tolerant grasses on contaminated soil. Some of the grass seeds that land in the area will have alleles which result in their being better adapted to these conditions. They can survive on soils that are heavily contaminated with copper. They germinate, grow and reproduce, passing on these alleles to the next generation. This will produce a change in the proportion of alleles in the **population**. It is thought that a large number of small changes like this could eventually result in the evolution of a completely new species. The flow chart overleaf summarises this process.

M

> Contaminated soil colonised by grass seeds. These seeds vary. A few of them will be copper-tolerant

↓

> The seeds that cannot tolerate high concentrations of copper die; those that can survive

↓

> The tolerant grass plants have less competition. They grow and reproduce, passing on the alleles associated with copper tolerance

↓

> Selection has operated. The population has changed from grasses which could not tolerate copper to grasses which could tolerate it

micrometre (μm): one thousandth of a millimetre. **Bacteria** are approximately 1 μm in diameter. In practice, they are the smallest objects that can be seen with a light microscope. Animal and plant cells are larger. A human **red blood cell**, for example, is about 7.5 μm in diameter, while a palisade cell from the **mesophyll** of a leaf may be 200 μm long.

micronutrient: a substance required by an organism in amounts that are very small when compared to the other nutrients which it requires. Micronutrients include **vitamins** and mineral ions. The table shows some examples of mineral ions that are required as micronutrients by plants and animals.

Micronutrient	Main functions
Plant micronutrients	
Iron	● Essential for the synthesis of **chlorophyll**
	● Contained in **cytochromes**
Molybdenum	● Needed for the **enzyme** nitrate reductase to function. This enzyme reduces nitrates during the synthesis of **amino acids**
Animal micronutrients	
Iodine	● Contained in **thyroxine**, a **hormone** that controls the **basal metabolic rate (BMR)**. Thyroxine is secreted by the **thyroid gland**
Iron	● Contained in cytochromes, which are substances that form part of the **electron transport chain**. A cytochrome is a protein that is combined with another chemical group containing iron or copper molecules
	● Part of **haemoglobin**
	● Necessary for enzymes like catalase to work

microorganism: any organism that is very small and can only be seen with a microscope. The three main groups of microorganisms – the **viruses**, the **bacteria** and the fungi – are different in their basic structure. Although they are responsible for many plant and animal diseases, they are also particularly useful in industrial processes. This is mainly because they have simple nutritional requirements, a very rapid growth rate and either produce or, with *genetic engineering* can be altered to produce, a whole range of useful substances.

microphage: a small phagocytic white blood cell.

micropropagation: a process by which large numbers of genetically identical plants can be produced from the growth region of a single parent. The growth region or *meristem* is carefully dissected from the apical bud and transferred to a tube containing nutrient *agar* to which *plant growth substances* have been added. The meristem grows rapidly into a mass of cells called a callus. After a short time, leafy shoots begin to grow from the callus tissue. These are separated and transferred to a different medium on which they grow roots and develop into complete plants. This process is summarised in the flow chart.

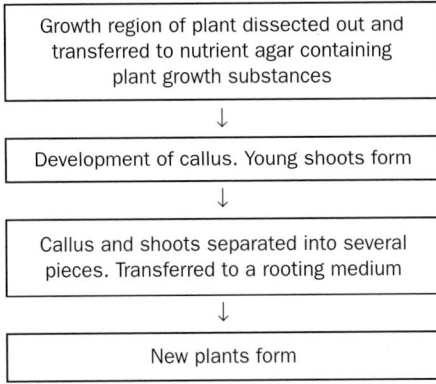

The great advantages of this technique are that it requires relatively little space, heat and light, and the plants that are produced are disease-free.

microsatellite repeat sequence (MRS): one way in which the sequence of *bases* in *DNA* varies between individuals that belong to the same species. It involves repeated sequences, between one and six *base pairs* in length. Scientists have used microsatellite repeat sequences to show genetic similarities and differences between different populations.

microtubule: part of the *cytoskeleton* of a cell. Microtubules have many functions. They are components of *cilia* and *flagella* and aid in their movement. They are also involved in the movement of *chromosomes* during *mitosis*.

microvilli: tiny, finger-like projections (singular microvillus) from the *cell-surface membrane* of an animal cell. Microvilli are found in large numbers on the epithelial cells that line the *small intestine*. They are also found on the cells that form the wall of the *first convoluted tubule* in a *nephron*. Microvilli increase the surface area over which *diffusion* and *active transport* take place. You should take care to distinguish a microvillus from a villus. Villi are much larger structures. Those in the small intestine, for example, are made up of many cells.

middle lamella: the layer that cements together the *cell walls* of neighbouring cells.

migration: the movement of a population of organisms from one place to another. The timescale over which this takes place can vary enormously. Small organisms in the plankton move up towards the surface waters during the daytime and move back into the depths at night. This is a daily movement. Swallows, on the other hand, undergo an annual migration. They leave the temperate regions of Europe in the autumn, spend the winter in Southern Africa and return to Europe the following spring. In addition, migration can involve many return journeys during the lifetime of the animal, a single return journey or even a one-way trip.

mitochondria: the organelles (singular mitochondrion) in which the biochemical reactions associated with **aerobic respiration** take place. Mitochondria vary in size and shape, but they are often small, elongated structures about a micrometre in length. Each mitochondrion is surrounded by an envelope made up of two **plasma membranes**. The inner one is folded to form structures called **cristae**. These cristae stick out into the matrix of the mitochondrion. The matrix contains the **enzymes** associated with the **Krebs cycle** while the **electron transport chain** is located on the cristae. Numbers of mitochondria vary considerably from cell to cell. Cells that contain particularly large numbers of these organelles usually require large amounts of **ATP**. This may be for one of the following reasons:

- **active transport**. This process requires ATP to move substances against a concentration gradient. Active transport is important in the **absorption** of the products of **digestion** in the **small intestine**, reabsorption by the cells of the **first convoluted tubule** in a **nephron**, and in loading sugar into the sieve tubes (see **sieve tube element**) in the phloem. Cells associated with these processes contain large numbers of mitochondria
- movement. The contraction of muscles, the separation of chromosomes in dividing cells, and the swimming of sperms all require energy. Once again large numbers of mitochondria are associated with the cells concerned
- chemical reactions involved in producing new molecules also require energy. Examples of cells in which such reactions take place include **liver** cells and the **rod cells** in the **retina** of the eye.

mitosis: a type of nuclear division that results in each daughter cell having the same number of chromosomes as the parent cell. Although mitosis is a continuous process, it is convenient to divide it into four stages. These are:

- prophase. The chromosomes become shorter and thicker and show up clearly. Each one consists of a pair of **chromatids** which are joined together by a **centromere**. The nuclear envelope starts to break down and the **spindle** forms
- metaphase. The chromosomes are now arranged across the equator of the cell. They are attached at their centromeres to the spindle fibres
- **anaphase**. The pair of chromatids which make up each chromosome are pulled apart by the spindle fibres. One chromatid from each pair goes to one pole of the cell and the other goes to the other pole. Each chromatid has now become a chromosome
- telophase. The chromosomes form two groups, one at each pole. They gradually lose their form and can no longer be recognised as distinct structures. A new nuclear envelope forms round each group of chromosomes.

monoamine oxidase (MAOA): an **enzyme** that breaks down **neurotransmitters** such as **adrenaline** and **noradrenaline**. Too much or too little monoamine oxidase has been associated with various aspects of behaviour including depression, substance abuse and attention deficit disorder. Scientists in New Zealand found that maltreated children with a mutant form of the gene for monoamine oxidase were very likely to develop antisocial behaviour. They explained the results of their investigation by suggesting that genetic susceptibility to a condition varies with environmental exposure.

monoclonal antibodies: antibodies produced by the descendants of a single B cell. Monoclonal antibodies are normally made in the laboratory by the following technique.

```
┌─────────────────────────────────────┐
│  A mouse or other small mammal is    │
│  injected with a specific antigen. B cells are │
│  then obtained from the animal       │
└─────────────────────────────────────┘
                   ↓
┌─────────────────────────────────────┐
│  The B cells are fused with other cells so │
│  that they will effectively grow forever in │
│  cell culture                        │
└─────────────────────────────────────┘
                   ↓
┌─────────────────────────────────────┐
│  The fused cells are diluted and placed │
│  individually in the wells on a culture plate │
└─────────────────────────────────────┘
                   ↓
┌─────────────────────────────────────┐
│  Each individual cell is allowed to multiply so │
│  that a pure cell line or clone is established. │
│  This clone will only produce a single type of │
│  antibody, a monoclonal antibody     │
└─────────────────────────────────────┘
```

Monoclonal antibodies are extremely useful in *biotechnology* because they bind tightly to specific molecules. Because of this, they may be used to distinguish between similar molecules. They may be used, for example, to distinguish between human *proteins* and chimpanzee proteins or between molecules on the surface of cancer cells and molecules on the surface of healthy cells.

monocotyledon: a member of the group of *flowering* plants characterised by having a single *cotyledon* in each of its seeds.

monoculture: a type of farming in which a large area of land is devoted to the growth of a single crop. In contrast to this, more primitive methods of agriculture, such as subsistence farming, involve the growth of many different species of plant on a relatively small amount of land. Monoculture has a number of distinct advantages over agricultural techniques like subsistence farming:

- Monoculture is much less labour-intensive than subsistence farming. It relies on the use of machinery. Much greater profits may be made from monoculture.
- Because large areas are devoted to a single crop, it is much easier to use specialised machinery for crop management. Combine harvesters, for example, can only be used for harvesting wheat if it is grown in large fields.
- Different crops require different soil conditions. Such features as soil pH and the correct balance of nutrients can be managed so they are at an optimum for the crop concerned.

On the other hand, monoculture has some disadvantages:

- Large areas are devoted to the growth of single species of plants. This encourages the development of large pest populations. At present, the only effective way of controlling these pests is with the repeated use of chemical *pesticides*.

179

- The search for more and more productive crops results in little genetic variation. A lack of genetic *diversity* may create problems if circumstances were to change in the future.
- Devoting larger and larger areas to the growth of a single crop encourages farmers to remove hedges and make bigger fields. This removes *habitats* for wildlife and may lead to soil erosion.

monocyte: a type of white blood cell. Monocytes are large cells which are made in the *bone* marrow. They engulf bacteria and other *microorganisms* by *phagocytosis*.

monohybrid cross: a genetic cross between individuals in which *alleles* at a single *locus* are considered. For an example of how to set out a genetic cross, see *dihybrid cross*.

monomer: one of the small similar molecules that join together to form a *polymer*. There are three important biological monomers: *amino acids*, *monosaccharides* and *nucleotides*.

mononucleotide: see *nucleotide*.

monosaccharide: a carbohydrate with molecules that consist of a single sugar unit. Monosaccharides are small molecules and dissolve readily in water. They are classified according to how many carbon *atoms* each molecule contains:

- Trioses contain three carbon atoms in each molecule. They are important intermediate compounds in biochemical pathways, such as those of *respiration* and *photosynthesis*.
- Pentoses are five-carbon sugars. Ribose and *deoxyribose* are important constituents of *RNA* and *DNA*. *Ribulose bisphosphate* is the substance that combines with carbon dioxide in the *light-independent reaction* of *photosynthesis*.
- *Hexoses* are six-carbon sugars such as *glucose* and *fructose*.

motor neurone: a cell that carries *nerve impulses* away from the *central nervous system* to an *effector*, such as gland or a muscle.

motor neurone disease: a disease that involves degeneration of the *motor neurones* in the *brain* and *spinal cord*. This leads to muscle wastage and weakness. The disease gets gradually worse and the patient often dies 3 to 5 years after the initial diagnosis.

MRI: see *magnetic resonance imaging (MRI)*.

mRNA: messenger ribonucleic acid or mRNA, a nucleic acid that acts as a messenger. It takes a copy of the *genetic code* from the nucleus into the *cytoplasm* during *protein synthesis*. It is a type of RNA and consists of a single *polynucleotide* chain with a backbone built up of alternating ribose sugars and phosphate groups. One of four bases – *adenine*, *cytosine*, *guanine* or *uracil* – is attached to each ribose sugar. In the process of *transcription*, part of the *DNA* in the nucleus unwinds and acts as a template for the formation of mRNA. The mRNA then takes the code for the gene concerned from the nucleus into the cytoplasm, enabling *amino acids* to be assembled in the correct sequence for the required protein. A second type of RNA, transfer RNA or *tRNA*, is found in the cytoplasm. Transfer RNA is important in assembling amino acids in the correct position on the mRNA during the process of *translation*.

MRS: see *microsatellite repeat sequence (MRS)*.

MRSA: methicillin-resistant *Staphylococcus aureus*, a strain of the bacterium *Staphylococcus aureus* that is resistant to the *antibiotic* methicillin. It is not only resistant to methicillin but also to many other commonly used antibiotics. Therefore, infections caused by MRSA are very difficult to treat. MRSA is a particular problem in hospitals where it may infect patients who have weakened immune systems or who are recovering from surgery.

mucosa: the layer of tissue that surrounds the *lumen* of the *small intestine*. It is specialised for secretion and *absorption*.

mucous membrane: a thin layer of tissue that lines many tubular organs and cavities in the body. It contains *glands* that secrete *mucus*.

mucus: a slimy substance produced by *goblet cells*. One of its main components is a *glycoprotein* called mucin. Mucus secreted in the *lungs* traps particles in the air that is breathed in. *Cilia* on the surface of the epithelial cells that line the airways waft the mucus and trapped particles away from the gas-exchange surface. Mucus also acts as a lubricant such as, for example, in the *alimentary canal*. Invertebrate animals also produce mucus. Filter-feeding worms and molluscs have tentacles or *gills*. Mucus secreted by cells on these structures traps small food particles. These are drawn towards the mouth by cilia.

multiple alleles: a gene that has more than two forms. An example of a gene that has multiple *alleles* is the one which controls the inheritance of the ABO *blood groups* in humans. There are four blood groups in this system: A, B, AB and O. Their presence is determined by a single gene with three alleles: I^A, I^B and I^O. The alleles I^A and I^B are *codominant*. The allele I^O is *recessive* to both I^A and I^B. Each gene occupies a particular place on a chromosome called its *locus*. Since chromosomes are found in homologous pairs, it follows that there will be two places that can be occupied by alleles of this gene, one on one chromosome of the pair and the other at the corresponding place on the other chromosome. It is only possible, then, to have two blood group alleles. They may be the same as each other or they may be different. The table shows the various combinations that it is possible to have and the resulting blood groups.

Possible genotypes	Blood group
$I^A I^A$ or $I^A I^O$	A
$I^B I^B$ or $I^B I^O$	B
$I^A I^B$	AB
$I^O I^O$	O

multiple birth: this ccurs when a woman gives birth to more than one child at the end of a pregnancy. In other words, more than one fetus is carried to full term. Where twins are produced, they may be either monozygotic or dizygotic. Monozygotic twins are produced from a single fertilised egg and are genetically identical. Dizygotic twins are produced from two fertilised eggs. They are not genetically identical.

multiple pregnancy: a woman who is pregnant and has more than one *embryo* in her uterus. Where *in vitro fertilisation (IVF)* has been used, more than one fertilised egg may be inserted into the uterus. This increases the probability of a fetus going to full term and the woman giving birth to a child at the end of the pregnancy. Fertility drugs also increase the probability of a multiple pregnancy.

multiple repeat: *DNA* that does not code for a *polypeptide* and is found between *genes*. A lot of the DNA in a *eukaryotic* cell such as a human cell does not code for polypeptides. Some of this non-coding DNA is found within genes. Non-coding regions of DNA within a gene are called **introns**. Non-coding DNA may also be found between genes. Some of this DNA consists of short sequence of **nucleotides** that are repeated a number of times.

muscle: an **effector** that contracts and brings about movement in animals. There are three types of muscle tissue found in mammals:

- **Cardiac muscle** is found only in the heart.
- **Skeletal muscle** or striated muscle is attached to **bones** in the skeleton and brings about limb movements. It also helps to maintain posture.
- **Involuntary muscle** is involved with the movement of organs that are not under conscious control. **Peristalsis** in the gut, the change in diameter of **blood vessels** and the alteration in the size of the pupil of the eye in response to changes in light intensity are controlled by involuntary muscle.

muscularis externa: the muscle coat in the wall of the **small intestine**. The muscularis externa consists of an inner layer of circular muscle fibres and an outer layer of longitudinal fibres. Contraction of the muscle fibres in these layers is responsible for movements of the gut such as **peristalsis**.

muscularis mucosa: a layer of smooth muscle in the wall of the **small intestine**. It separates the **mucosa** from the **submucosa**. Contraction of this muscle results in movement of the **villi** and improves their contact with the products of **digestion**.

mutagen: an environmental factor that increases the rate of **mutation**. There are many different mutagens. They include certain types of radiation, many organic substances and some viruses. Mutations occur randomly, and under natural conditions the rate at which mutation occurs is slow. Mutagens simply increase this rate. Exposure to mutagens in the environment is medically important because of the link between mutations and cancer (see **tumour**). Substances in tobacco smoke that act as mutagens also increase the likelihood of developing cancer.

mutation: a change in either the amount or the arrangement of the genetic material in a cell. Mutation can take place in any cell. If it occurs in a **gamete**, the resulting characteristic can be inherited. There are mutations, however, which occur in the normal body cells. These are called somatic mutations. They are not passed on to the next generation. Mutations differ in their effects. Chromosome mutations involve large-scale changes and are concerned with parts of, or even whole, chromosomes. **Gene mutations**, on the other hand, are on a much smaller scale and are concerned with changes to relatively small numbers of bases in the genetic material. The rate at which mutation takes place can be increased by exposure to various **mutagens**, such as ultraviolet radiation, **X-rays** and a wide range of organic substances.

mutualism: a relationship between two different species of organism where both gain a nutritional advantage. In explaining examples of mutualism, you should identify nutritional rather than more general advantages, such as the provision of warmth and shelter. A good example is provided by the **microorganisms** that live in the **rumen** of cattle. The mammal provides a constant supply of food rich in **cellulose** for the microorganisms. In turn, the microorganisms provide the cattle with **fatty acids** obtained from the **digestion** of cellulose.

mycelium: the mass of thread-like **hyphae** that make up the body mass of many fungi.

Mycobacterium tuberculosis: the bacterium that causes **tuberculosis (TB)**.

myelin: a fatty substance produced by **Schwann cells**. It forms a layer round the **axons** in many mammalian nerves. As it is an insulator, myelin only allows the ion movements that result in the next **action potential** to occur at the gaps where no myelin is present. These gaps are called the **nodes of Ranvier**. **Nerve impulses** therefore travel along in a series of jumps from one node to the next. This is known as **saltatory conduction** and leads to impulses moving much more rapidly in myelinated than in non-myelinated nerves.

myocardial infarction: another name for a heart attack. The blood supply to the heart muscle may be interrupted by, for example, a blood clot in one of the coronary arteries. This results in the death of a large part of the muscle concerned. The patient experiences severe pains in the chest. The main danger is that a myocardial infarction may cause the heart to beat in an uncontrolled way without actually pumping any blood. This rapidly leads to death. With appropriate treatment, however, most patients can return to a normal, active life.

myofibril: one of the many small fibres that are arranged parallel to each other in a **skeletal muscle fibre**. When seen with a light microscope, each myofibril is made up of alternating light and dark bands. This is due to a regular arrangement of filaments made from the **proteins actin** and **myosin**. The diagram shows the structure of a myofibril.

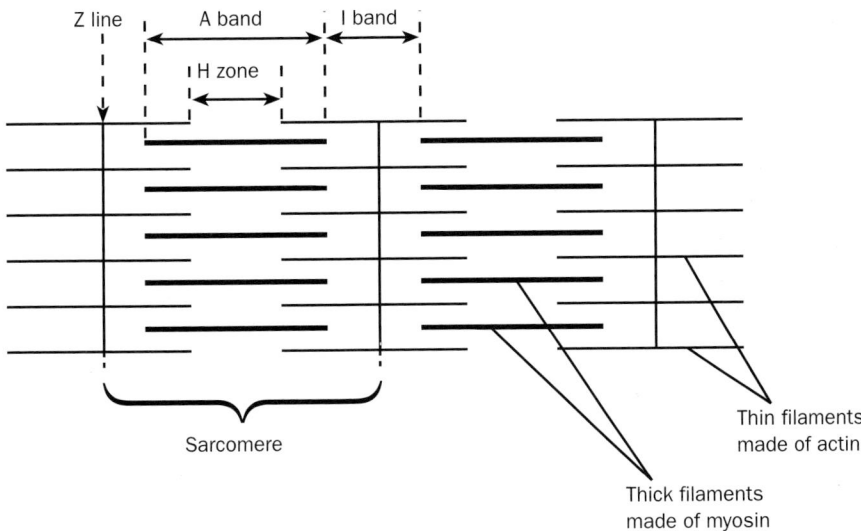

The structure of a myofibril

myogenic: heart muscle is described as myogenic because it contracts automatically. Most of the other muscle in the body is neurogenic and contracts only when stimulated by a **nerve impulse**. The automatic contraction of heart muscle is particularly important in the **sinoatrial node (SAN)** where it gives rise to the impulse that controls the heart beat.

myoglobin: an oxygen-storing pigment found in muscle. Myoglobin is similar to **haemoglobin** as it is made up of a polypeptide chain associated with an **iron**-containing **haem** group. It is different, however, because it only contains one of these units, not four like haemoglobin. Myoglobin is found in the muscles of many animals, but there are particu- larly high concentrations in diving mammals such as seals and whales. The myoglobin acts as an oxygen store. It is saturated with oxygen at relatively low partial pressures and is able

M

to hold on to this oxygen, only releasing it when the concentration in the tissues falls to a very low level.

myometrium: the outer layer of the *uterus*. It is mainly made of *involuntary muscle*.

myosin: the protein of which the thick filaments in the *myofibrils* of *skeletal muscle* are made. Each myosin molecule consists of two parts. It looks rather like a golf club with a long, rod-shaped part and a globular head. The myosin heads go through a cycle of attaching to, changing their angle to the rest of the molecule and detaching from points on the *actin* filaments. This lets the two types of filament slide between each other and the muscle to shorten. This is the basis of the *sliding-filament model* that explains how skeletal muscle contracts.

A–Z Online

Log on to A–Z Online to search the database of terms, print revision lists and much more. Go to **www.philipallan.co.uk/a-zonline** to get started.

NAD: a coenzyme that is important in respiration. ***Aerobic respiration*** involves a series of reactions in which ***glucose*** is oxidised step by step to carbon dioxide and water. NAD is reduced as a result. The reduced NAD then transfers ***electrons*** and protons to the ***electron transfer chain***. As the electrons are passed from one molecule to the next along this chain, energy is released. This energy is used to form ***ATP***. For each molecule of glucose that is oxidised, ten molecules of reduced NAD are produced.

NADP: a coenzyme that is important in ***photosynthesis***. In the ***light-dependent reaction*** of photosynthesis, light is absorbed by ***chlorophyll*** molecules. This causes some of the ***electrons*** in the chlorophyll to become excited. They reduce NADP. Reduced NADP is involved in converting carbon dioxide to carbohydrate in the ***light-independent reaction***.

nail-patella syndrome: an inherited condition in which affected people have poorly developed nails and knee-caps. It is caused by the inheritance of a ***dominant allele***. The ***gene*** for nail-patella syndrome is carried on the same chromosome as the gene that controls the ABO ***blood groups***. These genes are linked.

natural fertiliser: any form of ***fertiliser*** that is not an ***artificial fertiliser*** and is therefore not made either from naturally occurring rocks or by industrial processes. Natural fertilisers are produced from organic material that is broken down to provide a supply of plant nutrients. They include manure and ***compost***. Natural or organic fertilisers:

- are mixtures of substances and may contain trace elements. These are substances that are important to plants in small amounts
- add organic matter to the crop. This may improve soil structure and water-holding properties and reduce erosion
- release the nutrients they contain over a long period of time.

natural selection: see ***selection***.

negative feedback: many substances or systems in living organisms have a set concentration or level. Negative feedback is the process in which a departure from this set level results in changes that lead to a return to the original value. The process is important in ***homeostasis***. A good example of negative feedback is provided by ***temperature control*** in a mammal. Humans, for example, maintain a body temperature within a degree or so of 37°C. If the temperature rises too high, the resulting increase in blood temperature is detected by receptors in the ***hypothalamus*** of the ***brain***. As a result, the heat loss centre, also in the hypothalamus, sends ***nerve impulses*** to structures such as ***arterioles*** and sweat ***glands*** in the skin. They bring about a fall in temperature. Cold conditions, on the other hand, are detected by receptors in the skin. Impulses are sent to the hypothalamus. This time, the heat conservation centre triggers mechanisms which conserve the body's heat, or generate more heat by actions such as shivering. The outcome, again, is a return to the normal level. This is summarised in the diagram.

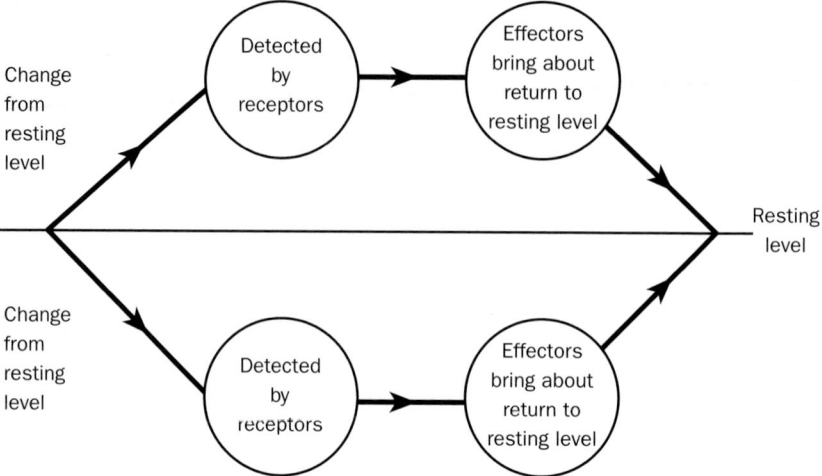

Negative feedback and body temperature

There are many other examples of negative feedback in mammals. Blood *glucose* concentration, respiration rate and the concentrations of some of the *hormones* which regulate reproduction are all controlled by negative feedback. You should distinguish negative feedback from *positive feedback*.

nephron: one of the thousands of tubules that are present in the *kidney* and produce *urine*. Each nephron is divided into five main parts. In the *renal capsule*, *ultrafiltration* takes place. As the filtrate produced by this process passes down the *first convoluted tubule*, useful substances are reabsorbed. Finally, the contents are concentrated in the *loop of Henle*, the second convoluted tubule and the *collecting duct*. The different parts of the nephron and their main functions are summarised in the diagram.

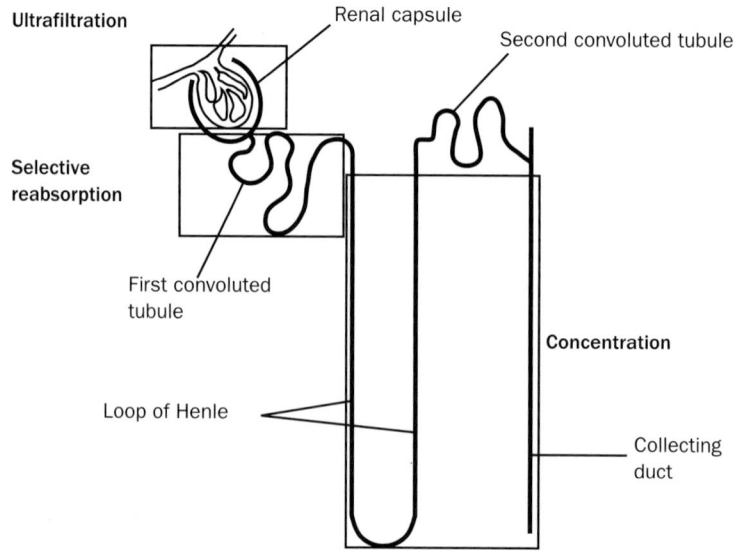

The structure and functions of a nephron

nerve: a bundle of neurones or nerve cells enclosed in a sheath of **connective tissue**. Nerves may contain **sensory neurones**, which conduct impulses from sense organs; **motor neurones**, which conduct impulses to **effectors**; or they may be mixed, containing both sensory and motor neurones.

nerve conduction velocity test: a test used to measure the speed at which a **nerve impulse** passes along a nerve. Electrodes are placed over the nerve at a specific point. These stimulate the nerve. The resulting impulse is then recorded by other electrodes. The distance between the stimulating electrodes and the recording electrodes and the time taken are used to measure the speed of the impulse. The test can be used to help diagnose conditions in which damage has occurred to nerves.

nerve impulse: the way in which information is transmitted along a neurone or nerve cell. The inside of the cell is negatively charged when compared with the outside, so there is a potential difference across its **cell-surface membrane**. This is called the **resting potential**. When an impulse travels along the nerve, a wave of electrical activity called an **action potential** passes along. The charge is temporarily reversed and the inside of the nerve cell becomes positive. A nerve impulse is an **all or nothing** action. Either a full nerve impulse is generated or no nerve impulse at all. This is important as it means that the only way that information about the strength of a stimulus can be transmitted is by varying the frequency of nerve impulses.

net productivity: the rate at which new substances accumulate in an organism. **Photosynthesis** produces organic substances in plants. In animals, food is digested and the products are absorbed. The term gross production is used to refer to all the organic matter gained, either from photosynthesis in plants or by **absorption** in animals. Some of this, however, is used for respiration, so not all the material produced is available for growth. Net productivity is the amount of organic material that accumulates in the organism during the process of growth. It can be defined by the equation:

net productivity = gross productivity − respiration

neuromuscular junction: a **synapse** between a neurone or nerve cell and a muscle fibre. Neurones stimulate **skeletal muscle** fibres to contract. A **nerve impulse** travels down the **axon** to the point where it ends close to the muscle fibre. The events that follow are similar to those which occur in a synapse between two nerve cells. **Acetylcholine (ACh)** is released from small vesicles in the axon terminal. This diffuses across the gap between the axon and the **cell-surface membrane** of the muscle fibre. As a result, there is a change in the membrane's permeability to sodium ions. This causes the muscle fibre to contract.

neurone: a nerve cell. A neurone has:
- a cell body containing a nucleus
- many fine branches known as dendrites that extend from the cell body. These dendrites form **synapses** which connect to neighbouring nerve cells
- one or more longer processes called **axons**. Axons transmit **nerve impulses** away from the cell body. The axons of many neurones in mammals are enclosed in a sheath of a fatty material called **myelin**. Nerve impulses travel faster in myelinated axons than in non-myelinated ones.

Neurones that carry impulses from **receptors** towards the **central nervous system** are called **sensory neurones**. Those which conduct impulses away from the central nervous

z

system towards an *effector*, such as a gland or muscle, are *motor neurones*. Intermediate or relay neurones provide connections between sensory and motor neurones.

neurotransmitter: a substance that transmits information across a *synapse*. Two important neurotransmitters in mammals are *acetylcholine (ACh)* and *noradrenaline*, but there are many others. The molecules of a neurotransmitter have a particular shape. This enables them to fit into receptor molecules on the postsynaptic membrane of the synapse. Since other molecules, often used as drugs, have similar shapes, they can interfere with synaptic transmission. They may either mimic the action of the neurotransmitter or they may block its action.

neutrophil: a type of white blood cell or *leucocyte*. It has a lobed nucleus and small granules in its *cytoplasm*. Neutrophils form part of the immune system and destroy bacteria by *phagocytosis*. They are formed from *stem cells* in *bone* marrow.

niche: a description of the way in which an organism fits into its *environment*. A definition of an organism's niche should refer to where it lives and what it does there. The two-spot ladybird is a small red and black beetle that is common on many plants where it feeds on *aphids*. Its niche would be described in terms of the *abiotic* aspects of its *habitat*, such as the temperature range it can tolerate and the height above ground at which it feeds. A full description would also refer to the biological aspects of its ecology such as the particular species of plant on which it is found and the size of the prey that it normally eats.

An important ecological principle is that no two different species have exactly the same ecological niche. The two-spot ladybird and the seven-spot ladybird both feed on aphids. The seven-spot ladybird, however, is found on a much wider range of plants where it eats slightly larger aphids. The two species do not share the same niche.

nitrification: an important stage in the *nitrogen cycle* in which ammonium compounds are converted to nitrites and nitrates. The bacteria that do this are *chemoautotrophs*. Nitrification is an *oxidation* reaction and releases energy. Nitrifying bacteria use this energy to produce organic compounds from simple inorganic substances.

Nitrobacter: a *genus* of rod-shaped bacteria important in the *nitrogen cycle*. Two groups of soil-dwelling bacteria are involved in *nitrification* reactions. *Nitrobacter* oxidises nitrites to nitrates. This reaction requires oxygen and produces *ATP*.

nitrogen: a biologically important chemical element. Nitrogen forms part of the molecules that make up *amino acids* and *proteins*, *nucleic acids* and many other substances found in living organisms. The *nitrogen cycle* describes how nitrogen circulates or cycles in an *ecosystem*.

nitrogen cycle: the way in which the element nitrogen circulates or cycles in an *ecosystem*. The basic principles of the nitrogen cycle are exactly the same as for other *nutrient cycles*.

The particular points to note in this cycle are:
- Nitrogen from animals and plants is made available both when they die and through excretory products such as *urea*.
- *Decomposers* break down organic nitrogen-containing substances such as protein. This produces ammonia, which is then oxidised to nitrites and nitrates in the process of *nitrification*.
- Some nitrates may be converted to nitrogen gas. This is *denitrification*.
- Nitrogen gas can be made available to plants by *nitrogen fixation*.

nitrogen fixation: the conversion of *nitrogen* gas present in the atmosphere into nitrogen-containing compounds. There are several ways in which this can occur. All of them require a considerable amount of energy. This is because the *atoms* in a nitrogen molecule have to be separated before fixation can occur.

- During thunderstorms, nitrogen and oxygen in the atmosphere combine to produce various oxides of nitrogen. This process uses electrical energy in lightning. The oxides are washed into the soil by the rain and taken up by plants as nitrates. In those parts of the world where violent electrical storms are common, this is an important method of nitrogen fixation.
- *Microorganisms* living free in the soil also fix nitrogen. Some soil bacteria produce an *enzyme* called nitrogenase. With this enzyme, they convert nitrogen in air pockets in the soil into ammonium compounds. This requires energy from *ATP*. They can then make the organic nitrogen-containing substances that they require.
- Nitrogen is fixed by microorganisms that live in the *root nodules* of *leguminous plants*. This is a similar process to the one described above.
- Industrial processes make large amounts of ammonia each year. Nitrogen and hydrogen are combined directly. This needs high temperatures and pressures and the presence of an inorganic catalyst. Much of this ammonia is then used to produce inorganic *fertilisers*, which are added to the soil. The nitrogen-containing ions in the fertiliser are taken up directly by plants.

Nitrosomonas: a *genus* of rod-shaped bacteria important in the *nitrogen cycle*. Two groups of soil-dwelling bacteria are involved in *nitrification* reactions. *Nitrosomonas* oxidises ammonia to nitrites. This reaction requires oxygen and produces *ATP*.

node: the part of a plant stem from which a leaf comes.

nodes of Ranvier: gaps in the *myelin* sheath round an *axon*. *Nerve impulses* travel in a series of jumps from one node to the next. This is *saltatory conduction*. Saltatory conduction is responsible for impulses moving much faster in myelinated than in non-myelinated nerves.

non-competitive inhibition: this occurs when the rate of an *enzyme*-controlled reaction is slowed down or stopped. The rate of reaction is changed because of an inhibitor. This inhibitor affects the enzyme but does not combine with its *active site*. It attaches somewhere else on the enzyme molecule and, in doing so, changes the shape of the active site. As a result, the *substrate* can no longer fit and form an *enzyme–substrate complex*. (See also *competitive inhibition*.)

non-cyclic photophosphorylation: photophosphorylation is the process in which energy from light is used to produce *ATP* in the *light-dependent reaction* of *photosynthesis*. In non-cyclic photophosphorylation, the *electrons* that drive this process pass along an *electron transport chain*, but they do not return to the *chlorophyll* molecule from which they came.

non-disjunction: the failure of chromosomes to separate properly during *meiosis*. This may occur either in the first division or in the second division. It results in *gametes* that contain an incorrect number of chromosomes. In humans, non-disjunction of chromosome 21 is the cause of *Down syndrome*. During the first division of meiosis, the chromosomes belonging to pair number 21 fail to separate. This results in gametes that contain an extra chromosome. When *fertilisation* takes place, a person is produced with 47 chromosomes in each body cell, instead of the usual 46. There will be three copies of chromosome 21.

non-infectious disease: a disease not caused by a pathogen such as a bacterium or a virus. Non-infectious diseases may result from the genes that we inherited from our parents or they may be linked with our lifestyle and associated with factors such as diet, smoking and lack of exercise.

non-reducing sugar: a sugar that does not produce a positive result when heated with Benedict's solution. A *reducing sugar* contains either a free aldehyde group or a free ketone group. It is the presence of one or other of these chemical groups that enables reducing sugars to reduce Benedict's solution. All *monosaccharides* and most *disaccharides* are reducing sugars. Sucrose is an exception. It is a non-reducing sugar. In order to test for the presence of a non-reducing sugar, a test solution is first hydrolysed and broken down into reducing sugars. The test involves the following steps:

- confirm that there are no reducing sugars present by using *Benedict's test*. You should obtain a negative result and the solution should stay blue in colour
- hydrolyse the sample by heating with dilute hydrochloric acid
- neutralise with sodium hydrogencarbonate
- carry out Benedict's test again. A positive result will now be obtained and an orange-red precipitate formed.

non-reproductive cloning: cloning used to produce cells, tissue or organs rather than a new organism. Biologists can grow *stem cells* in cultures. In appropriate conditions, these stem cells can differentiate into specialised cells. This is an example of non-reproductive cloning.

non-shivering thermogenesis: the production of body heat by a method other than shivering. When a mammal moves into a colder environment, it maintains its body temperature. It does this either by limiting heat loss or by producing heat. There are three basic ways in which it can produce heat:

- from metabolic reactions
- from muscle activity
- from the processes involved in the *digestion* and *absorption* of food.

Non-shivering thermogenesis is the production of extra heat from metabolic reactions in the body. An important source of the heat produced this way is found in infants and hibernating mammals. This is from a special sort of *fat* known as brown fat. Brown fat acts rather like a radiator. It is respired, but the energy released is in the form of heat rather than as *ATP*. There is a lot of brown fat around the main arteries in the chest so the heat produced can be distributed rapidly round the body by the blood.

noradrenaline: a substance similar in structure and function to *adrenaline*. It is a *neurotransmitter* involved in some of the *synapses* in the *sympathetic nervous system*. These synapses are known as *adrenergic* synapses.

normal distribution: data which when plotted on a graph produce a bell-shaped curve.

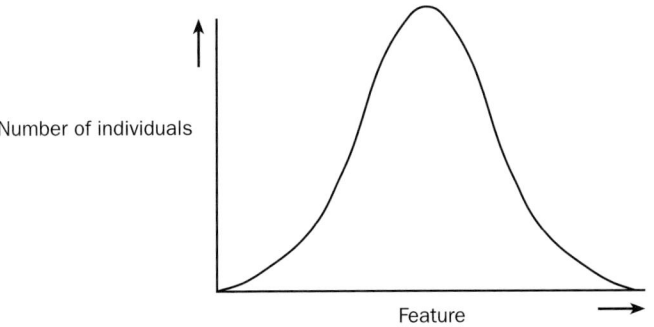

A normal distribution curve

Measurements may be made of the variation shown by a particular feature in a population of mature animals or plants. When these measurements are plotted on a suitable graph, they often show a normal distribution. A normal distribution curve has a number of mathematical properties:

- The average value is the commonest value. In other words, the mean and the mode are the same.
- The curve is symmetrical. Half of all values will be greater than the mean and half will be less.
- The way in which the results are spread out is the same for all normal distribution curves. Approximately 95% of all measurements will be within two **standard deviations** of the mean.

Norplant: a contraceptive in which small rods impregnated with **progesterone** are implanted under a woman's skin. The implant may remain effective for up to 5 years.

notifiable disease: an **infectious disease** that must be reported to the appropriate health authority. This provides information that helps with prevention and control. Notifiable diseases vary from country to country. In the UK, they include measles, mumps and **tuberculosis (TB)**.

nuclear envelope: the membrane layers that surround the nucleus in a **eukaryotic** cell. The nuclear envelope consists of two **plasma membranes**. There are pores called nuclear pores which penetrate the nuclear envelope. The nucleus communicates with the **cytoplasm** through these pores. Molecules of messenger RNA, for example, pass through the nuclear pores during **protein synthesis**.

nucleic acid: a substance that is important in carrying genetic information or producing **proteins** from this. There are two different sorts of nucleic acid. **DNA** acts as the store of genetic information and carries the genes which determine the characteristics of the organism. Different sorts of **RNA** molecules are involved in the production of proteins from the information coded on the DNA. Nucleic acids are **polynucleotides** made from **nucleotides** linked together by **condensation**.

nucleolus: a dark-staining area in the nucleus of a cell. It contains **DNA** with the copies of genes that code for the ribosomal RNA. This is a type of RNA found in **ribosomes**. Ribosomal RNA is produced in the nucleolus.

N

nucleotide: the basic unit or *monomer* from which *nucleic acids* are formed. The diagram shows a nucleotide.

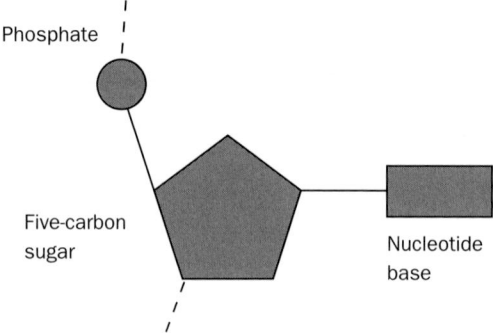

Phosphate

Five-carbon sugar

Nucleotide base

The structure of a nucleotide

It is made up of:

- a five-carbon or pentose sugar. This is either ribose, in the case of RNA, or *deoxyribose* in *DNA*
- a phosphate group
- a nucleotide base. There are five different bases. These are *adenine*, *cytosine*, *guanine*, *thymine* and *uracil*, which are usually abbreviated to A, C, G, T and U. The nucleotides that make up DNA contain one of the nucleotides, adenine, cytosine, guanine or thymine. In RNA, the thymine is replaced by uracil.

These three components are linked by means of *condensation* reactions to make a polynucleotide.

nucleotide base: see *base*.

nucleus: a large organelle (plural nuclei) that carries a cell's genetic material and controls its activities. The *prokaryotic* cells of bacteria do not have nuclei. Nuclei are present in the *eukaryotic* cells of animals and plants. There are some eukaryotic cells, however, that do not have nuclei when they are mature. These include animal cells such as *red blood cells* and some of the plant cells found in phloem. Eukaryotic cells without nuclei, however, always develop from cells with nuclei.

A nucleus is surrounded by a nuclear envelope. The nuclear envelope consists of two membrane layers and has a large number of pores in it. The nucleus communicates with the *cytoplasm* through these pores. Molecules of messenger RNA, for example, pass through the nuclear pores during *protein synthesis*. Within the nucleus is the genetic material of the cell. This is *DNA*. The DNA is bound together with *proteins* to form *chromosomes*. The chromosomes in a cell can only be seen clearly when the cell divides. Between divisions, this genetic material is much more spread out, so individual chromosomes are not visible. Within the nucleus is a dark-staining structure known as the *nucleolus*. This is involved in the production of RNA.

null hypothesis: the starting point in using a statistical test to analyse the results of a scientific investigation. A null hypothesis is based on the assumption that there is no significant difference between sets of observations. Some examples of null hypotheses are:

- there is no difference in the mean surface area of leaves from nettle plants growing in the shade and from those growing in the sunlight
- temperature has no effect on the proportion of black ladybirds in a population
- there is no difference between the number of female chaffinches and the number of male chaffinches visiting a bird table in winter.

Statistical tests can be used to test these hypotheses. From the results of such a test, a decision can be made as to whether to accept or reject the null hypothesis. If we accept the null hypothesis, it suggests that any difference is likely to be due to chance. If we reject the null hypothesis, it indicates that the results are likely to be biologically significant.

nutrient cycle: the way in which chemical elements pass through *ecosystems*. Whatever the element involved, the basic principles are the same. These are summarised in the diagram.

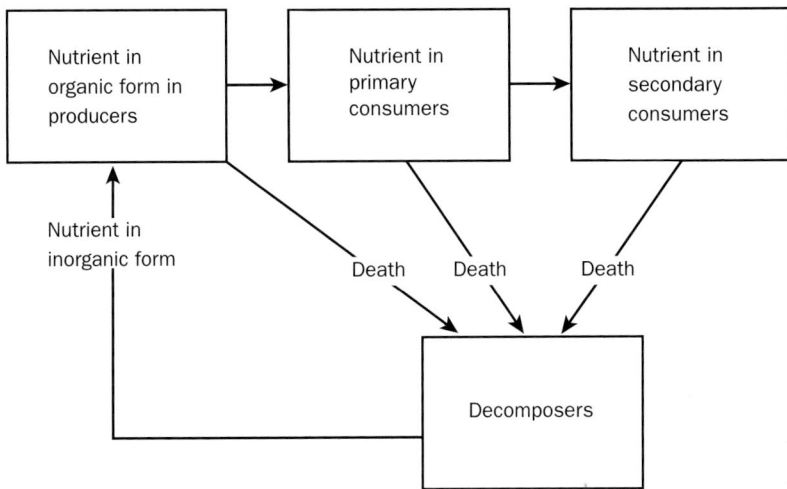

The element is taken up in an inorganic form and incorporated into organic molecules in plants. From the plant or *producer*, it is passed along the *food chain* to the *primary consumer* and secondary consumer. Death of any of these organisms results in their decomposition by *microorganisms* in the soil. The element is released in inorganic form once again and the cycle is complete. The *carbon cycle* and *nitrogen cycle* are specific examples of nutrient cycles.

nutrition: how organisms obtain the raw materials from which they build up their organic compounds. There are two main methods of nutrition:

- In *autotrophic nutrition*, an organism synthesises the organic substances that it requires from simple inorganic substances such as carbon dioxide and water. To do this, an energy source is necessary. Autotrophic nutrition is characteristic of plants and other producers.
- In *heterotrophic nutrition*, organisms gain their nutrients from complex organic substances. These substances are broken down to simple, soluble substances by digestive *enzymes*, and then built up again to form the organic substances which the organism requires.

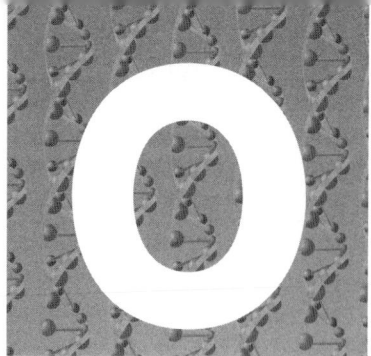

obesity: a condition in which a person's *body mass index (BMI)* is greater than 30. Obesity involves the accumulation of subcutaneous *fat*. This results when energy intake from food is greater than expenditure during exercise. Obesity is the commonest nutritional disorder in affluent societies and increases the risk of *diabetes* and cardiovascular disease.

oedema: accumulation of tissue fluid resulting in swelling, often of the ankles and feet. It may follow long periods of standing or immobility such as on long-haul flights. It may also be a symptom of a more serious condition such as starvation, or heart or *kidney* disease.

oesophagus: the part of the *alimentary canal* of a mammal that takes food from the mouth to the *stomach*. Food is usually chewed in the mouth and mixed with *saliva*. It is then swallowed and a wave of muscular contraction or *peristalsis* pushes it down into the stomach. Layers of circular and longitudinal muscle in the wall assist with peristalsis.

oestrogen: a female sex *hormone* produced by cells in the *ovary*. It is responsible for changes such as the development of pubic hair, the growth of breast tissue and the broadening of the pelvic girdle that take place in the female body during puberty.

During the first half of the *menstrual cycle*, oestrogen is secreted by the cells of the developing follicle. It has several effects on the reproductive system but, in particular, it brings about growth of the lining of the uterus. This becomes thicker and more glandular as a result. Oestrogen has a *negative feedback* effect on the concentration of another hormone, *FSH*. At the start of the cycle, there is a rise in the concentration of FSH secreted by the anterior lobe of the *pituitary gland*. FSH stimulates follicle development and oestrogen is secreted by the follicle cells. The increase in the concentration of oestrogen in the blood inhibits the release of more FSH. This is summarised in the diagram opposite.

This negative feedback effect results in oestrogen being used as an *oral contraceptive*, either on its own or with *progesterone* in a combined pill.

The females of many mammals show a distinctive pattern of behaviour around the time of ovulation. This is known as *oestrus* and is triggered by the peak in oestrogen secretion that occurs just before ovulation.

oestrous cycle: the reproductive cycle in a female *mammal*. Different mammals have reproductive cycles with different patterns and different lengths. Cattle, rats and humans have cycles which follow, one after another, for the entire reproductive part of the animal's life. Dogs and sheep are only reproductively active at a particular time or times during the

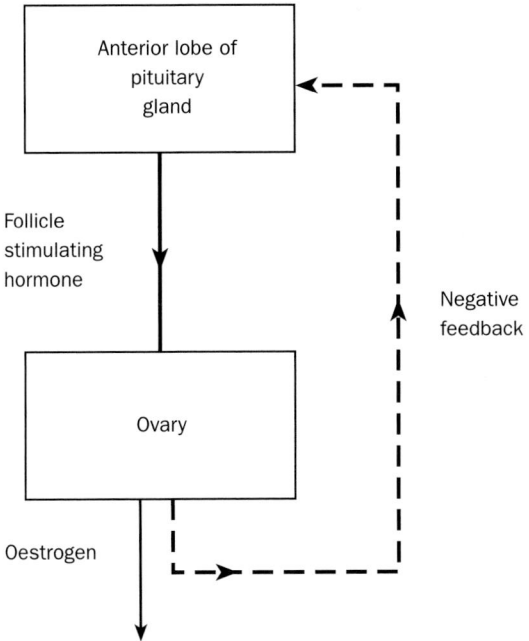

Negative feedback effect of oestrogen

year. In all these animals, however, the oestrous cycle is controlled by the **hormones FSH** and **LH**. These hormones are produced by the anterior lobe of the **pituitary gland**. Under their control the **ovaries** and other reproductive organs undergo a series of changes leading to ovulation and preparation for possible pregnancy. If pregnancy does not occur, there is a short recovery period and then the cycle starts again. These events are summarised in the diagram.

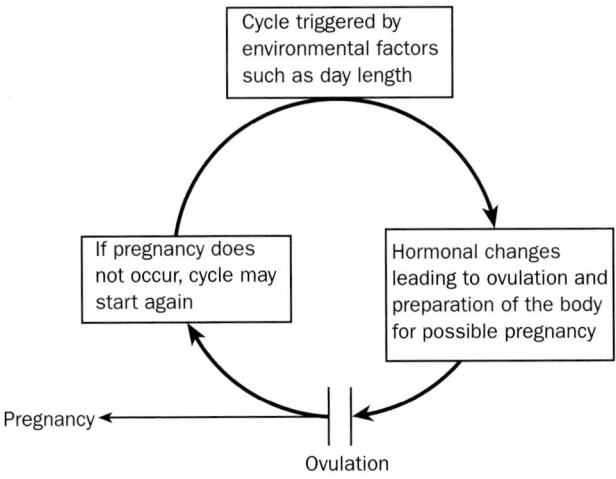

Oestrous cycle

oestrus: the distinctive pattern of behaviour shown by some mammals when they ovulate. Many female mammals live solitary lives and are only fertile for a short period of time. If they are to conceive, it is essential that they advertise their reproductive condition to potential mates. This is the purpose of oestrus. Oestrus may involve:

- changes in the normal behaviour pattern. Cattle, for example, are generally restless, walk about more and feed less
- the intense coloration and swelling of patches of skin, such as those on the chest and around the reproductive area in some monkeys
- the production of chemical signals or **pheromones**, such as those produced by a female dog.

oil: a *triglyceride* that is liquid at temperatures below 20°C. This is because it contains a low proportion of saturated *fatty acids*. Generally speaking, the triglycerides found in plants are oils. Those found in animals are *fats* and are solid at a temperature of 20°C.

omnivore: an animal that eats both plant and animal material. Omnivores feed at different *trophic levels* in a *food web*. Because they eat plant material they are *primary consumers*. They are also secondary or tertiary consumers because they eat animal material.

oncogene: a *gene* that may cause cancer (see *tumour*). Proto-oncogenes are found in healthy cells. They are important in regulating cell division. A proto-oncogene may mutate (see *mutation*) and become an oncogene. Oncogenes may stimulate cells to divide in an uncontrolled way and cause cancer to develop.

oocyst: a thick-walled spore produced by some parasitic organisms. Oocysts can survive outside their hosts for long periods.

oocyte: an immature egg cell. It is a cell that is undergoing *meiosis* and will eventually become a female *gamete*.

oogenesis: the process in which the female *gametes* are formed in an animal. In the *ovaries* of a female mammal, there is a layer of cells called the germinal *epithelium*. These cells will develop into female gametes. The cells of the germinal epithelium first divide by *mitosis*. This produces many immature follicles, each consisting of an immature gamete surrounded by follicle cells. **Hormones** produced during the **oestrous cycle** cause one or more of these immature follicles to develop and become a mature ovarian follicle. At the same time, the immature gamete that it contains increases in size and divides by *meiosis*. The basic sequence of mitosis, growth and meiosis is exactly the same as in the process of *spermatogenesis* in a male. The differences are in the events that take place during meiosis:

- At the first division of meiosis, the nuclear material in the immature gamete divides equally in two. The surrounding *cytoplasm*, however, almost all goes with one of the resulting sets of chromosomes. This gives a large cell which will become the female gamete and a small one called a polar body. The polar body eventually disappears.
- Ovulation now takes place and the female gamete, which is still immature, travels down the oviduct.
- The second division of meiosis now follows. Again, the nuclear material divides equally, but all the cytoplasm goes with the now mature, female gamete. The remaining nuclear material goes to form another polar body.

open circulatory system: see *blood system*.

operant conditioning: a form of learned behaviour in which a reward or a punishment modifies the original behaviour pattern. A rat is put in a cage that has a lever on its wall. The rat runs round the cage and treads on the lever. Treading on the lever releases a pellet of food. This reinforces the behaviour and the rat treads more frequently on the lever as a result.

optic chiasma: crossing over of the *optic nerves* on the underside of the *brain*.

optic nerve: the nerve that takes *nerve impulses* from the *retina* to the *brain*.

opsin: a light-sensitive protein. *Rhodopsin* is found in the *rod cells* of the eye. A molecule of rhodopsin consists of two parts: opsin and *retinal*. Retinal is a substance derived from vitamin A. When light strikes a molecule of rhodopsin, it is broken down into its two parts. A *nerve impulse* is produced and this conveys sensory information along the *optic nerve* to the *brain*. Before it can be stimulated again, rhodopsin must be regenerated from opsin and retinal. Another sort of opsin is found in some photosynthetic bacteria. These bacteria do not contain *chlorophyll*. They use opsin to trap the light energy that they use for *carbon fixation*.

oral contraceptive: pills consisting of one or more female sex *hormones* taken by women to prevent pregnancy. Most oral contraceptives are combined pills that contain:
- *oestrogen*. This prevents the development of a mature ovarian follicle by inhibiting the secretion of *FSH* from the *pituitary gland*. Without a mature follicle, ovulation will not take place and pregnancy cannot occur
- *progesterone*. This also blocks the action of hormones controlling the *menstrual cycle* but it has other contraceptive effects as well. It affects the lining of the uterus and the mucus around the cervix. If ovulation does occur, this makes it much less likely that *implantation* will be successful.

Contraceptives containing progesterone only are called minipills. There is a slightly greater risk of pregnancy with them but the absence of oestrogen means that there are fewer side-effects.

oral rehydration solution (ORS): a solution used to rehydrate a person who has become dehydrated because of diarrhoea. The contents of the *small intestine* in a healthy person are liquid. This is a result of the secretions which are continually pouring into it. Further down the intestine, soluble molecules and ions are absorbed. This produces a *water potential* gradient. The water potential of the epithelial cells lining the intestine is lower than that of the gut contents. Water is therefore removed from the intestine contents by *osmosis* and firm, dry *faeces* are produced. In patients with diarrhoea, more water goes into the intestine than can be reabsorbed. Oral rehydration solutions give the person concerned a balanced solution of salts and *glucose*, which stimulates the mechanism leading to reabsorption of water. Faeces will then return to their normal consistency. Oral rehydration solutions were originally developed to combat *cholera*, but they have proved useful in treating diarrhoea resulting from other causes.

order: a level of *classification*.

organ: a structure made up of different *tissues* and that has a specific function. The *stomach* is an example of an organ. *Epithelial tissue* lines it and secretes *gastric juice*. Surrounding this is a wall that has three layers of muscle tissue. Contraction of this muscle results in the stomach contents being continually mixed and churned. The functions of the various tissues are coordinated by nerve tissue, while nutrients are supplied by the blood. Blood is a *connective tissue*. The stomach is therefore an organ made up of four different tissues. Its function is the storage and *digestion* of food. A number of organs may work together to form a *system*.

organ system: see *system*.

organelle: a structure found inside a cell. It has a specific function. Many organelles, such as *mitochondria* and *chloroplasts*, are surrounded by *plasma membranes*. These are called membrane-bounded organelles. Other organelles, such as ribosomes, are not surrounded by a plasma membrane. The table summarises some of the properties of the main organelles found in plant and animal cells.

Organelle	Main features	Function
Chloroplast	A large organelle, found only in cells in the green tissues of plants	The site of *photosynthesis*
Golgi apparatus	Found in the *cytoplasm* of both animal and plant cells. It is made up of a stack of flattened membranes	Has several functions concerned mainly with packaging and processing molecules produced by the cell
Lysosome	A small sac surrounded by a membrane and containing digestive *enzymes*	Concerned with the breakdown of molecules and structures inside the cell
Mitochondrion	Found in all cells in animals and plants in which *aerobic respiration* takes place	The site of the reactions associated with aerobic respiration
Nucleus	Largest cell organelle, a single nucleus is found in most living cells. Some cells such as the red blood cells in a mammal do not have nuclei	Contains the cell's *DNA*. This is the material which forms the genes. These control all the activities of the cell
Ribosomes	Found in the cytoplasm of plant and animal cells and in other organelles such as chloroplasts and mitochondria	Site of *protein synthesis*

Biologists use various methods to investigate the structure and function of the organelles that are found in a cell. Apart from chloroplasts and nuclei, most organelles are much too small to be seen with a light microscope. *Electron microscopes* allow their detailed structure to be examined. It is possible to use *cell fractionation* to get a suspension containing a single type of organelle. Different techniques can then be used to investigate the structure and functions of the organelle concerned.

organophosphate pesticide: a phosphate-containing substance that is used as an insecticide. Organophosphates inhibit the *enzyme acetylcholinesterase*. This enzyme is found in *synapses* where it breaks down *acetylcholine (ACh)* released in transmission across *cholinergic synapses*.

ornithine cycle: the biochemical pathway by which the *liver* converts ammonia to *urea*. Excess *amino acids* cannot be stored in the body. Instead they undergo *deamination*. In this process, the amino group is separated from the rest of the amino acid and used to form ammonia. Ammonia is a very toxic substance and, in most animals, has to be converted into a less poisonous substance before it can be excreted safely. In mammals, this substance is urea. The ornithine cycle is a cycle of reactions in which ornithine takes up ammonia and produces urea.

osmoreceptor: a receptor that detects changes in *water potential*. Osmoreceptors in the *hypothalamus* detect a decrease in the water potential of the blood. *Nerve impulses* from these receptors cause the release of the *antidiuretic hormone (ADH)* from the *pituitary gland*. This results in more water being reabsorbed in the *kidney* and the water potential of the blood returning to its normal level.

osmoregulation: the ability of an organism to regulate the concentration of its body fluids. The overall concentration of ions in sea water is approximately the same as that in the body fluids of marine organisms. The simplest marine organisms do not have osmoregulatory mechanisms.

In most other *habitats*, the situation is very different. In fresh water, the total ion concentration is much lower. The *water potential* of the surrounding medium will therefore be higher and water will enter the organism by *osmosis*. If this is not controlled in some way, the cells concerned would rupture. This would clearly have disastrous consequences.

Terrestrial animals have different osmoregulatory problems. They live in an environment where water is constantly being lost by evaporation from gas exchange surfaces and in the removal of excretory products from the body. Water intake and loss need to be balanced and regulated. Different organisms have different mechanisms which allow them to do this. These range from the *contractile vacuoles* found in single-celled organisms like amoeba to the *kidneys* of vertebrates.

osmosis: a special case of *diffusion* involving the movement of water molecules. The diagram shows two solutions separated by a *partially permeable membrane*.

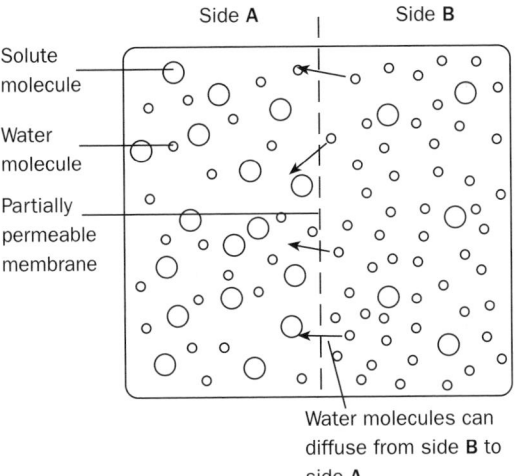

A simple model to show osmosis

Water molecules are able to pass freely through the membrane, but the solute molecules are too large and cannot pass through. The solution in side A has a greater concentration of solute molecules and a lower concentration of water molecules than the solution in side B. Since the membrane will only allow the water molecules to pass through, they will diffuse from side B to side A. This is one way of defining osmosis: the net movement of water molecules through a partially permeable membrane from where they are in a higher concentration to where they are in a lower concentration. Since the concentration of water molecules is greater in side B, this solution has a greater **water potential** than the solution in side A. We can also define osmosis in terms of water potential. It is the net movement of water molecules through a partially permeable membrane from a solution of higher water potential to a solution of lower water potential.

osteoarthritis: a disease affecting joints, particularly the knee and the hip. The **cartilage** is lost and this may result in damage to the underlying **bone**. The condition is painful and affected people experience difficulty in moving the joint concerned. Osteoarthritis commonly affects older people and may result from past injury or physical stress.

osteoporosis: a condition that accounts for a high proportion of **bone** fractures among the elderly. As a person grows, bone tissue gets denser as more bone substance is formed. After the age of about 40, however, there is a gradual decrease in the amount of bone substance present. Once this falls below a critical level, bones become brittle and there is an increased risk of fracture. This is part of the normal ageing process. The situation is, however, more complicated in women. One of the female sex **hormones**, **oestrogen**, plays an important part in maintaining bone density. Once a woman passes through **menopause**, she stops secreting oestrogen. This can lead to a sharp decrease in bone density and a greatly increased risk of arm, hip and leg fractures. People with osteoporosis also suffer from back pain resulting from fractures of the vertebrae in the spine.

ovarian follicle: a structure found in a mammalian ovary. At the start of an **oestrous cycle** in a female mammal, the **hormone FSH** is secreted by the **pituitary gland**. This hormone stimulates one or more primary follicles found in the ovaries to develop into a mature ovarian follicle. Each mature follicle consists of an immature female **gamete** surrounded by a mass of follicle cells. The role of the follicle cells is to produce the hormone **oestrogen**. Oestrogen has various effects on the reproductive system. In particular, it brings about the growth of the lining of the uterus. At ovulation, the developing gamete leaves the ovary and passes into the oviduct, while the remaining cells of the follicle develop into a structure known as the **corpus luteum**.

ovary (animal): the organ in a female animal that produces the **gametes** or sex cells. In the ovaries of a female mammal, there is a layer of cells called the germinal **epithelium**. These cells will develop into female gametes. The cells of the germinal epithelium first divide by **mitosis**. This produces many immature follicles, each consisting of an immature gamete surrounded by follicle cells. **Hormones** produced during the **oestrous cycle** cause one or more of these immature follicles to develop and become a mature ovarian follicle. When ovulation takes place, the developing gamete leaves the ovary and passes into the oviduct, while the remaining cells of the follicle develop into a structure known as the **corpus luteum**. The process by which female gametes are formed is called **oogenesis**. In addition to gamete formation, the mammalian ovary has another important function. It secretes **oestrogen** and **progesterone**. These hormones are responsible for the changes that take place in the female body during puberty and for controlling many of the processes associated with the oestrous cycle and pregnancy.

ovary (plant): part of the *flower* that contains an ovule or ovules. The female part of the flower is known as the gynoecium. It is made up of one or more units called carpels. Each carpel has an ovary and a *stigma* which receives the pollen. The stigma is joined to the ovary by a *style*. After *fertilisation*, the ovary develops into a *fruit*, while the ovules that it contains become the *seeds*.

oviduct: a tube that takes the female *gametes* from the ovary. In a mammal there are two oviducts, one from each *ovary*. These lead to the single *uterus* in humans or to paired uteri in many other species. *Fertilisation* usually takes place near the top of the oviduct. The fertilised egg is then moved down into the uterus by the beating action of *cilia* on the epithelial cells which line the oviduct, and contraction of muscle in its walls.

oxaloacetate: a four-carbon compound produced in the *Krebs cycle*. Acetylcoenzyme A is synthesised in the *link reaction*. It contains two carbon *atoms* which are involved in the process of respiration. Acetylcoenzyme A is fed into the Krebs cycle. It combines with oxaloacetate to form a six-carbon compound. This six-carbon compound is then broken down via intermediate compounds to produce oxaloacetate again.

oxidation: a chemical reaction involving loss of *electrons*. As the electrons that are lost must go somewhere, oxidation of one substance is always accompanied by *reduction* of another. In biological systems, oxidation and reduction are controlled by *enzymes* called *oxidoreductases*. Oxidation of organic molecules is the basis of the process of *respiration*.

oxidative phosphorylation: the part of the respiratory pathway in which energy released in the *electron transfer chain* is used in the production of *ATP*. In living organisms, there are several ways in which ATP is produced. Two of the most important are oxidative phosphorylation and *photophosphorylation*. Oxidative phosphorylation and photophosphorylation are similar because:

- they both involve the passage of *electrons* from one molecule to another through a system of carriers called an electron transport chain
- energy released in this process is used to combine *ADP* with inorganic phosphate to produce ATP.

There are also some important differences between these two processes. These are summarised in the table.

Difference	Oxidative phosphorylation	Photophosphorylation
Source of electrons	Chemical reactions during *glycolysis* and the *Krebs cycle*	Excited electrons produced as a result of light energy being absorbed by *chlorophyll*
Electron acceptor	Usually *NAD*	*NADP*
Location of process	Inner membranes of mitochondria	Granal membranes in *chloroplasts*
Uses of ATP made in this process	Providing energy for many cell processes and reactions	Providing energy for the *light-independent* reactions of *photosynthesis*

oxidoreductase: a type of *enzyme* that is responsible for catalysing a reaction involving *oxidation* and *reduction*, a redox reaction. Many important oxidoreductases are associated with the reactions that take place in *respiration*. Some, however, catalyse other biochemical reactions. The reason why *fruits* such as apples and bananas turn brown when cut open is because they contain certain substances that are oxidised on exposure to air. The enzyme involved in this process is an oxidoreductase.

oxygen debt: the amount of oxygen required to oxidise the *lactate* produced during *anaerobic respiration*. Muscles in the body of a mammal rely on the presence of oxygen in order to respire. During periods of vigorous exercise, however, there is insufficient oxygen available to meet their respiratory requirements and they respire anaerobically. Lactate is produced as a result and accumulates in the body tissues. When activity returns to normal, this lactate is removed. It is either converted to pyruvate or to carbohydrate in the *liver*. The oxygen needed for removing the lactate is the oxygen debt.

oxygen dissociation curve: a curve showing the amount of oxygen carried by the *haemoglobin* in the blood plotted against the concentration of oxygen present. The graph shows an oxygen dissociation curve for human haemoglobin.

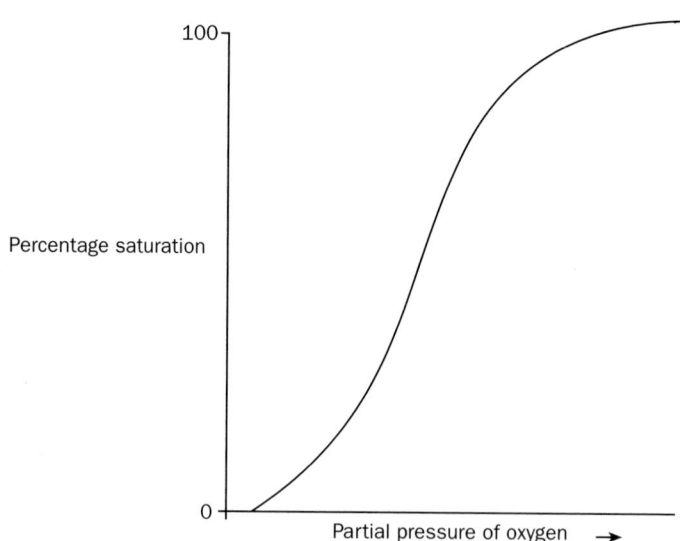

You should note that:

- the haemoglobin is almost completely saturated with oxygen at the concentration normally found in the **lungs**
- the steep fall in the curve means that blood will rapidly give up oxygen to respiring tissues.

There are many different types of haemoglobin. These result from differences in the sequence of **amino acids** that make up their polypeptide chains. Each has slightly different oxygen-carrying properties. These can be related to the **environment** in which the animal lives. The graph shows oxygen dissociation curves for oxygen-carrying pigments of some different animals.

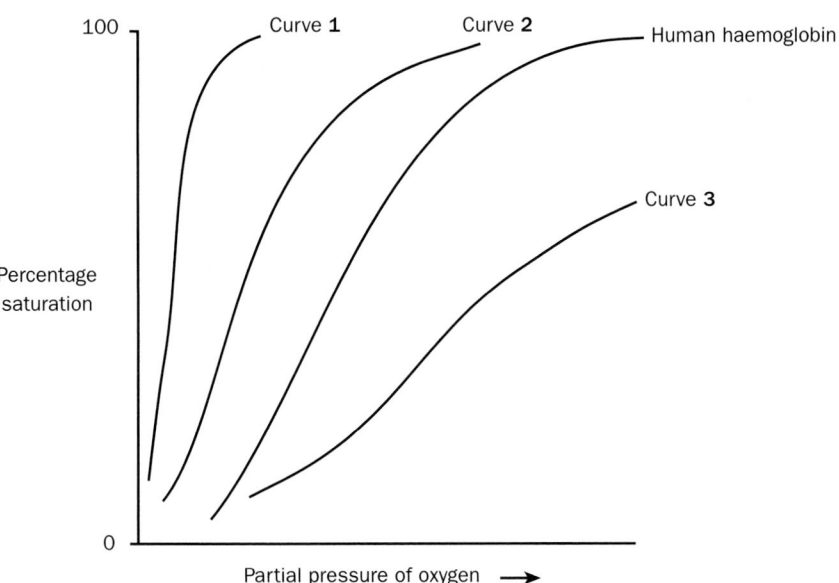

- Curve 1. Pigments which have dissociation curves of this type usually act as oxygen stores. They are saturated with oxygen at relatively low partial pressures. They hold on to this oxygen, releasing it when the concentration in the tissues falls to a very low level. The pigment **myoglobin** found in the muscles of seals and whales has a curve of this shape.
- Curve 2. Found in animals that live in an environment which has relatively little oxygen present. Examples include the llama which lives high in the Andean mountains of South America, the prairie dog which spends much of its time in burrows deep underground and the mammalian fetus in the uterus of its mother.
- Curve 3. Haemoglobin with a dissociation curve like this gives up its oxygen readily. It is found in animals which have a very high respiration rate such as birds and small mammals like shrews.

oxyhaemoglobin: the compound formed when oxygen combines with *haemoglobin*.

oxytocin: a *hormone* released by the posterior lobe of the *pituitary gland*. It causes contraction of the *involuntary muscle* in the wall of the uterus during birth. It also brings about the release of milk from the mammary *glands* by causing muscle fibres in these glands to contract.

ozone layer: a layer in the Earth's atmosphere that contains a relatively high concentration of ozone. This layer absorbs a lot of harmful ultraviolet light which would otherwise damage living organisms. The thickness of the ozone layer varies worldwide. It is thinner at the equator and thicker in polar regions but this thickness is subject to seasonal variation. As a result, seasonal ozone holes have appeared over polar regions. Gases such as the chlorofluorocarbons (*CFCs*) used in aerosols cause the thinning of this ozone layer.

Do you need revision help and advice?

Go to pages 306–314 for a range of revision appendices that include plenty of exam advice and tips.

pacemaker: see *sinoatrial node (SAN)*.

Pacinian corpuscle: a type of pressure receptor. In humans, Pacinian corpuscles are most numerous in the skin of the fingers and toes. If a longitudinal section is cut through one of these receptors it looks rather like a slice through an onion bulb. A nerve ending, consisting of a *dendron* from a *sensory neurone*, runs through the middle. This is surrounded by a *capsule* formed of layers of *connective tissue* separated from each other by a thick jelly-like substance. When pressure is applied to the capsule, these membranes are distorted. This affects the permeability of *ion-channel proteins*. Sodium ions flow into the sensory neurone and produce a *nerve impulse*.

packed red cells: when blood has been centrifuged (see *centrifuge*), this is the volume of *red blood cells*. It is given as a fraction of the total volume of blood in the sample.

palisade cell: tall, narrow, photosynthetic cell found in the *mesophyll* of a leaf. This shape enables palisade cells to be packed closely together and provides an excellent arrangement for absorption of light. Palisade cells contain many *chloroplasts* that are able to move within the *cytoplasm*. The chloroplasts can, therefore, absorb light efficiently in low light intensities but are protected from bleaching in bright light.

pancreas: a gland found in the abdomen of a mammal. It is an *endocrine gland* and secretes the *hormones insulin* and *glucagon*, which are important in the regulation of blood *glucose*. It is also an *exocrine gland*, secreting pancreatic juice along the pancreatic duct into the first part of the *small intestine*. Pancreatic juice contains many digestive *enzymes*, in particular *amylase*, *lipase* and *trypsin*, in a liquid made alkaline by the presence of hydrogencarbonate ions.

pancreatic enzyme replacement therapy (PERT): treatment given to a patient to improve the *digestion* of food when his or her *pancreas* fails to produce enough digestive *enzymes*. This may result from a number of medical conditions including pancreatic cancer, *pancreatitis* and *cystic fibrosis*.

pancreatitis: *inflammation* of the *pancreas*. Digestive *enzymes* secreted by the pancreas become active inside the organ and break down its tissues. In acute pancreatitis, the disease develops rapidly but lasts a relatively short time (see *acute disease*). In chronic pancreatitis, it is long lasting and changes in the condition only take place slowly (see *chronic disease*). Pancreatitis may be diagnosed by testing the blood for the presence of enzymes from damaged pancreas cells, or by monitoring the concentration of pancreatic enzymes in *faeces*.

pandemic: an outbreak of disease that affects large numbers of people in many different countries. It is an *epidemic* on a global scale. Examples of pandemics include the Black Death, which spread through Asia and Europe in the fourteenth century, and the *influenza* outbreak that followed the 1914–18 war. It is considered by many that AIDS should also be described as pandemic.

papillary muscles: small, cone-shaped muscles in the heart. When these muscles contract, they tighten the tendons attached to the flaps of the *atrioventricular valves*. This prevents the valves from turning inside out and blood flowing back into the atria.

Paneth cell: a cell found in the wall of the *small intestine*. Paneth cells are thought to play a role in protecting the gut wall from pathogenic bacteria.

parasite: an organism that lives in or on a host organism. The parasite gains a nutritional advantage from this relationship, while the host suffers a disadvantage. Parasites are adapted in different ways to living inside or on their hosts but these adaptations frequently involve:
- a means of attachment to the host. This depends on the parasite concerned but may involve hooks and suckers in tapeworms or the strongly curved claws of ticks and lice
- methods of resisting the defence mechanisms of the host. Parasites must avoid the digestive *enzymes* of the host's gut or destruction in its blood by the immune system
- development of reproductive organs and enormous powers of reproduction. Only by producing very large numbers of young can parasites successfully complete their life cycles
- a complex life cycle. This is often an adaptation that enables infection of another individual to occur
- reduction of body systems other than those concerned with reproduction.

parasympathetic nervous system: a division of the *autonomic nervous system* that generally controls the resting functions of the body. It is more important when the body is at rest. Some features of the parasympathetic system are:
- *synapses* in this system secrete acetylcholine as a *neurotransmitter*
- its effects are usually inhibitory. For example, parasympathetic stimulation decreases the breathing rate, and decreases the heart rate and *stroke volume*.

parenchyma: relatively unspecialised cells found in plants. Their main characteristics are that they have thin, permeable, *cell walls* of *cellulose* and therefore have living contents. In addition, they are not as elongated as the cells in many other plant tissues. Parenchyma is often referred to as packing tissue, since it fills the spaces between other types of cells. This is misleading since it suggests that it does not have any real function. Parenchyma cells are extremely important in providing support to young plants and they photosynthesise and store substances as well.

Parkinson's disease: a *degenerative disease* of the *brain* most common in elderly people. The commonest symptom is a tremor. This tremor often at first affects one of the hands. It gradually spreads to the other limbs. Parkinson's disease is caused by a deficiency of *dopamine*. This is a *neurotransmitter* produced in the brain.

partial pressure: a measure of the amount of gas present in a mixture of gases. It is a term frequently used when describing the transport of oxygen by *haemoglobin*.

partially permeable membrane: a membrane that only lets small molecules, such as those of water, pass through. When a solution containing a solute such as sucrose is separated from distilled water by a partially permeable membrane, there will be a higher concentration of water molecules in the distilled water. Since only the water molecules can pass through the membrane, this results in their **diffusion** from where they were in a higher concentration to where they are in a lower concentration. This special case of diffusion is known as **osmosis**.

passage cell: a cell in the **endodermis** of a root that allows substances to pass freely from the outer part of the root to the xylem. The endodermis consists of a ring of cells between the outer part of the root or the **cortex** and the vascular tissue in the centre. A band of water-proof material called the **Casparian strip** runs round the walls of each of these endodermal cells. It prevents substances from moving through the **cell walls** along the **apoplastic pathway**. Passage cells do not have this band of material present and therefore allow water and other substances to pass more readily through the endodermis.

passive immunity: when **antibodies** made by one organism enter the body of another organism. In mammals, this process occurs naturally as some antibodies are passed from the mother to the **fetus** through the **placenta**. In addition, the secretion produced by the mammary **glands** immediately after birth is not milk but a liquid called **colostrum**. Colostrum is rich in antibodies. Injections of antibodies are useful in circumstances where the body would not have enough time to produce antibodies of its own by **active immunity**. They are used to treat snake bites and diseases such as tetanus. Antibodies produced by one organism are foreign **proteins** when they are introduced into the body of another. They act as **antigens** and are usually destroyed rapidly. Protection by passive immunity is therefore only short-lived.

passive transport: the movement of a substance from where it is more concentrated to where it is less concentrated; in other words, it is the movement of a substance down a concentration gradient. In liquids and gases, the kinetic energy that molecules possess results in them continually moving about. This movement is random, so equilibrium is eventually reached and the molecules are spread out evenly. Because energy from **ATP** is not required, **diffusion** is described as passive transport.

paternal chromosome: the member of the pair of **homologous chromosomes** that originally came from the sperm cell produced by the male parent.

pathogen: a **microorganism** that is a **parasite** of an animal or plant and causes a **disease**. There are many different microorganisms found in and on the surface of the human body but not all of them are pathogens. The nineteenth-century biologist, Robert Koch, put forward a series of principles called **Koch's postulates** which can be used to confirm that a particular microorganism is a pathogen and causes a particular disease.

PCR: see **polymerase chain reaction (PCR)**.

peak expiratory flow rate (PEFR): a measure of the maximum rate at which air can be exhaled. **Asthma** is a condition in which the smooth muscle in the walls of the **bronchi** contracts. Together with the production of large amounts of **mucus**, this leads to a narrowing of the airways and the formation of a blockage which limits the rate at which air can enter and leave the **alveoli**. By using a peak flow meter it is possible to monitor this rate and control asthmatic attacks with the timely use of appropriate drugs.

peak flow: see **peak expiratory flow rate**.

pectin: a carbohydrate found in and between plant **cell walls**, where it helps to cement **cellulose** fibres together. Pectins are commercially important in the extraction of **fruit** juices.

When fruit ripens, the pectin molecules start to break down. Part of the molecule is insoluble and remains attached to the cell wall, while the part of it which is soluble extends into the cell, mixing with the juice. The result of this is that the juice is thick and difficult to squeeze out. The addition of pectinases to fruit such as blackcurrants considerably increases the amount of juice that can be extracted. The cloudiness of some extracted fruit juices is also due to pectin. In this case, treatment with pectinase *enzymes* helps to clarify it. Many *microorganisms* produce pectinases and the enzymes that are used commercially come from this source.

Pediculus: the *genus* to which human head lice belong.

PEFR: see *peak expiratory flow rate (PEFR)*.

pentose: a sugar that has five carbon *atoms* in each of its molecules. Pentoses are *monosaccharides*. Biologically important pentose sugars include *deoxyribose* and ribose, which form part of the structure of the *nucleotides* which make up *DNA* and *RNA*. Ribulose is a pentose which, when combined with two phosphate groups, becomes ribulose bisphosphate. This substance acts as the carbon dioxide acceptor in the *light-independent reaction* of *photosynthesis*.

penicillin: an *antibiotic* that prevents the formation of bacterial *cell walls*. When bacteria enter a person's body they are in an environment that has a higher *water potential* than that of their own *cytoplasm*. Water therefore enters the bacterial cell by osmosis. Without a strong cell wall, the bacterial cells swell and burst. This is called osmotic *lysis*.

pepsin: a protein-digesting *enzyme* produced in the *stomach*. Pepsin is secreted by peptic cells in the lining of the stomach wall. They produce the enzyme as inactive pepsinogen. This is activated on contact with hydrochloric acid in the stomach. Pepsin is an *endopeptidase*. It breaks the peptide bonds between certain *amino acids* only when they are in the middle of a polypeptide chain. It does not break them down if they are at the ends of the chain. As a result, the products of *digestion* are various small polypeptides. Pepsin works best in acid conditions and has an optimum pH of between 1.6 and 3.2. It can, therefore, only function in the stomach. Once the contents of the stomach enter the *small intestine*, they are neutralised by the alkaline secretion produced by the *pancreas*. Under these conditions, pepsin rapidly stops working.

pepsinogen: the inactive form in which the enzyme *pepsin* is secreted by cells in the lining of the stomach wall. Once in the *stomach*, some of the *peptide bonds* in the pepsinogen are hydrolysed by hydrochloric acid. The pepsin produced as a result can now begin to digest *proteins* in the food. Once some pepsin has been produced, it can activate more pepsinogen, so the activation process may be thought of as being self-catalysing.

peptic ulcer: an open sore on the inside wall of the *stomach*. Infection with the bacterium *Helicobacter pylori* may cause the stomach wall to become inflamed. This leads to a condition called gastritis and reduces the resistance of the stomach wall to damage by acid and *pepsin*. As a result, an ulcer may form.

peptide: a compound formed by joining a small number of *amino acids* together with peptide bonds. Peptides are difficult to distinguish from *polypeptides* and the two terms are often used imprecisely. In general, a peptide consists of a few amino acids and is therefore shorter in length than a polypeptide. Peptides and smaller polypeptides result from the *digestion* of *proteins* by *endopeptidases*, such as *pepsin* and *trypsin*. Some *hormones* are peptides. Among these are the hormones known as release factors which are produced by the *hypothalamus* in the *brain*. These are taken in the blood to the anterior lobe of the *pituitary gland*, where they stimulate the secretion of other hormones, such as *growth hormone* and follicle stimulating hormone, *FSH*.

peptide bond: the chemical bond formed when *amino acids* are linked together. The diagram shows how two amino acids are joined together by a peptide bond.

Removal of molecule of
water to form a peptide bond

This is **condensation** and involves the removal of a molecule of water. A peptide bond may be broken or hydrolysed by the addition of a water molecule.

percentage cover: the proportion of ground in an area being sampled that is covered by a particular species of plant. Percentage cover can be found in one of two ways. It may be estimated from inspecting the area enclosed by a *quadrat* frame. An alternative method is to use *point quadrats*. In this case the proportion of pins hitting the species concerned is recorded and expressed as a percentage.

Percentage cover has advantages over methods which rely on counting the number of individual plants present. It probably gives a better idea of the ecological importance of a particular plant. A single large plant, for example, may have a much greater percentage cover than ten very small ones. A large plant will also exert a much greater influence in the *community*. Another advantage is that it is often extremely difficult to identify individual specimens of plants such as grasses. By using percentage cover, there is no need to do so.

peripheral nervous system: nerves carrying *nerve impulses* towards or away from the *brain* and *spinal cord*.

peristalsis: the wave of contraction of *involuntary muscle* by which food is pushed along the gut. After a meal, the increase in the contents of the gut stretch the muscle in the gut wall. This initial stretching causes the circular muscles to contract. The *lumen* of the gut narrows as a result and squeezes the contents. As the muscle contraction moves in a wave along the gut, this squeezing pushes the food along in front of it.

peritoneal dialysis: a method of removing *urea* and other substances from the blood of someone with *kidney* failure using the membrane that naturally encloses the abdominal cavity. Dialysing fluid consists of a suitable solution of salts. It is inserted into the abdominal cavity through a small tube called a catheter. The fluid is kept apart from the blood by the membrane that surrounds the cavity. The small soluble molecules of urea pass out of the blood, through the membrane and into the fluid. Larger **protein** molecules and blood cells are unable to do so and remain in the blood. The waste fluid is then removed from the abdominal cavity through the catheter.

peroxidase: one of a group of **enzymes** that catalyse reactions in which substances are oxidised by a peroxide such as hydrogen peroxide. Peroxidases are found in many plants but they are also found in white blood cells.

PERT: see **pancreatic enzyme replacement therapy (PERT)**.

pest: an organism that has features that bring it into conflict with human interests. It may have characteristics that are injurious, or it may simply be unwanted in a particular area. Pests often cause damage to agricultural crops or to domestic livestock. Other than this, pests may:

- damage natural **ecosystems**. Himalayan balsam and Japanese knotweed are introduced species of plant that have outcompeted native plants in some areas
- carry disease, often from one human to another. Mosquitoes are pests because the females of some species transmit **malaria**.

pesticide: a substance that is used to control pests. An ideal pesticide should be:

- quick acting, rapidly killing the pest against which it has been used
- specific in its action and harmless to other organisms such as humans and other mammals, beneficial insects, soil bacteria and the crop on which it is used
- effective at low doses
- **biodegradable** so that, as soon as it has had the required effect, it is broken down and does not produce a long-lasting impact on the **ecosystem**.

Frequent use of pesticides may mean that pest populations become resistant, so there is a constant need to develop new, more effective substances. The most important agricultural pesticides are **herbicides** or weedkillers, **insecticides** and fungicides.

PET scan: a technique used to produce a three-dimensional image of processes that take place in the body. The patient is given a biologically active substance that has a small amount of a radioactive tracer attached to it. The radioactive tracer has a short half life and as it decays an image can be produced. A substance similar to **glucose** is commonly used to produce an image that shows up the metabolic activity in different organs.

petal: one of an outer ring of structures in a **flower** which are often brightly coloured and together make up the **corolla**.

pH: a scale that measures the concentration of hydrogen ions in a solution. Solutions with pH values below 7 contain high concentrations of hydrogen ions and are acidic. Solutions with values above 7 have low concentrations of hydrogen ions and are alkaline. The pH of the surrounding medium can have a marked effect on protein molecules and hence on reactions controlled by **enzymes**. Soil and water pH are also important ecological factors which help to determine the distribution of living organisms.

phage: a type of virus that infects bacteria and fungi. The T2 bacteriophage attacks bacteria. It has been studied in detail and has contributed much to our present knowledge of the life cycle of viruses. Bacteriophages are important in **genetic engineering**, as they can be used as vectors to introduce genes into bacterial cells. With the growth of **biotechnology**, large numbers of bacteria are cultured in order to produce useful products. If a culture becomes infected with bacteriophages, it will be rapidly destroyed.

phagocyte: a cell from the *immune system* that engulfs debris or *microorganisms*. The ingested material is enclosed in a *vacuole* and then digested by *enzymes* or broken down by other substances produced by the cell. This process is called *phagocytosis*.

phagocytosis: the transport of large particles or cells through the *cell-surface membrane* into the *cytoplasm*. The membrane surrounds the particles concerned. A *vesicle* or *vacuole* is formed which is pinched off and moves into the cytoplasm. Although phagocytosis takes place in many cells, two particularly good examples are provided by amoeba and certain kinds of white blood cell. Amoeba changes its shape as it moves by extending its cytoplasm to form processes called pseudopodia. These cytoplasmic processes engulf prey so that it becomes enclosed within a *food vacuole* in the cytoplasm. *Enzymes* are secreted into this vacuole and digest the prey. Various kinds of phagocytic white blood cell engulf pathogenic bacteria. Again, enzymes are released into the resulting vacuoles and the bacteria are digested.

phenotype: the characteristics of an organism that result both from the genes the organism possesses and from the environment in which it has developed. The Himalayan rabbit provides a good example. These rabbits have white fur but black feet, ears and tail. Crossing two pure-bred Himalayan rabbits always results in animals with these characteristics. However, the black pigment will only develop on parts of the body that are at a low temperature. Himalayan rabbits possess the *gene* for making black fur, but only in the right environment will the typical phenotype develop.

pheromone: a substance that is produced and released into the environment by one organism and affects the behaviour of another organism of the same species. Pheromones have many different functions. For example, they are present on *faeces* used to mark *territory* in badgers; they provide information about sex and reproductive state in dogs, and can delay or hasten sexual maturity in mice. Pheromones should be distinguished from *hormones*.

phloem: the plant tissue that transports the products of *photosynthesis* away from the leaves to other parts of the plant. It has a number of cell types, of which the most important are sieve tubes (see *sieve tube element*) and *companion cells*. Sieve tubes are made up from individual cells arranged end to end. The walls between these cells are perforated and form the characteristic sieve plates. Sieve tubes do not contain nuclei and have only a little *cytoplasm*. Associated with the sieve tubes are companion cells, which have densely stained cytoplasm. The method by which phloem transports sugars through the plant is not well understood. Although most biologists consider that substances travel by a system of mass flow, there is still some debate over the mechanism involved.

phosphocreatine: a substance that breaks down to give creatine. This releases energy which can be used in muscles to produce *ATP*. At rest, there is only a little ATP present in a muscle and this is used up rapidly when the muscle contracts. The function of phosphocreatine is to replace and provide a supply of ATP for immediate use. Phosphocreatine can be replenished from the ATP produced in respiration. The relationship between phosphocreatine and ATP is summarised in the diagram overleaf.

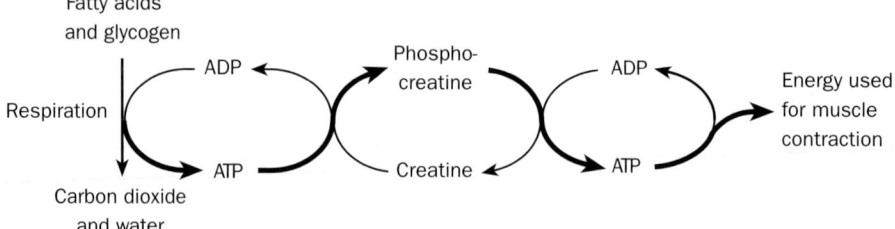

The relationship between phosphocreatine and ATP

phospholipid: a type of *lipid* that contains a phosphate group. The diagram shows a phospholipid molecule.

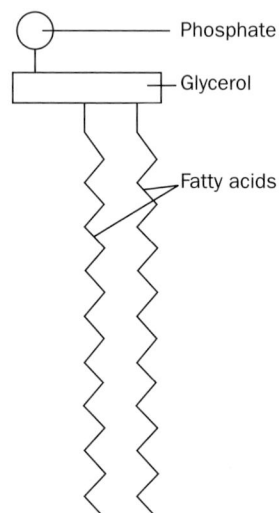

Phosphate

Glycerol

Fatty acids

*A simplified diagram of
a phospholipid molecule*

You can see that the glycerol has two **fatty acids** joined to it. In a **triglyceride** molecule there would be a third fatty acid. In a phospholipid molecule this third fatty acid is replaced by a phosphate group. The phosphate-containing head of the molecule is water soluble, unlike the fatty acid tail. This property is important in the structure of a **plasma membrane**.

photoautrotroph: an **autotroph** that uses light energy to convert simple inorganic molecules into organic ones. All green plants are photoautotrophs, as are a number of organisms belonging to the kingdoms **Prokaryotae** and **Protoctista**. Photoautotrophs are **producers** and form the basis of most **food chains** and **food webs**.

photolysis: the breakdown of water molecules that takes place during the **light-dependent reaction** of **photosynthesis**. Water molecules break down to produce protons, **electrons** and oxygen. This is summarised by the equation:

$$2H_2O \quad \rightarrow \quad 4H^+ \quad + \quad 4e^- \quad + \quad O_2$$

water protons electrons oxygen

The electrons replace those lost when light strikes the **chlorophyll** molecule and oxygen is given off as a waste product. The protons combine with more electrons to reduce **NADP**.

photoperiod: the ratio of light to dark over a 24-hour period. The photoperiod controls many biological responses, such as the timing of nesting behaviour in birds and the loss of leaves from *deciduous* trees in the autumn. *Flowering* in many plants is also a photoperiodic response.

photophosphorylation: the part of the *light-dependent reaction* of *photosynthesis* in which energy released in the *electron transfer chain* is used to produce *ATP*. Light striking a *chlorophyll* molecule causes some of its *electrons* to become excited. These are lost from the chlorophyll molecule and pass through a series of electron acceptors. As a result, energy is progressively lost and some of this is used to make ATP. This is the process of photophosphorylation.

photosynthesis: the process by which plants and other *photoautotrophs* use light energy to convert carbon dioxide into *carbohydrates*. The diagram summarises the main features of the biochemical pathways involved in photosynthesis. (See also *light-dependent reaction* and *light-independent reaction*.)

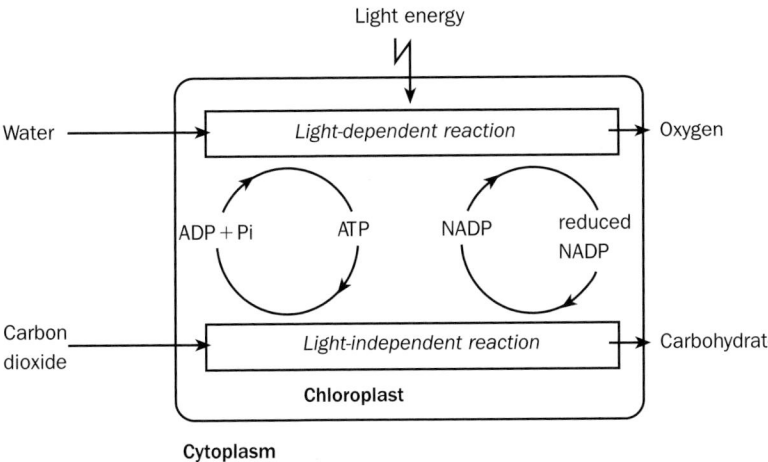

A summary of the biochemical pathways involved in photosynthesis

Photosynthesis is often thought of as just producing carbohydrates, but it is the starting point for the synthesis of all the organic molecules in a plant. Mineral ions are absorbed through the roots and added to carbon-containing substances produced in photosynthesis, to make *amino acids*, *lipids* and *nucleic acids*.

photosynthetic pigment: a pigment found in *chloroplasts* and in many photosynthetic bacteria. Photosynthetic pigments trap the light energy used in *photosynthesis*. *Chlorophyll* is the best-known photosynthetic pigment. Others such as the orange-coloured carotene and the yellow xanthophyll are also found in chloroplasts. Some photosynthetic bacteria do not contain chlorophyll. They use a light-sensitive protein called *opsin* to trap the light energy that they use for *carbon fixation*.

photosystem: a group of 300 or so *chlorophyll* molecules found on the membranes in a *chloroplast*. A photosystem functions as a sort of light funnel. It absorbs light and passes the energy from molecule to molecule until it reaches one particular chlorophyll molecule called the reaction centre. When the light energy reaches the reaction centre, some of its *electrons* become excited. These are lost from the reaction centre molecule and pass through a series of electron acceptors to produce *ATP*.

phototropism: a growth movement of a whole plant organ in response to light. Stems are generally positively phototropic and grow towards the direction of the light. Roots are usually negatively phototropic and grow away from the light. (See also see *tropism*.)

phrenic nerve: a nerve that goes from the *brain* to the *diaphragm*. Regular bursts of *nerve impulses* pass from the respiratory centre in the *medulla* down the phrenic nerve. They cause the diaphragm to contract and flatten when a mammal breathes in.

phylogeny: the evolutionary relationship between organisms. Originally systems used to classify living organisms were based on their appearance. These systems did not always reflect evolutionary relationships. More recently, scientists have used different approaches to classification. They have compared amino acid sequences in *proteins* and base sequences in *DNA*. These approaches have clarified the phylogenetic relationships between the organisms and have led to changes in the classification of some organisms.

phylum: a level of *classification*. Each kingdom is divided into a number of *phyla*.

physiology: the study of the way in which the organs and systems in a living organism function.

phytochrome: a light-absorbing pigment found in plants. It exists in two forms. One form is particularly sensitive to red light and is referred to as P_R. The other form, which is more sensitive to the far-red light that is just outside the normal visible spectrum, is called P_{FR}. The diagram shows the relationship between these two forms of phytochrome.

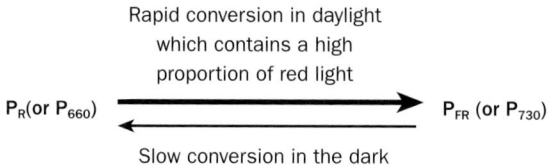

Rapid conversion in daylight which contains a high proportion of red light

P_R(or P_{660}) → P_{FR} (or P_{730})

Slow conversion in the dark

The relationship between the two forms of phytochrome

After a period in the daylight, most of the phytochrome will be in the form P_{FR} because the P_R will have had time to absorb red light and be converted into P_{FR}. However, the amount of P_{FR} will decline if the plant is left in the dark. The P_{FR} appears to be the active form and promotes activities such as *flowering* in the plant. It may also stimulate *germination*.

phyto-oestrogen: an *oestrogen*-like substance produced by plants. Phyto-oestrogens occur naturally in many plants that are eaten by humans. These include soya beans, pulses and some vegetables. Biologists think that many of these substances do not have any specific function in the plant and may just be metabolic by-products. Some people suggest that food rich in phyto-oestrogens:

- can relieve some of the more unpleasant symptoms of *menopause*
- lower the risk of some cancers.

pickling: a method of preventing food spoilage by storing in acid conditions. Acid conditions may be achieved by *fermentation* to produce *lactic acid* or by adding vinegar. A pH below 4.6 kills most bacteria.

pigment: a coloured substance. Pigments have many different functions in living organisms. The table shows some examples.

Pigment	Function
Anthocyanins	Pigments found in *cell sap*. They produce blue and red colours in *flowers*. Slight changes to the basic molecule will change the colour of the pigment. The addition of a methyl (CH_3) group, for example, will produce a redder colour
Chlorophylls	The group of pigments responsible for absorbing light in *photosynthesis*
Carotenoids	Responsible for the red, brown and yellow colours of plants. The yellow colour of daffodils and the red of tomato *fruits* are due to the presence of these pigments
Rhodopsin	The light-sensitive pigment found in the *rod cells* in the eyes
Respiratory pigments	Molecules such as the red *iron*-containing *haemoglobin* and the blue-green copper-containing pigment, haemocyanin, found in the blood of many invertebrates, are involved in the transport of oxygen

pinocytosis: the way in which a cell takes fluids through the *cell-surface membrane* into the *cytoplasm*. The membrane folds round a tiny drop of the solution or suspension of particles concerned and forms a small vesicle. This vesicle is pinched off and moves into the cytoplasm.

pioneer species: a species of organism that is among the first to colonise a new *habitat*. Sand dunes are common in many coastal areas. The sand is very unstable and is constantly being blown by the wind. It dries out easily and there are also very low concentrations of soil nutrients. This is a very harsh environment and one of the first plants to colonise it is marram grass. Marram grass is a pioneer species in sand dune succession. It has adaptations for growing on sand dunes:
- The leaves of marram grass grow from a vertical underground stem. They remain above the sand that the wind blows on to the plant.
- Marram grass leaves are rolled inwards and they have a thick waxy *cuticle*. These are adaptations that reduce water loss.

pit: a small depression in a plant *cell wall*. Newly formed plant cells have cell walls made of *cellulose* cemented together with other substances. This is called the primary cell wall. At a later stage a secondary cell wall may be added. This is made of *lignin* and makes the wall very strong. It also makes it impermeable. Pits usually only have primary cell walls. Because there is no secondary wall, they allow the passage of substances through the cell wall from one cell to another.

pituitary gland: an *endocrine gland* attached to the underside of the *hypothalamus*. It consists of two separate *glands*: the anterior lobe and the posterior lobe. The anterior lobe is connected to the hypothalamus by a *portal vein*. It secretes a number of different *hormones* and their secretion is under the control of the hypothalamus. This is summarised in the diagram overleaf.

P

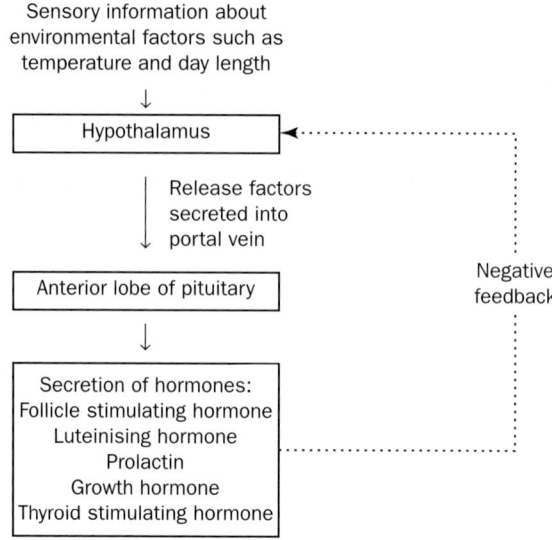

Sensory information about environmental factors such as temperature and day length

↓

Hypothalamus

Release factors secreted into portal vein

↓

Anterior lobe of pituitary

↓

Secretion of hormones:
Follicle stimulating hormone
Luteinising hormone
Prolactin
Growth hormone
Thyroid stimulating hormone

Negative feedback

The posterior lobe does not actually produce any hormones. It stores and releases *antidiuretic hormone (ADH)*, which has been produced in special cells in the hypothalamus.

placenta: an organ formed partly from membranes surrounding the developing mammalian fetus and partly from the lining of the *uterus*. It is an exchange organ. It is where oxygen and nutrients enter the fetal blood system from the blood of the mother. Waste products such as *urea* and carbon dioxide are also removed from the blood of the fetus to that of the mother via the placenta. As with all exchange organs, it has a number of adaptations (see *Fick's law*):

- large surface area. The exchange surface of the placenta is folded into villi and microvilli (see *villi* and *microvilli*)
- thin membrane. There is a reduction in the number of cell layers between the blood of the foetus and that of the mother
- maximum difference in concentration on either side of the exchange surface. *Fetal haemoglobin* has a much higher affinity for oxygen than adult haemoglobin. The blood of the mother more or less flows in the opposite direction to that of the fetus, forming a *counter-current* system

In addition to acting as an exchange organ, the placenta has an important endocrine function and secretes a number of *hormones*, including *progesterone* and *oestrogen*.

plankton: the name given to the microscopic organisms that float in the surface layers of open water. Phytoplankton consists of photosynthetic organisms which, when conditions are right, can multiply and increase in numbers extremely rapidly. They are mainly small unicellular algae. Zooplankton, on the other hand, are the animals of the plankton. They feed both on the phytoplankton and on other zooplanktonic organisms. Many different organisms make up the zooplankton. They include small arthropods and the larval stages of many larger animals. Marine plankton is extremely important ecologically. It forms the food of many marine animals, including some of the largest fish and many whales. It also plays an extremely important part in the *carbon cycle* and is probably one of the most important ways in which the carbon in carbon dioxide can be converted into organic carbon compounds.

plant growth substance: a substance produced by a plant that is involved in the control of various aspects of its growth and development. Plant growth substances used to be called plant *hormones* but they differ from the hormones produced by animals in a number of ways. In particular, they are generally produced close to the site of action rather than being released from one organ and transported to another. There are five main groups of plant growth substance – *auxins*, *cytokinins*, *gibberellins*, *abscisic acid (ABA)* and *ethene* – and they have a wide range of effects on plant tissues. Different concentrations of the same substance may have different effects on different tissues. They may even have different effects on the same tissue. Additionally, the effect of one substance may be modified by the presence of another. Plant growth substances have widespread commercial applications.

Plantae: the *kingdom* containing plants (see *classification*). This kingdom contains not only the *flowering* plants but also others such as mosses and ferns. Plants share the following features:
- they have *cell walls* containing *cellulose*
- they can photosynthesise and their *chlorophyll* is found in special organelles known as *chloroplasts*
- they have a life cycle that shows *alternation of generations* involving a sexually reproducing generation and an *asexual* generation.

plasma: the liquid part of the blood. It consists mainly of water but about 10% is made up of dissolved substances. Many of these dissolved substances, such as *glucose*, are maintained at approximately constant concentrations by the body's homeostatic mechanisms; others vary according to circumstances.

plasma cells: produced when *B cells* in the blood are exposed to an infection by a *pathogen*. B cells are *lymphocytes* and divide rapidly by mitosis to form plasma cells. The plasma cells produce and secrete *antibodies*.

plasma membrane: a membrane found either on the outside of a cell or inside it. Each cell in a eukaryotic organism is surrounded by a *cell-surface membrane* and its *cytoplasm* also contains membranes and membrane-bounded organelles such as *mitochondria* and *chloroplasts*. Plasma membranes are extremely thin. They are only about 7 nm thick and so cannot be seen with a light microscope. A *transmission electron microscope*, however, shows a plasma membrane as consisting of three lines forming a sandwich. The two outer lines are dark in colour while there is a lighter one in between. Plasma membranes are made mainly of *lipids* and *proteins* but the actual amount of these substances differs from cell to cell. As it is impossible, even with an *electron microscope*, to see how the actual molecules are arranged in a cell membrane, scientists have produced a model to explain the membrane's properties. The most accurate model of membrane structure that has been developed is the *fluid mosaic model* which describes most of the properties of a plasma membrane.

Plasma membranes play an important part in the biology of cells. They regulate the movement of substances into and within cells and they also have important receptor molecules on their surfaces that enable cells to respond to substances such as *hormones*.

plasma proteins: the *proteins* that are normally present in the liquid part of the blood, the *plasma*. These proteins have a number of functions. They are *buffers* and help to maintain a constant blood pH. They also contribute to the *water potential* of the plasma and play an important part in the formation of tissue fluid and its return to the blood. There are three main types of plasma protein:

- The albumins are the commonest. Among their functions is the transport of other substances in the plasma.
- **Globulins** are the proteins that include **antibodies**.
- **Fibrinogen** is a protein that is extremely important in **blood clotting**.

Most plasma proteins are made in the **liver**. They circulate in the plasma and are eventually broken down again in the liver.

plasmid: a small circular piece of **DNA** found in bacteria and some other organisms. In bacteria, most of the DNA in the cell is joined to make a loop. This is the largest piece of DNA in the cell but most bacteria also carry plasmids. Plasmids often contain genes that code for **resistance** to **antibiotics** and they can be passed from one bacterium to another by a variety of processes. These two features give them considerable importance. Overexposure to antibiotics may result in otherwise harmless bacteria, such as those found in the gut, being selected for antibiotic resistance. This resistance can be passed via plasmid transfer to disease-causing bacteria, which then become more difficult to control. In addition, the fact that plasmids are readily accepted by bacteria makes them useful as **vectors** in **genetic engineering**.

plasmodesmata: thin strands of protoplasm that penetrate the **cell walls** of plants and connect the **cytoplasm** of one cell to that of another. They are part of the **symplastic pathway**, by which substances are able to go from one cell to the next without having to pass through cell walls and **cell-surface membranes**. Most mineral ions move through a plant in this way.

Plasmodium: the parasite that causes **malaria**.

plasmolysis: the shrinkage of **cytoplasm** away from the **cell wall** of a plant cell as a result of water being lost from the cell by osmosis. When a plant cell is put in a concentrated sugar solution, the concentration of water molecules inside the cell is higher than that outside. Water moves by **osmosis** from the higher **water potential** inside the cell to the lower water potential outside. The cell contents shrink away from the cell wall and the cell is plasmolysed. At the point where the water potential outside is exactly the same as that inside, the cytoplasm is just shrinking away from the cell wall. This is **incipient plasmolysis**.

The relationship between water potential, pressure potential and **solute potential** is given by the equation:

water potential = solute potential + pressure potential

At incipient plasmolysis, the cell wall is not pressing on the **cytoplasm** so the pressure potential will be zero. If this is the case:

water potential = solute potential + 0

or:

water potential = solute potential

Therefore, finding the water potential by this method also enables the solute potential of the cell to be determined.

plastid: an **organelle**, found in the **cytoplasm**, that is characteristic of plant cells. **Chloroplasts** and **amyloplasts** are plastids, and there are other types as well as these. Plastids contain an internal system of membranes and pigments or storage substances.

platelets: small cell fragments found in the blood that play an important part in *blood clotting*. They are formed as a result of the breaking up of large cells found in *bone* marrow. When the body is wounded and a blood clot forms, soluble *fibrinogen* is converted to insoluble *fibrin*. This forms a mesh of protein fibres over the surface of the wound that traps red blood cells and forms a clot. A complex mechanism controls the process of blood clotting. Fibrinogen can only be converted into fibrin in the presence of the *enzyme*, *thrombin*. Thrombin is normally present in the blood as inactive prothrombin. When the blood comes into contact with air or damaged tissue, the platelets break down and produce thromboplastin. Along with other substances, thromboplastin converts the inactive prothrombin to thrombin.

Atherosclerosis is a disease of the arteries caused by *atheroma* or fatty deposits in the walls. These can result in rough patches to which platelets may stick and produce thromboplastin. If this happens, there is a risk that a blood clot may form and block an *artery*. This is thrombosis and it can cause the death of tissues which are supplied by the particular artery. When atherosclerosis affects the *coronary arteries* it may lead to a heart attack or *myocardial infarction*.

Platelets are also called thrombocytes.

Platyhelminthes: the animal phylum containing the *flatworms* and two important groups of *parasite*, the tapeworms and the flukes. Platyhelminthes share the following features:
- they are dorso-ventrally flattened, unsegmented worms
- they do not have a *coelom*.

In those species that have a gut, it is branched and has only a single opening.

plumule: part of an embryo plant found in a seed. When *germination* occurs, the plumule grows upwards towards the light. It is the shoot of the embryo plant.

pluripotent: a cell that can give rise to many of the cell types that make up the organism. Human embryonic *stem cells* may be removed from an *embryo* that has been donated for research purposes and grown on dishes containing a coating of embryonic mouse skin cells. The stem cells in this culture are pluripotent. When they are cultured in appropriate conditions they can develop into most of the cell types found in an adult human.

podocyte: a cell from the inner layer of the *renal capsule*. Podocytes have processes sticking out from them. Fluid entering the renal capsule from the blood goes through the gaps between these processes. This makes *ultrafiltration* more efficient as the gaps do not provide a barrier to the movement of fluid from the blood into the *kidney* tubule.

point mutation: an alternative name for *gene mutation*.

point quadrat: a *quadrat* reduced in size to a pinpoint. A point frame consisting of a group of 10 pins set about 5 cm apart is normally used. The pins are lowered in turn and the number of hits or misses on the particular species of plant under consideration is recorded. From this the *percentage cover* can be estimated.

polar body: nuclei with a thin covering of *cytoplasm* that are produced during *oogenesis*. At the first division of *meiosis*, the nuclear material in the immature *gamete* divides equally in two. The surrounding cytoplasm, however, almost all goes with one of the resulting sets of chromosomes. This gives a large cell which will become the female gamete and a small one called a polar body. The polar body eventually disappears. During the second division of meiosis the nuclear material again divides equally, but all the cytoplasm goes with the now mature, female gamete. The remaining nuclear material forms another polar body.

pollen: the stage in the *flowering* plant life cycle that contains the male *gametes*. Pollen mother cells in the *anther* divide by *meiosis* to produce four immature pollen grains. The nucleus in each of these immature pollen grains then divides by *mitosis* to produce a pollen grain which, when it is mature, contains three *haploid* nuclei.

One of these nuclei is the tube nucleus which controls the *pollen tube*. The other two will be the male gametes. The outer coat of the pollen grain is tough and very resistant to decay. This and the fact that the pollen grains from different species are very different in appearance can give us information about past climates.

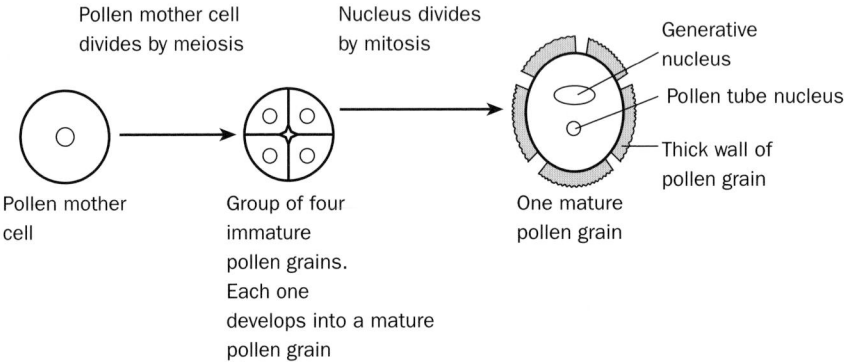

The development of a pollen grain

pollen tube: a structure produced by a germinating pollen grain. It grows away from oxygen and towards moisture, down through the *style* to the ovules. It usually enters the ovule through a small pore called the micropyle. The male *gametes* travel down this pollen tube to the ovule.

pollination: the transfer of pollen from the *anther* to the *stigma*. You should take care to distinguish pollination from *fertilisation*, which involves the actual fusion of male and female *gametes*, and from seed dispersal which, as its name suggests, involves the dispersal of seeds. Although there are a variety of different methods by which pollen can be transferred, insect and wind pollination are the most frequent in temperate regions. The table shows some differences between the pollen of insect- and wind-pollinated plants.

Pollen from insect-pollinated plants	Pollen from wind-pollinated plants
Pollen grains larger in size, usually over 25 μm in diameter	Pollen grains smaller in size, usually between 10 and 25 μm in diameter
Produced in smaller amounts	Produced in larger amounts: a single hazel catkin, for example, may produce up to 4 000 000 pollen grains
Pollen grains with sticky or rough surface	Pollen grains with smooth, dry surface

polluter pays principle: the principle by which those who are responsible for pollution must pay for the resulting damage to the environment.

pollution: the contamination of the environment with harmful substances or in other ways such as with excessive heat, light or noise.

polyculture: a type of farming in which different crops are grown on an area of land. Polyculture may be compared with *monoculture*. This is a type of farming in which a large area of land is devoted to the growth of a single crop.

polygenic inheritance: where more than one gene *locus* is responsible for controlling a characteristic. Mendel's work showed that height in peas is controlled by a single gene with two *alleles*, the allele for tall plants, T, being *dominant* to that for short plants, t. As a result, pea plants are either tall or dwarf. They show *discontinuous variation*. Human height, however, shows *continuous variation*, with a complete range from the shortest person to the tallest person in a population. Height in humans is an example of polygenic inheritance. Continuous variation reflects the fact that many different combinations of alleles are possible.

polymer: a large molecule that is built up from a number of similar smaller molecules or *monomers*. There are three biologically important groups of polymers found in living organisms. Some of their characteristics are summarised in the table.

Polymers	Property		
	Monomers from which they are built up	Chemical elements they contain	Relative molecular mass
Nucleic acid	*Nucleotides*	Carbon Hydrogen Oxygen Nitrogen Phosphorus	Between 10^4 and 10^{10}
Polysaccharide	*Monosaccharides*	Carbon Hydrogen Oxygen	Between 10^4 and 10^6
Proteins	*Amino acids*	Carbon Hydrogen Oxygen Nitrogen Sulphur (in some amino acids)	Between 10^4 and 10^6

Biological polymers are all formed by *condensation*. This involves linking the monomers by removing a molecule of water. They can be broken down to the monomers from which they are built by *hydrolysis*. Hydrolysis involves the addition of a molecule of water. Condensation and hydrolysis are shown in the diagram overleaf.

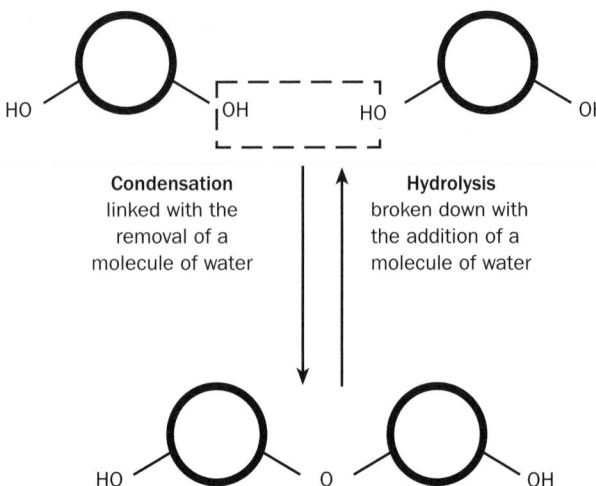

Condensation and hydrolysis

polymerase chain reaction (PCR): a process used by biologists to make large amounts of identical **DNA** from very small samples. It is explained in the diagram below.

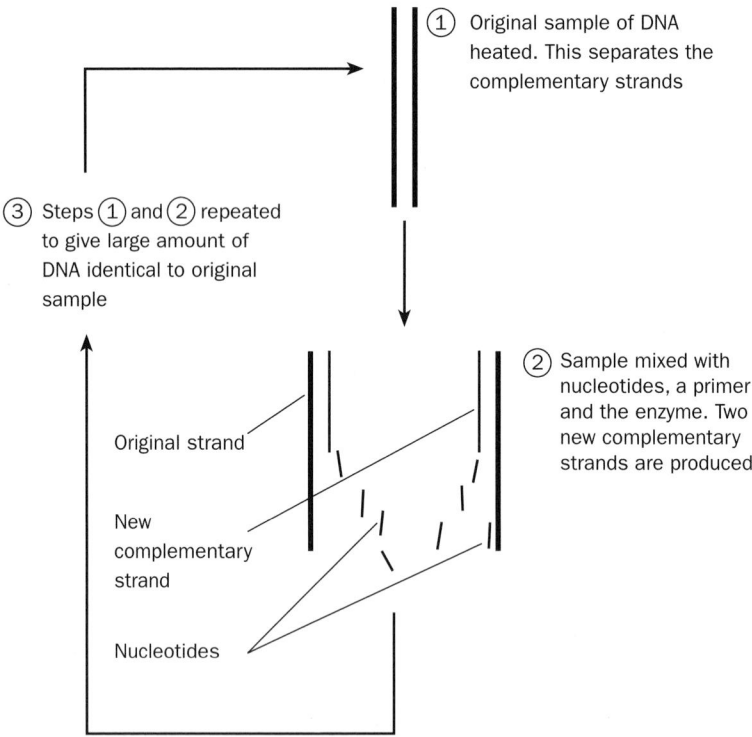

① Original sample of DNA heated. This separates the complementary strands

③ Steps ① and ② repeated to give large amount of DNA identical to original sample

② Sample mixed with nucleotides, a primer and the enzyme. Two new complementary strands are produced

Original strand

New complementary strand

Nucleotides

polymorphism: a type of variation between members of a **species**. Polymorphism refers to having different **alleles** of a particular **gene** in a population.

polynucleotide: a **polymer** made up from a number of **nucleotides** joined to each other by **condensation**. **DNA** consists of two polynucleotide strands twisted around each other to form a spiral or helix. There is one complete turn of the spiral for every ten nucleotides. *RNA* is a shorter molecule and consists of a single polynucleotide chain.

polypeptide: a *polymer* consisting of a chain of *amino acid* molecules. They are joined by peptide bonds formed by *condensation*. One or more polypeptides form a *protein*.

polyploid: a term that can refer to cells, organisms or stages in the life cycle in which the nuclei contain more than two copies of each *chromosome*. Many cultivated plants are polyploid. Bananas, for example, may be triploid (3n) or tetraploid (4n) and many modern varieties of wheat are hexaploid (6n). Triploid bananas grow faster and produce bigger *fruit* than diploid varieties. Because they have an odd number of sets of chromosomes, however, their chromosomes are unable to pair during *meiosis*. Triploid bananas, therefore, cannot reproduce sexually.

polysaccharide: a *carbohydrate* which is a *polymer*. Polysaccharides are made up of many sugar units or *monosaccharides* joined by *condensation*. Since they have large molecules, they are insoluble. Their main functions in living organisms are to act as storage substances (*starch* and *glycogen*) or as structural substances (*cellulose*).

population: a group of organisms belonging to the same species. In ecological terms, this definition refers to a particular area, so it is possible to consider the population of frogs in a pond or the population of *aphids* on a sycamore tree. In evolutionary terms, however, it is necessary to add that members of a population are able to breed with each other.

population growth curve: a graph that shows the growth of a population of organisms over a period of time. The diagram shows a population growth curve for yeast cells growing in a sucrose solution.

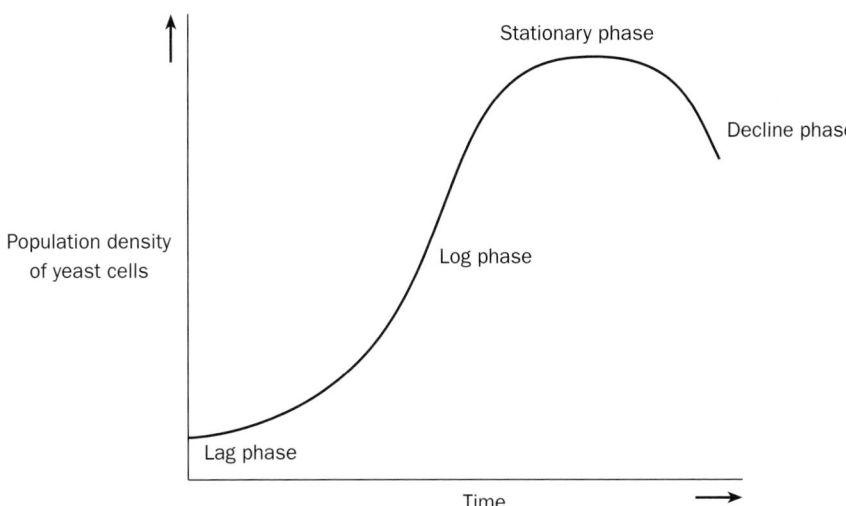

We can divide this curve into four separate stages or phases:
- *lag phase*. During this stage, the organisms do not increase in number; they are adapting to the medium in which they are growing and synthesising *enzymes*
- *log phase*. This is a period of rapid population growth. Nutrients are in plentiful supply and the amount of toxic waste products produced is very low
- *stationary phase*. The population remains more or less constant

- *decline phase*. A shortage of nutrients and a build-up of toxic waste products lead to the death of many organisms.

population growth rate: a statistic used to analyse data on human populations. Population growth rate can be calculated by subtracting death rate from the *birth rate*.

population pyramid: a way of showing the structure of a population graphically. A population pyramid shows the numbers of males and females in each age group in the population. Although most frequently used to describe human populations, it can also be used to represent the population structure of other organisms. The diagrams show the shapes of two typical pyramids. Diagram A represents a fast-expanding population in which there is a large number of young individuals. Diagram B shows a stable population in which there are relatively few immature individuals and more adults.

portal vein: a *blood vessel* that has a *capillary* network at both ends. The hepatic portal vein is a blood vessel that transports the products of *digestion* from the intestine to the *liver*. The blood in most other parts of the blood system is remarkably constant in composition. For example, the blood *glucose* concentration is usually around 90 mg per 100 cm^3 of blood. It is kept within narrow limits by the action of *hormones*, particularly *insulin* and *glucagon*. The glucose concentration in the hepatic portal vein, however, may fluctuate much more and may rise significantly higher than this after a meal.

positive feedback: many substances or systems have a set level or norm within a living organism. Positive feedback involves a departure from this set level bringing about changes which produce further departures. A good example is provided by what happens in *hypothermia*. The human body temperature has a set level, generally around 37°C. If the body temperature falls, then the *metabolic rate* of the body will fall. Less heat will be produced and the body temperature will fall further. Ultimately, this process may lead to the death of the person concerned. For another example see *action potential*.

postsynaptic membrane: the *cell-surface membrane* of one of the *neurones* that forms a *synapse*. A *nerve impulse* arriving at a synapse causes the release of a *neurotransmitter* from the presynaptic neurone. This diffuses across the synaptic cleft and produces another nerve impulse by depolarising the postsynaptic membrane.

potassium–argon dating: a method of dating fossil remains that has proved particularly useful for human fossils. It relies on the principle that radioactive potassium, or potassium-40, gradually breaks down into argon. Like all radioactive elements, potassium-40 breaks down at a steady rate; in this case it takes 1.3 billion years for half of the potassium-40 in a sample to break down. If the amount of argon is measured and the rate of breakdown of potassium-40 into argon is known, then the age of the sample can be calculated. There are some difficulties, however. Only volcanic rocks are suitable for this method of dating and they do not contain fossils. In addition, only very small amounts of argon are produced by decay of potassium-40, and it is easily contaminated by the argon that is normally present in the air.

potassium-movement hypothesis: a mechanism which has been suggested to account for the opening and closing of *stomata*. It is outlined in the diagram.

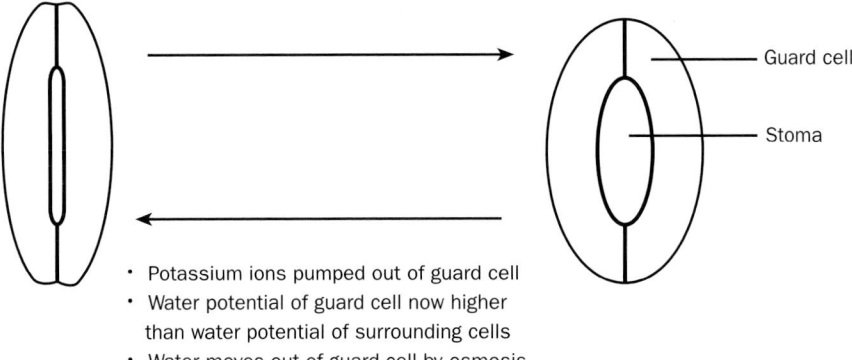

- Potassium ions pumped into guard cell
- Water potential of guard cell falls
- Water moves into guard cell by osmosis
- Thin outer walls of guard cell get longer
- Stoma opens

Guard cell

Stoma

- Potassium ions pumped out of guard cell
- Water potential of guard cell now higher than water potential of surrounding cells
- Water moves out of guard cell by osmosis
- Stoma closes

potometer: a piece of apparatus used to estimate the rate of transpiration from a plant shoot. At its simplest, a potometer consists of a capillary tube with a piece of rubber tubing at one end. The apparatus is filled with water and a leafy shoot is cut and fitted into the rubber tubing. When the shoot transpires, it draws water up through the capillary tube. By measuring the rate of movement of an air bubble introduced into the capillary tube, we can measure the rate of transpiration. When we use a potometer we assume that the water taken up by the shoot equals the amount of water lost in transpiration. This is not always the case, so a potometer only gives us an estimate of the rate of transpiration.

Prader-Willi syndrome: a rare genetic disorder in which affected people are short, have small hands and feet and may be mentally retarded. In people with this condition, 7 of the genes on chromosome 15 are not expressed. These are always the genes on the paternal chromosome. Prader-Willi syndrome is an example of *epigenetic imprinting*. This occurs where heritable changes in the phenotype of an organism are not accompanied by changes in its *genotype*.

predator: an organism that feeds on another organism, killing it before eating it. Most familiar examples of predators, such as lions and ladybirds, are animals and it is usual to think of prey as animals as well. However, predators and their prey can also be found in other kingdoms such as the *Protoctista*.

predator strip: areas of rough grass left undisturbed at the edges of fields. They encourage natural predators that assist in pest control.

preimplantation genetic diagnosis: a procedure in which the genes of an early *embryo* are examined before *implantation* in the uterus. In *in vitro fertilisation (IVF)*, fertilised eggs are at first incubated outside the mother's body. The fertilised egg divides to become an embryo. Cells may be removed from this embryo. This does not harm the embryo in any way. The *DNA* in these cells is analysed and this information is used to find out whether the embryo is carrying an *allele* that might cause it to be seriously disabled. Doctors can then be sure that the embryo that is put into the *uterus* does not have this allele.

pre-mRNA: formed during *protein synthesis* in a *eukaryotic* cell. During *transcription*, the entire base sequence of a *gene* is transcribed to produce pre-mRNA. This includes the non-coding sequences (*introns*) as well as the coding sequences (*exons*). Before it leaves the nucleus, the pre-mRNA is edited. The non-coding sections are removed. The coding sections are then spliced together to produce *mRNA* that only carries the coding sections of the gene.

preservation: maintaining areas that are largely untouched by human activity in their present condition. This term is often confused with *conservation*. Conservation involves management for the sustainable use of resources.

pressure potential: the pressure produced by the *cell wall* pushing against the *cell-surface membrane*. In the *cytoplasm*, water molecules move around at random. Some collide with the membrane that surrounds them. They exert a pressure on this membrane. This is the *water potential*, the symbol for which is the Greek letter psi, ψ. The pressure produced by these randomly moving water molecules pushes against the membrane from the inside, while the pressure potential is pushing against the membrane from the outside. They are acting in the opposite direction to each other. Since all values of water potential are negative, values of pressure potential will be positive. The relationship between pressure potential, ψ_p, water potential, ψ, and *solute potential*, ψ_s, is summarised by the equation:

$$\psi_p = \psi - \psi_s$$

presynaptic membrane: the *cell-surface membrane* of one of the *neurones* that forms a *synapse*. A *nerve impulse* arriving at a synapse causes the release of a *neurotransmitter* from the presynaptic neurone. Vesicles containing neurotransmitter fuse with the presynaptic membrane. The neurotransmitter then diffuses across the synaptic cleft and produces another nerve impulse in the postsynaptic neurone.

prevalence: the total number of cases of a particular disease in the population at any one time.

primary consumer: an organism that feeds on plants. In the *food chain*

nettle plant → large nettle aphid → two-spot ladybird

the nettle plant is the *producer*. It converts the energy in sunlight into energy in the organic molecules in the plant. The large nettle *aphid* feeds on the nettle and is the primary consumer.

primary immune response: the response shown by the immune system when a person is first exposed to a particular *antigen*. Suppose a person is infected with the virus that causes measles. The virus acts as an antigen. When an appropriate B cell comes into contact with a measles virus, it divides rapidly and produces two sorts of cells, *plasma cells* and *memory cells*.

The plasma cells produce and secrete **antibodies**. The primary immune response is a rise in the concentration of specific antibodies after the first exposure to a particular antigen. In this case it is the rise in anti-measles antibodies after a person is first infected with measles viruses. The memory cells remain in the blood system. They respond rapidly if re-infection with the same antigen occurs. The memory cells multiply and produce more plasma cells. The response to a second infection, the **secondary immune response**, is much more rapid.

primary metabolite: a substance that is produced by **microorganisms** when they are still actively growing. When microorganisms are grown in culture, they soon enter a phase of rapid growth. During this stage, **enzyme**-controlled reactions produce the substances required for growth. These substances are called primary metabolites. **Secondary metabolites** are substances that are usually produced later in the life cycle, as the microorganisms age.

primary structure (protein): the sequence of **amino acids** that make up a **protein** molecule. A particular protein has a specific number of amino acids in each of its molecules. Typically, this is somewhere between 100 and 650. In addition to this, there are about 20 different amino acids which may occur in a protein molecule. They may be arranged in a variety of different ways. Variation in the number and arrangement of the amino acids in a molecule means that it is possible to have an enormous number of different **proteins**.

primary succession: **succession** in a place where plants have not grown before. Primary succession may take place on bare rock, sand dunes or on areas left bare after retreating glaciers.

primate: a member of the order of mammals that includes lemurs, monkeys and apes. Humans also belong to this order. Most primates are tree-dwellers and many of the characteristics associated with the group are adaptations to living in trees.

- The forelimbs have considerable freedom of movement. This is shown with the freely movable shoulder joint; an elbow joint that allows rotation of the forearm; and fingers and thumbs which are, to some extent, opposable. This should be compared with the degree of movement shown in the front limbs of mammals such as horses and dogs. The claws are flattened and form finger- and toe-nails.
- The eyes are extremely well developed; primates can distinguish colours and have **binocular vision**.
- The **brain** is relatively large in proportion to the size of the body.
- Young are usually born singly and the female has a single pair of mammary **glands**. These are on her chest.

primer: a short single-stranded piece of **DNA** that is used to mark the section of DNA to be copied in the **polymerase chain reaction (PCR)**. Two different primers are produced. Each binds to one end of one of the strands of the DNA that is to be copied and forms a double stranded section. The primers are synthesised artificially so they can be designed to match exactly the bases at the start of the strand to be copied. Once the double-stranded section has been marked, the **enzyme DNA polymerase** replicates the DNA.

probability: a mathematical way of expressing the likelihood of a particular event occurring. In humans, a single gene with two **alleles** affects the appearance of the ear lobes. The allele for free ear lobes, E, is **dominant** to that for fixed ear lobes, e. If two people who are heterozygous for this gene have children, the probability of their having a child with fixed ear lobes is one in four, or 0.25.

When two or more factors are considered, the probabilities of the individual events must be multiplied. To use the same example, the probability of this couple having a child who is a girl with fixed ear lobes can be calculated:

- The probability of any child being a girl is one out of two, or 0.5.
- The probability of the child having fixed ear lobes is 0.25.
- Therefore, the probability of both events occurring is $0.5 \times 0.25 = 0.125$, or one in eight.

The results obtained in many experimental investigations may have a biological explanation. They may also be due to *chance*. Statistical tests can be used to determine the probability of results being due to chance.

probiotic bacteria: bacteria that live in the gut and are thought to have a beneficial effect on human health. People think that probiotic bacteria are beneficial because:

- they compete with pathogenic bacteria and prevent them attaching to the gut wall
- they bind to *carcinogens* that may be present and therefore lower the risk of some types of cancer
- they produce useful substances such as *vitamin K*.

producer: an *autotroph*. All *food webs* and *food chains* ultimately depend on producers. These are organisms that produce organic molecules from simple inorganic ones with the aid of an additional energy source. Most terrestrial food webs are based on *photoautotrophs*, organisms that depend on *photosynthesis*. There are some deep-sea *ecosystems*, however, that have developed around volcanic areas. Here, the producers are chemoautotrophic bacteria (see *chemoautotroph*) that use the energy released from chemical reactions to produce organic substances.

progesterone: a female sex *hormone* produced during the second half of the reproductive cycle and during pregnancy. In the *menstrual cycle*, it is secreted by the *corpus luteum*. During the first weeks of pregnancy, the corpus luteum continues to secrete progesterone, but in the later stages it is secreted by the *placenta*. Progesterone has several effects on the reproductive system. It maintains the lining of the uterus and inhibits contraction of the uterine muscle. It also exerts a *negative feedback* effect on the release of two more hormones, *FSH* and *LH*, from the anterior lobe of the *pituitary gland*. Progesterone can be used as an *oral contraceptive,* either on its own or combined with *oestrogen*.

programmed cell death: see *apoptosis*.

Prokaryotae: the *kingdom* containing *prokaryotic* organisms such as bacteria (see *classification*). Members of the Prokaryotae share the following features:

- they are single-celled organisms. Their small cells do not have nuclei
- their DNA is in circular strands. It is found in the *cytoplasm*
- they have relatively few organelles in their cytoplasm. None of these organelles is surrounded by a *plasma membrane*.

Prokaryotes have *cell walls*. The cell wall is made of peptidoglycan and other substances. It is not made of *cellulose*, as in plants. In some prokaryotes, the cell wall is surrounded by a *capsule*.

prokaryotic: a term used to describe the cells of organisms such as bacteria that do not contain a nucleus. Cells from animals and plants are eukaryotic. Prokaryotic cells differ from eukaryotic cells in a number of ways. These differences are summarised in the table.

Prokaryotic cells	Eukaryotic cells
Cells small with a mean diameter under 5 µm	Large cells up to 50 µm in diameter
Circular strands of **DNA** not associated with **proteins** and found in the **cytoplasm**. No nucleus present	DNA linear and associated with proteins to form **chromosomes**. The chromosomes are found in a nucleus
Few **organelles** present. None is surrounded by a **plasma membrane**	Many membrane-surrounded organelles such as mitochondria present

prolactin: a **hormone** secreted by the anterior lobe of the **pituitary gland**. It stimulates the production of milk by the mammary **glands**.

Propionibacterium acnes: a species of bacterium that has been linked with the skin condition **acne vulgaris**. It normally lives on **fatty acids** in the **sebaceous glands**. The ducts of these glands are often blocked. When they are blocked, the bacteria increase in number and break down the wall of the duct. This may lead to other bacteria entering the gland. They produce substances that lead to **inflammation**.

prostaglandin: a hormone-like substance that acts as a local chemical mediator, affecting cells in the immediate vicinity. Prostaglandins are secreted by most of the organs in the body and have many actions. They:
- cause contraction or relaxation of smooth muscle in the walls of **blood vessels** and help to regulate blood flow
- cause contraction of muscle in the walls of the uterus
- dilate or constrict airways in the **lungs**.

prostate gland: a male sex gland that opens into the **urethra**. During ejaculation, it produces a fluid that forms part of the semen. Cancer of the prostate gland is a common form of cancer in older men.

prosthetic group: a non-protein group that is attached to a **protein** molecule to form a **conjugated protein**. **Haemoglobin** has an **iron**-containing prosthetic group called haem. **Carbohydrates** and **lipids** form prosthetic groups in **glycoproteins** and **lipoproteins**.

protandry: a condition in which the male part of a flower matures first. It is one of the ways in which **self-fertilisation** can be prevented in flowers which possess both male and female organs. If the flowers of a rose-bay willow-herb are examined, for example, it will be seen that the **anthers** mature and shed their pollen before the lobes of the **stigma** open.

protease: an **enzyme** that hydrolyses protein molecules by breaking the bonds which join the **amino acids** together. Proteases are useful in **biotechnology**. They are used in biological washing powders and in solutions such as those used for cleaning protein from contact lenses. Other proteases are used in the food industry for tenderising meat. Some proteases have a general action and hydrolyse the protein molecule at various points along its length, but most are very specific. They only break **peptide bonds** between particular amino acids. Proteases can be divided into two categories, **endopeptidases** and **exopeptidases**, according to the point in the protein molecule at which they act.

protein: a **polymer** built from **amino acids** joined by peptide bonds. The resulting chain of amino acids (called a polypeptide) is then folded in different ways and to different extents. This is shown in the diagram overleaf.

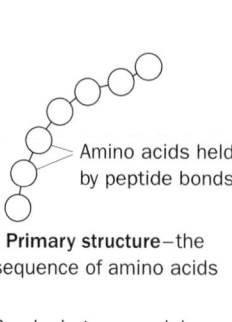

Amino acids held
by peptide bonds

Primary structure–the
sequence of amino acids

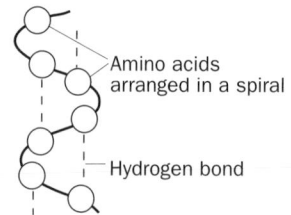

Amino acids
arranged in a spiral

Hydrogen bond

Secondary structure–the coiling of
an amino acid chain into a helix.
This is only one example of
secondary structure

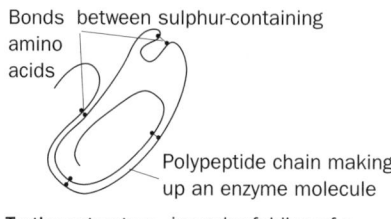

Bonds between sulphur-containing
amino
acids

Polypeptide chain making
up an enzyme molecule

Tertiary structure–irregular folding of a
polypeptide chain into a globular shape

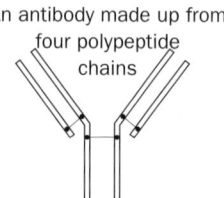

An antibody made up from
four polypeptide
chains

Quaternary structure–association
of more than one polypeptide chain

Many of the properties of **proteins** can be explained in terms of their shape. Structural **proteins** are made from long parallel chains which form fibres or sheets. In **globular proteins** the tertiary structure is important and gives them their shape. This shape is important in determining how they function as, for example, **enzymes** or **antibodies**.

protein synthesis: the process by which **proteins** are built up from **amino acids** using information carried by the **genetic code** on the cell's **DNA**. The diagram summarises the main features of the process in a **eukaryotic** cell.

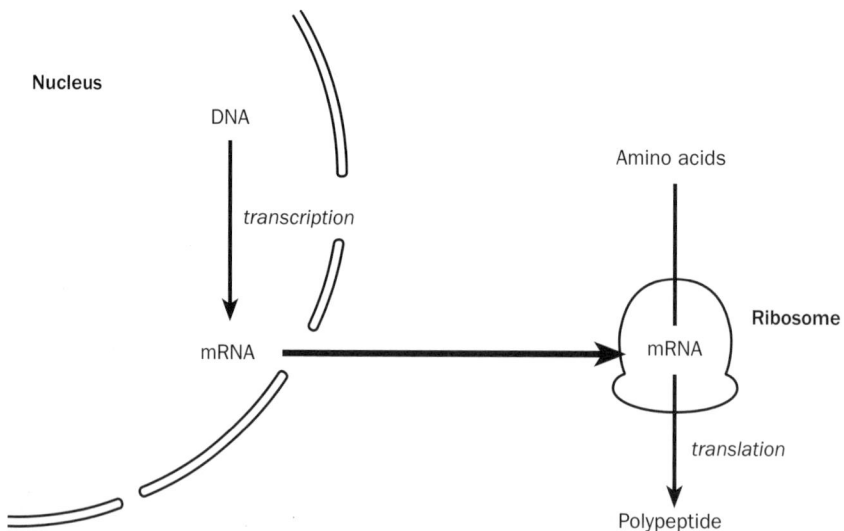

The main features of protein synthesis

The base sequence of a gene is first copied on to messenger RNA (**mRNA**) in the nucleus of the cell. This mRNA moves out into the **cytoplasm** of the cell. Here the code on the mRNA molecule is used to control the production of a polypeptide chain by a ribosome. (See also *genetic code*, *transcription* and *translation*.)

proteome: all the *proteins* found in an organism. *Proteomics* is the study of proteomes.

proteomics: the study of the structure and function of the *proteins* in an organism. Proteomics is a complex subject because proteins in an organism differ with time and from one cell to another.

prothallus: a structure that forms the *gametophyte* generation of a fern plant.

prothrombin: a substance involved in *blood clotting*. Prothrombin is an inactive form of the *enzyme thrombin*. It is converted to thrombin in the presence of a number of substances or factors, some of which are only produced when tissue is damaged.

Protoctista: the *kingdom* containing the *eukaryotic* organisms that cannot be classified as members of the *Fungi*, *Animalia* or *Plantae* (see *classification*). It contains a variety of different organisms ranging from those that are one-celled to algae such as seaweeds. It is not possible to give a list of features shared by all members of this kingdom. All that can really be done is to say that a eukaryotic organism belongs to the Protoctista if it does not belong to one of the other kingdoms.

protogyny: this occurs when the female part of a flower matures first. It is one of the ways in which *self-fertilisation* can be prevented in flowers that have both male and female organs. In buttercup flowers, for example, the *stigmas* are receptive before the *anthers* mature and pollen is shed.

proto-oncogene: a gene found in healthy cells that is important in regulating cell division. A proto-oncogene may mutate (see *mutation*) and become an *oncogene*. Oncogenes may stimulate cells to divide in an uncontrolled way and cause cancer to develop.

protoplast: the actively metabolising part of a cell. In animal cells, the protoplast consists of the whole cell. In plant cells it is the part of the cell that is surrounded by the *cell wall*.

protoxylem: the first *xylem* to be formed in a young plant. Its walls are thickened with *lignin* that is laid down in rings and spirals. This allows protoxylem to bend and stretch, important because at this stage the plant is still growing rapidly.

protozoa: a group of single-celled organisms that belong to the kingdom *Protoctista*. It includes free-living forms like amoeba as well as a number of important parasites such as *Plasmodium*, which causes *malaria*, and *Trypanosoma*, responsible for sleeping sickness.

proximal convoluted tubule: see *first convoluted tubule*.

proximal tubule: see *first convoluted tubule*.

pteridophyte: a member of the plant phylum that contains the ferns. Pteridophytes share the following features:
- Members of this phylum show an *alternation of generations*. The dominant stage in the life cycle is a *sporophyte*, but there is also a free-living *gametophyte* called a prothallus.
- The young leaves of ferns are tightly coiled. As they grow, they uncoil.
- Mature leaves usually have spore-bearing structures called sporangia on the undersurface.

puberty: the stage in human development when the sex organs mature and first produce *gametes*. Because of difficulties in recognising the first occurrence of these events, it is usually considered as involving the time over which the changes in the body that lead to sexual maturity take place. Puberty is controlled by **hormones**. There is a gradual increase in the amount of luteinising hormone, **LH**, secreted by the anterior lobe of the **pituitary gland**.

In females, the increase in LH brings about an increase in the **oestrogen** secreted by the **ovaries**. The increase in oestrogen leads to development of **secondary sexual characteristics** such as the growth of breast tissue, the broadening of the pelvis and the growth of pubic hair.

In males, the increase in LH leads to increased **testosterone** secretion by the testes. This stimulates the development of secondary sexual characteristics such as the deepening of the voice and growth of facial and body hair.

Puberty takes place later in the life of humans than in that of any other mammal. This may have been important in human evolution as it prolongs the period of childhood dependency and allows greater opportunity for learning from the parents. In humans, there has been a recent trend towards earlier puberty. This is particularly apparent in more developed countries and can probably be explained by improved nutrition.

pulmonary artery: the **blood vessel** that takes the blood from the right **ventricle** of the heart to the **lungs**. There is a difference between the pulmonary **artery** and other arteries in the body because it contains deoxygenated blood.

pulmonary vein: the **blood vessel** that takes blood from the **lungs** to the left **atrium** of the heart. It is the only vein in the body that contains oxygenated blood.

pulmonary ventilation: the volume of air that a person breathes in in one minute. It is also called the minute ventilation rate. To work out pulmonary ventilation, we need to know how much air is taken in with each breath when a person is breathing steadily. This is the tidal volume. We also need to know how many breaths are taken each minute. This is the ventilation rate. Pulmonary ventilation can then be calculated using the equation:

$$\text{pulmonary ventilation} = \text{tidal volume} \times \text{ventilation rate}$$

pulse: the surges in pressure that are caused by the **ventricles** contracting and pumping blood through an **artery**. The pulse is easily felt where an artery runs close to the surface of the body.

pupil: the opening in the **iris** that allows light to enter the eye. In bright light, circular muscle round the iris contracts. This makes the pupil smaller and reduces the amount of light passing through. In dim light, radial muscles contract. This makes the pupil larger and increases the amount of light entering the eye.

purine: one of the two chemical families of nucleotide bases (see **base**) found in a **nucleic acid** molecule. **Adenine** and **guanine** are the purines found in both **DNA** and **RNA** molecules. A purine contains two rings of **atoms**.

Purkinje tissue: see **Purkyne tissue**.

Purkyne tissue: the specialised heart tissue that carries the **nerve impulse** that triggers contraction of the **ventricles**. The **bundle of His** runs through the wall between the right and left ventricles, from the **atrioventricular node (AVN)** to the base of the ventricle. This bundle splits into finer and finer branches. These form the Purkyne tissue, which conducts the impulse to the rest of the ventricle muscle.

pyramids of number, biomass and energy: various quantitative ways of describing the relationships between different *trophic levels* in an *ecosystem*. As the names suggest, they refer to the numbers of individual organisms, the *biomass* and the energy transferred through the different trophic levels. If sufficient information is provided, it is possible to construct them to scale; in most cases, however, it is sufficient to indicate the relative proportions. The chart shows the shape of these pyramids for three simple *food chains*.

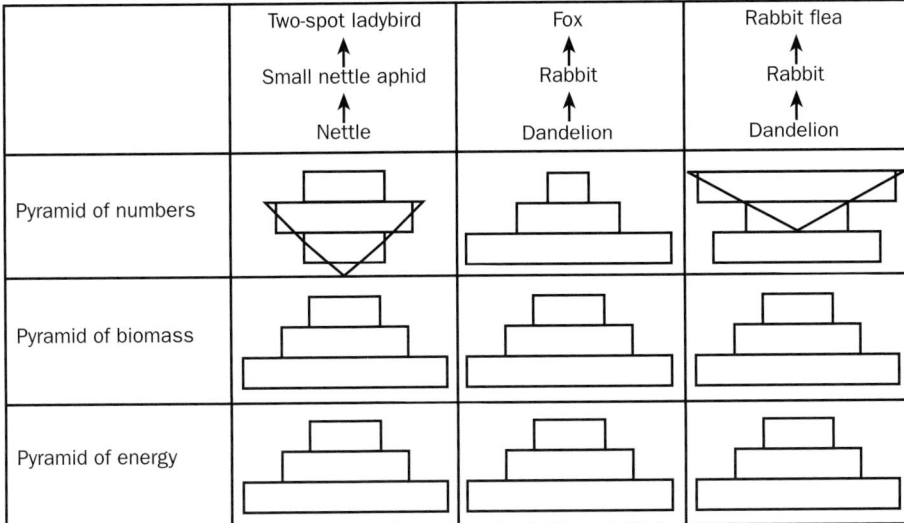

The shapes of pyramids for three simple food chains

You should note the exceptions to the pyramid shape with pyramids of numbers. The shape is inverted where the producer is large in size compared to the *primary consumers* that feed on it. It is also inverted where small parasites are feeding on a larger host.

A pyramid of numbers and a pyramid of biomass refer to the organisms that are present at a particular time. A pyramid of energy, on the other hand, refers to the total amount of energy that is transferred through the different trophic levels over a period of time. It is because of this that pyramids of energy are always pyramid shaped.

pyrimidine: one of the two chemical families of *nucleotide bases* found in a *nucleic acid* molecule. *Cytosine* and *thymine* are the pyrimidines found in a molecule of *DNA*. Those found in *RNA* are cytosine and *uracil*. A pyrimidine contains a single ring of *atoms*.

pyruvate: a three-carbon substance formed in *glycolysis*. Glycolysis is the first part of the respiratory pathway. In glycolysis, a molecule of *glucose* is broken into two three-carbon pyruvate groups. When oxygen is present, pyruvate is converted to acetylcoenzyme A in the *link reaction*. In *anaerobic respiration* pyruvate is converted either to *lactate* or to ethanol and carbon dioxide.

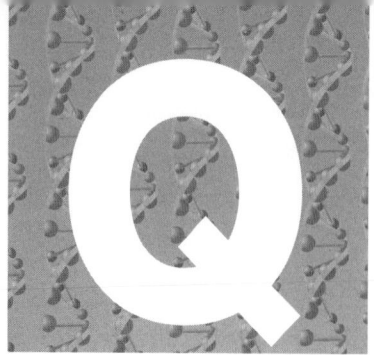

Q_{10} (temperature coefficient): the increase that occurs in the rate of a process when the temperature is increased by 10°C. At temperatures between 0°C and approximately 40°C, most **enzyme**-controlled reactions will have a temperature coefficient of approximately two. In simple terms, this means that a 10° rise in temperature will cause the rate of an enzyme-controlled reaction to double. A simple formula allows the temperature coefficient to be calculated from appropriate information.

> A solution containing the enzyme catalase and hydrogen peroxide gave off 63 cm³ of oxygen in one minute at 25°C. At 15°C, a second solution containing an identical mixture of catalase and hydrogen peroxide produced 29 cm³ of oxygen. Calculate the Q_{10} of this reaction.
>
> $$Q_{10} = \frac{\text{rate of reaction at } (T + 10)°C}{\text{rate of reaction at } T°C}$$
>
> $$= \frac{63}{29}$$
>
> $$= 2.1$$

Processes that do not rely on enzymes also take place in living organisms. These also increase in rate with an increase in temperature. However, the increase is not as much and the Q_{10} has a value of about 1.2 for processes such as **diffusion**.

quadrat: a sample area marked out to study the organisms that it contains. Quadrats are used in ecological investigations to obtain quantitative information about the distribution of animals and plants. They are frequently used to record the vegetation in a fairly uniform area. In using, the first step is to place quadrats at random so that the results will be representative of the area studied. Once the quadrat is in place, information can be collected concerning the species of organism present. **Frequency** and **percentage cover** are commonly used measures. Other studies, such as those involving the investigation of **succession** or the effects of grazing, make use of permanent quadrats that may stay in place for many years.

quaternary structure (protein): involves a **protein** being made up from more than one polypeptide chain. Two examples of **proteins** with quaternary structures are the **haemoglobin** and **antibody** molecules found in the blood of a mammal. Both of these molecules contain four polypeptide chains. Some **enzyme** molecules are even more complex. One of those involved in respiration, for example, contains 42 separate polypeptide chains.

quota: the total amount of fish that may be taken from a particular fishery in a given time. Quotas are one of the ways that are used to conserve stocks of fish.

r selected species: species that live in *habitats* which are short-lived and unpredictable. Because of this, mortality rates are often high. An r selected species:

- produces many young, each of which has a low probability of surviving to become an adult
- matures early and has a short generation time
- disperses its young widely.

Examples of r selected species are insects, weeds and small mammals such as mice.

rabies: a viral disease that affects the *central nervous system* of mammals including humans. If a person has been bitten by an infected animal, daily injections of rabies anti-serum may prevent the disease from developing. The anti-serum contains antibodies against the rabies virus. It is, therefore, an example of *passive immunity*.

radial symmetry: organisms that could be cut in more than one plane to produce a mirror image. This is because their parts are arranged in a circle around a line which goes down through the centre. Most radially symmetrical animals are *sessile*. They do not move from one place to another unless they are simply being carried by water currents. It is an advantage for such animals to be radially symmetrical as the stimuli to which they respond may come from any direction. Members of the phylum *Cnidaria* are sessile, radially symmetrical animals. In addition to animals, plants are also sessile and show radial symmetry to some extent.

radicle: part of an embryo plant found in a seed. When *germination* occurs, the radicle usually emerges first and grows downwards. It is the root of the embryo plant.

Ramapithecus: an extinct *primate*. When scientists first studied the fossilised *bones* of *Ramapithecus* they suggested that it was more closely related to modern humans than were gorillas and chimpanzees. This led to the suggestion that humans split away from gorillas and chimpanzees over 9 million years ago. Later, other scientists used different techniques to study the relationships between modern humans, gorillas and chimpanzees. Their evidence suggested that the split took place about 5 million years ago. This led to them looking more carefully at the fossil evidence. The work with *Ramapithecus* shows some of the problems of interpreting the fossil evidence for human evolution.

random assortment: see *independent segregation*.

random sampling: the taking of samples in a way that avoids bias. Random sampling is particularly important in ecology when placing *quadrats*. If samples are not taken at random there is always a danger that the results will be affected by deliberate choice. Throwing quadrats, whether with the eyes shut or not, does not result in a genuinely random distribution. You should use a method such as that shown in the diagram overleaf.

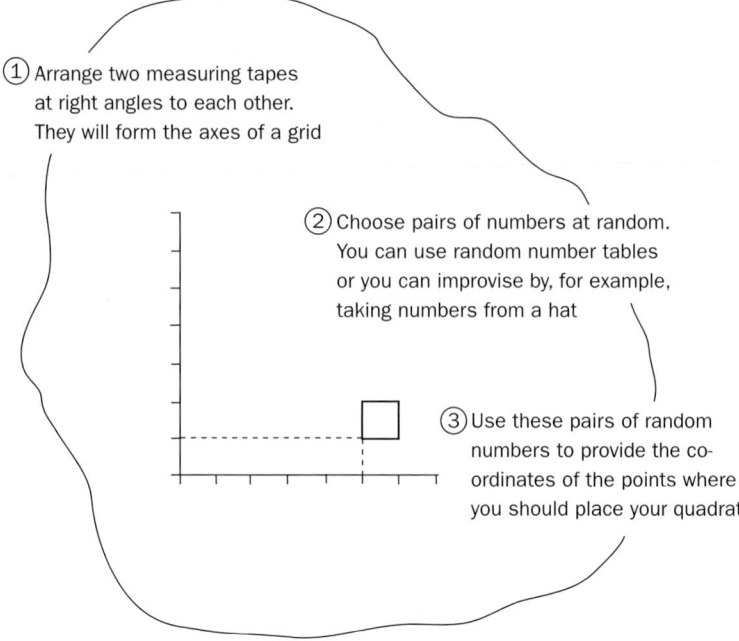

① Arrange two measuring tapes at right angles to each other. They will form the axes of a grid

② Choose pairs of numbers at random. You can use random number tables or you can improvise by, for example, taking numbers from a hat

③ Use these pairs of random numbers to provide the co-ordinates of the points where you should place your quadrat

Random sampling

receptor: a cell that detects changes in its *environment*. There are many different types of receptor, but they have some similarities:

- they are specific. There are basically six different sorts of stimulus. These are produced by chemical substances, electrical fields, light, magnetic fields, mechanical movement and temperature. A particular receptor will only respond to one kind of stimulus. A temperature receptor in the skin, for example, will not respond to pressure or to light
- they are very sensitive. *Rod cells*, for example, can detect a single photon of light
- they are often grouped together to form complex sense organs that are able to magnify or amplify the stimulus. In the eye, for example, this is helped by the *lens* and the way in which a number of *rod cells* join to a single nerve cell in the *optic nerve*
- they convert sensory information into a form that can be transmitted to the *central nervous system*.

recessive: an *allele* is recessive if it is only expressed in the *phenotype* of a diploid organism when the other allele of the pair is identical. In peas, the allele for yellow pods, g, is recessive to that for green pods, G. As the allele for yellow pods is recessive, only pea plants with the *genotype* gg will have yellow pods.

recombinant: a cell that contains a different combination of *genes* or *DNA* from that found in its parent cell. Under natural conditions, recombinants often arise during *meiosis* as a result of *crossing over*. During the first stage of meiosis, the *chromosomes* come together in their homologous pairs, each chromosome consisting of a pair of *chromatids*. These chromatids may break and rejoin with other chromatids. This results in the genes on one part of a chromatid combining with the genes which were present on part of another chromatid. If one of the chromatids concerned originally came from the mother, while the other came from the father, there would be a new combination of genes produced and recombinants would be formed.

Recombination also occurs as a result of **genetic engineering**. A particular ge[ne is] isolated from one organism and a vector used to insert this gene into the DNA[of another] organism. This produces recombinant DNA.

rectum: the last part of the **alimentary canal**. The rectum stores **faeces** befo[re they are] passed out of the body through the **anus**.

red blood cell: a cell that transports respiratory gases in the blood. When seen with a light microscope, human red blood cells (or erythrocytes) are biconcave discs 7 to 8 μm in diameter. Although each red blood cell contains large amounts of the oxygen-carrying pigment **haemoglobin**, it has no nucleus or **mitochondria**. Red blood cells are found in large numbers. There are about 5 million in each cubic millimetre of human blood. Their actual number varies from one person to another and may be estimated with a specially marked slide called a **haemocytometer**.

Red blood cells are produced in the **bone** marrow and their formation is controlled by a **hormone** called erythropoietin, secreted by the **kidney**. Red blood cells do not contain either nuclei or ribosomes, so they are unable to make new **proteins**. Partly because of this, they have a limited life span. At the end of approximately 4 months they are broken down by the **liver** and other organs. Some of the material they contain, such as the **iron** found in the haemoglobin, is recycled. The rest is excreted and forms substances such as **bile pigments**. Red blood cells, together with the haemoglobin they contain, have three important functions:

- Oxygen combines reversibly with the haemoglobin in red blood cells to form oxyhaemoglobin. Each molecule of haemoglobin can transport up to four molecules of oxygen from the **lungs** to the respiring tissues.
- Red blood cells are also important in the transport of carbon dioxide. An **enzyme**, **carbonic anhydrase**, is found inside red blood cells. This catalyses the reaction between carbon dioxide and water that produces carbonic acid. Some carbon dioxide is able to combine with haemoglobin to produce carbaminohaemoglobin. Just over 20% of the carbon dioxide transported by the blood is carried in this form.
- Haemoglobin is an important **buffer** and prevents the accumulation of hydrogen ions in the blood. This keeps the pH of the blood constant.

reduction: a chemical reaction involving the gain of **electrons**. Since the electrons that are gained must come from somewhere, reduction of one substance is always accompanied by **oxidation** of another. Reduction is an important process, occurring during both respiration and **photosynthesis**.

reducing sugar: a sugar that produces a positive result, an orange-red precipitate, when heated with Benedict's solution. A reducing sugar contains either a free aldehyde group or a free ketone group. It is the presence of one or other of these chemical groups that enables it to reduce Benedict's solution. All **monosaccharides** and most **disaccharides** are reducing sugars. Sucrose is an exception. It is a **non-reducing sugar**.

reflex: a simple behaviour pattern involving a rapid, automatic response to a **stimulus**. If a person accidentally puts his hand on a hot object, he quickly pulls it away. This is a reflex action. Receptors detect the stimulus. They send **nerve impulses** along **sensory neurones** to the **spinal cord**. An impulse from the spinal cord brings about muscle contraction and

...emoval of the hand. Because the nervous pathway involved in a reflex involves few neurones, reflexes allow rapid responses to be made.

Although reflexes are automatic with the same stimulus always producing the same response, there are times when the conscious part of the nervous system can modify or even override a reflex. It is possible, for example, to grasp a hot object and maintain the grip even though it causes considerable pain.

Reflexes play an important part in:

- protecting the body and avoiding damage by removing the part concerned rapidly from the source of danger
- making rapid adjustments to external stimuli. The presence of food on the tongue, for example, stimulates the flow of *saliva*, and a bright light reduces the diameter of the pupil in the eye. Reflexes associated with balance also come into this category
- enabling homeostatic adjustments to be made. Reflexes that involve the blood and breathing systems bring about the adjustments which are continually required to enable these systems to adapt to the changing requirements of the body.

reflex arc: the nervous pathway associated with a *reflex*. The simplest type of reflex arc in a vertebrate involves just two *neurones*, a *sensory neurone* and a *motor neurone*. This pathway can be represented as:

stimulus → receptor → sensory neurone → motor neurone → effector → response

An example of a simple two-neurone or monosynaptic reflex is the knee-jerk reflex. This is important in helping to maintain posture.

Other reflexes involve the presence of additional neurones. The impulse is conveyed from the sensory neurone to the motor neurone by way of intermediate or relay neurones. An example of a reflex arc which has a pathway like this is the one involved in the rapid removal of the hand from a hot or sharp object.

refractory period: the short recovery period that occurs immediately after the passage of a *nerve impulse* along the *axon* of a nerve cell. The sodium *ion-channel proteins* in the membrane are closed during the refractory period and the negative *resting potential* is re-established inside the axon. During this stage, these ion-channel proteins cannot open again, even if a much bigger stimulus is given. In other words, the membrane cannot be depolarised and a new *action potential* cannot be initiated. This has two important consequences:

- The refractory period separates nerve impulses from each other. Because it takes from 1 to 10 milliseconds, it limits the number of impulses that can be transmitted along an axon in a given time.
- Nerve impulses only pass in one direction along an axon. They go from an active region to a resting region. They cannot go the other way because the membrane immediately behind the impulse cannot be depolarised again.

relative growth rate: see *growth rate*.

relay neurone: a nerve cell that carries *nerve impulses* from a *sensory neurone* to a *motor neurone*. A relay neurone is sometimes called an intermediate neurone.

renal: an adjective meaning 'to do with the *kidney*'. The renal *artery*, for example, is the *blood vessel* supplying oxygenated blood to the kidney. The renal vein returns blood from the kidney.

renal capsule: the first part of a *nephron* or *kidney* tubule. It is a funnel-like structure which surrounds a ball of *capillaries* called a *glomerulus*. It consists of an outer layer of cells that is separated by a space from an inner layer of specialised cells called *podocytes*. The glomerulus and renal capsule are concerned with the first stage in the process of *urine* formation. It is here that *ultrafiltration* takes place. The renal capsule is also called Bowman's capsule.

replication: see *DNA replication*.

reproduction: the production of offspring. It is a process in which genetic material is passed from the parent or parents to a new generation. There are two main types of reproduction. These are *asexual* reproduction, in which a single individual gives rise to genetically identical offspring, and *sexual* reproduction, which involves the fusion of genetic material, usually from different organisms.

reproductive cloning: cloning used to produce an animal that has the same nuclear *DNA* as another animal. *Somatic cell* nuclear transplantation involves removing the nucleus from a cell such as a skin cell. This nucleus is inserted into an egg cell from which the genetic material has been removed. When this egg cell develops into a new animal, the new animal will be genetically identical to that from which the somatic nucleus was taken.

reproductive isolation: two populations are reproductively isolated when they are prevented from breeding with each other or when the hybrid offspring they produce fail to survive. It is convenient to divide the mechanisms involved into two categories.

Those that prevent organisms breeding and *fertilisation* taking place are examples of prezygotic reproductive isolation. For example, the chiffchaff and the willow warbler are small insect-eating birds that spend the summer in Britain. They are so similar to each other in appearance that it is difficult to tell them apart. However, their songs, which form an important part of courtship behaviour, are completely different. Since successful mating will only result from the correct behaviour pattern, chiffchaffs and willow warblers are reproductively isolated.

Postzygotic reproductive isolation occurs when the offspring that result from fertilisation between individuals from the two parent populations are at a disadvantage. The two populations are reproductively isolated because the hybrid offspring do not survive as well. There are two common species of poppy in Britain: the common poppy and the long-headed poppy. Pollen from one species can be transferred to the other and fertile seeds result. However, these hybrid seeds do not have mechanisms that limit their *germination* to favourable conditions. Many germinate in unsuitable areas and die. The hybrids, therefore, do not survive, so the two species are reproductively isolated.

residual volume: the amount of air that cannot be expelled from the *lungs* however deeply a person breathes out. (See also *lung capacities*.)

resistance: the ability of a pest, for example, to withstand the action of a particular pesticide. In a population of houseflies, there will be variation. A few of the flies may possess *alleles* that make them resistant to a particular *insecticide*. If, however, that insecticide is not in use, there will be no selection involved, either for or against the allele producing resistance. It is therefore likely that the allele will remain in small numbers in the population. If the insecticide is now introduced and used against the flies, those that are resistant will be at an advantage. They will be more likely to survive and will pass on their alleles to the next generation. In this way, a population of insecticide-resistant flies may arise. This has happened in many parts of Europe. Other important examples of resistance include:

- resistance associated with the control of *malaria*. Not only are many species of malarial mosquito resistant to the insecticides that were once used to control them but, in many parts of the world, the malarial parasite is resistant to drugs used to prevent and control the disease
- the development of resistance in rats to the poison warfarin, which was once used very successfully in their control
- the emergence of *MRSA* and strains of the *tuberculosis (TB)* bacterium that are resistant to all *antibiotics* at present in use.

resolution: the ability to distinguish between points that are close together. Take the example of a photograph that shows a crowd of people at a football match. It is impossible to make out the detail of individual faces. If you want more detail, you might think that the solution would be to take a small area of the photograph and blow it up. Unfortunately, when you do this, all that happens is that, instead of having a small picture where you cannot see the detail, you have a large picture in which you cannot see the detail. Magnification is only of real use if it is accompanied by greater resolution allowing more detail to be seen. The main problem with a light microscope is that the wavelength of visible light limits the detail that can be seen. The resolving power of a good light microscope is about 0.2 µm so, if two objects are closer than this, they will only be seen as one. If greater detail is required, it is necessary to use an *electron microscope*. With this instrument, a beam of *electrons* is used. It produces much greater resolution. Modern electron microscopes have a resolution of about 1 nm.

respiration: the biochemical pathway that takes place in cells and results in the release of energy from organic molecules. This energy is usually used to form *ATP*, although some is always lost as heat. The process of respiration is often summarised by the equation:

$$C_6H_{12}O_6 + 6O_2 \rightarrow 6H_2O + 6CO_2 + energy$$

It is important to appreciate that this formula is an oversimplification of what actually happens. *Glucose* is only one of a number of different substances that can act as a *respiratory substrate*. The process can take place either in the presence of oxygen (*aerobic respiration*) or in its absence (*anaerobic respiration*). If different respiratory *substrates* are used, the waste products will also vary. The only feature common to all respiratory pathways is the release of energy.

respiratory arrest: cessation of breathing. Respiratory arrest may occur at the same time as *cardiac arrest*. Emergency action involving *cardiopulmonary resuscitation (CPR)* may be necessary to avoid permanent damage to the *brain* caused by a shortage of oxygen.

respiratory pigment: a substance found in animals that is involved in the transport of oxygen and, in some cases, carbon dioxide. **Haemoglobin** is an important respiratory pigment found in the blood of mammals and many other animals. Another respiratory pigment, **myoglobin**, is associated with muscle where it acts as an oxygen store.

respiratory quotient (RQ): the amount of carbon dioxide produced by an organism in a given time divided by the amount of oxygen used in the same time. Different **respiratory substrates** produce different values for the respiratory quotient, so its calculation gives an indication of the substrate being respired. The table shows the RQ for some important respiratory substrates.

Respiratory substrate	RQ
Triglyceride (fat)	0.7
Amino acids and proteins	0.9
Carbohydrates	1
Anaerobic respiration of carbohydrate	> 1

There are, however, several different interpretations of, for example, an RQ of 0.9. The organism might be respiring proteins, but a mixture of fats and carbohydrates could give the same value. Care must be taken in interpreting the figures obtained in a particular situation.

Calculating the respiratory quotient from a chemical formula.

The equation represents the respiration of a typical fat:

$$2C_{51}H_{98}O_6 + 145O_2 \rightarrow 102CO_2 + 90H_2O$$

$$RQ = \frac{\text{amount of carbon dioxide produced in a given time}}{\text{amount of oxygen used in the same time}}$$

or, we could write this in another way:

$$RQ = \frac{\text{number of molecules of carbon dioxide produced}}{\text{number of molecules of oxygen used}}$$

So, from the formula, this would be

$$RQ = \frac{102}{145}$$

$$= 0.7$$

respiratory substrate: the organic substance that is being used for **respiration**. Oxygen is not a respiratory **substrate**.

respiratory tree: the branching system of the airways in a lung. Air passes from the mouth and nose into the **lungs** through the **trachea** or windpipe. The trachea divides into two main bronchi. These then divide into smaller bronchi, then into **bronchioles** and finally into **alveoli**.

resting potential: the difference in electrical charge across the **cell-surface membrane** of a nerve cell. There is a higher concentration of K^+ ions and a lower concentration of Na^+ ions inside a nerve cell than outside. The cell-surface membrane allows K^+ ions to diffuse out very

241

readily. It is much more difficult, however, for Na⁺ ions to diffuse in. This means that a large number of K⁺ ions leave the cell but this is not matched by the number of Na⁺ ions entering. Consequently, there are fewer positive ions inside the cell than outside. In other words, in a resting nerve cell, there is a potential difference across the membrane of about −60 mV. This is the resting potential. During the passage of a *nerve impulse*, a small number of Na⁺ ions enter the cell. A sodium pump in the membrane actively transports these Na⁺ ions back out so that the resting potential is re-established. (See also *action potential*.)

restriction endonuclease: one of a group of *enzymes* that cut *DNA* molecules at specific points along their lengths. Restriction endonucleases are found naturally in many bacteria, where they are thought to help in protecting them from *viruses* by cutting up the DNA of the virus. Each restriction endonuclease recognises a specific base sequence in a DNA molecule and will only cut it at this point. These enzymes are important tools in *genetic engineering*. They can be used as molecular scissors to isolate, for example, an individual gene from a length of DNA. By using a particular restriction endonuclease, a cut can be made at exactly the right points.

GTT AAC	GTTAAC
CAA TTG	CAATTG

Restriction enzyme *Hpa* 1 recognises this base sequence and cuts it as shown by the dotted line

Restriction enzyme *Eco* R 1 recognises the same base sequence but cuts it in a different way

Two examples of restriction endonucleases

restriction enzyme: see *restriction endonuclease*.

restriction fragment length polymorphism (RFLP): one way in which the sequence of bases in *DNA* varies between individuals that belong to the same species. These polymorphisms can be detected by cutting DNA into short lengths with *restriction endonucleases* and analysing the results with *electrophoresis*. In some individuals, a particular fragment will be short. In others it contains base sequences repeated many times so it will be longer. Because restriction fragment repeat polymorphisms are inherited in the same way as other *alleles*, they can be used as marker genes to investigate inheritance.

restriction mapping: a technique that uses *restriction endonucleases* to find out more about the structure of a particular piece of *DNA*. Restriction endonucleases cut DNA at particular sequences of bases called recognition sites. Restriction mapping involves cutting the piece of DNA that is being investigated into pieces with a restriction endonuclease. The pieces are separated from each other by *electrophoresis*. The distance between the different recognition sites can now be measured and a restriction map drawn showing where the restriction sites appear on the piece of DNA. More information can be obtained by repeating this procedure with different restriction endonucleases.

retina: the light-sensitive layer on the inside of the eye. The retina contains two sorts of light-sensitive cell: *rod cells* and *cone cells*. Rod cells are arranged more or less uniformly over the entire surface of the retina, although there are few present in the region known as the *fovea*. They are very sensitive and respond to light of extremely low intensities. Rods cannot, however, distinguish colour. Cone cells are fewer in number and are concentrated at the fovea. They are not as sensitive to low light intensities as rod cells, but they are able to distinguish colour. The retina also contains many **sensory neurones** with which rods and cones form **synapses**. These sensory neurones allow some sorting out and processing of information from the rods and cones before it is sent via the **optic nerve** to the **brain**.

retinal: part of a *rhodopsin* molecule. A molecule of rhodopsin consists of a protein called **opsin** and retinal. Retinal is formed from vitamin A. There are two forms of retinal: cis retinal and trans retinal. When light strikes a molecule of rhodopsin, cis retinal is converted to trans retinal and the rhodopsin molecule is broken down into its two parts. The chemical events associated with this lead to **nerve impulses**. They carry sensory information along the **optic nerve** to the **brain**. Before a **rod cell** can be stimulated again, trans retinal is converted back to cis retinal and rhodopsin must be regenerated. This requires **ATP**.

retinal convergence: the way in which groups of **rod cells** form **synapses** with a single nerve cell in the **retina**. This arrangement helps to make the eye sensitive to low light intensity. The stimulus affecting a single rod cell may not be strong enough to trigger a **nerve impulse** but the combined effects of a weak stimulus affecting a number of rod cells add together to produce an impulse in the nerve cell. This arrangement contrasts with that in **cone cells**, where each cone cell synapses with an individual nerve cell. Cone cells are less sensitive to low light intensities but can discriminate much better between separate stimuli.

retrovirus: a virus that has **RNA** as its **nucleic acid**. When it infects a cell from its host, it is able to make a **DNA** copy of this RNA using an **enzyme** called **reverse transcriptase**. This DNA copy then becomes incorporated into the genetic material of the host. **HIV** is an example of a retrovirus. Others are thought to be involved in the development of certain cancers.

reverse transcriptase: an **enzyme** that produces a **DNA** molecule from the corresponding **mRNA**. Reverse transcriptase enzymes are found naturally in certain viruses called **retroviruses**. These are viruses in which the genetic information is carried on an **RNA** molecule. When one of these viruses infects a host cell, it uses this enzyme to make a complementary DNA (cDNA) copy of its genetic information which is then incorporated into the host DNA. Reverse transcriptase is one of the enzymes used in **genetic engineering**, where it is used to obtain a copy of a particular gene from the relevant mRNA. This has a number of advantages.

- A cell that is making a particular protein will only contain two copies of the relevant piece of DNA, one on each **chromosome**. These are transcribed to produce large amounts of mRNA. It is easier to locate the relevant mRNA than the DNA.
- The piece of DNA making up a gene is a lot longer than the piece of mRNA produced from it. This is because it contains some base sequences, called **introns**, which do not code for **amino acids**. When the mRNA is produced, it is edited so that these non-coding sequences are removed.

R_F value: a ratio used to identify substances separated from each other by **chromatography**. It is calculated by dividing the distance moved by the unknown substance by the distance moved by the solvent front:

$$R_F = \frac{\text{distance moved by the substance}}{\text{distance moved by the solvent front}}$$

The actual distance moved by a particular substance depends on how long it has been left. If the apparatus has only just been set up, the substance is unlikely to have travelled far. If it has been left for some time, it will have moved farther. Because of this we cannot use the distance moved by the substance to identify it. The R_F value, however, can be used because the ratio involved will be the same, however long the chromatogram has been running.

RFLP: see *restriction fragment length polymorphism (RFLP)*.

rhesus factor: an antigen that may be present in the *cell-surface membrane* of human red blood cells. It forms the basis of the rhesus blood group system. Most people are rhesus positive (Rh+). They have the antigen on their red blood cells. People who are rhesus negative (Rh−) do not have this antigen in their blood.

A woman who is rhesus negative becomes pregnant with her first child. Suppose that the fetus is rhesus positive. Some fetal blood may leak across the placenta into the woman's blood during birth. Because she has now been exposed to the rhesus antigen, she produces anti-rhesus *antibody*.

Suppose this woman becomes pregnant again and the fetus is again rhesus positive. Anti-rhesus antibody can now cross the placenta from the mother's blood to the blood of the foetus. These antibodies will destroy the red blood cells of the fetus. The consequences are obviously serious. A blood test early in pregnancy allows suitable precautions to be taken to ensure the safety of the fetus.

Rhizobium: a *genus* of nitrogen-fixing bacteria found free-living in the soil and in the *root nodules* of *leguminous plants*.

rhizoid: a thin, root-like structure found in primitive plants such as mosses and liverworts. The main function of rhizoids is to anchor the plant and, in this way, they are similar to roots. However, each rhizoid consists of a single cell or, in some cases, several cells. Rhizoids do not contain either *xylem* or *phloem*.

rhodopsin: a light-sensitive pigment found in the *rod cells* of the eye. A molecule of rhodopsin consists of two parts: *opsin*, which is a protein, and *retinal*, which is derived from vitamin A. When a photon of light strikes a molecule of rhodopsin, the pigment is broken down into its two parts in a process called bleaching. The chemical events associated with bleaching lead to *nerve impulses* that convey sensory information along the *optic nerve* to the *brain*. Before it can be stimulated again, rhodopsin must be regenerated from opsin and retinal. This requires *ATP*, which is produced in the *mitochondria* present in rod cells. In bright light, nearly all the rhodopsin in the eye has been bleached. The eye is said to be light adapted. If a person moves from brightly lit conditions into dim light, little can be seen as there is not enough rhodopsin present. Only after about half an hour in dim light conditions will enough rhodopsin have been generated for things to be seen. The eye is now said to be dark adapted.

ribonucleic acid: see *RNA*.

ribose: the five-carbon sugar or pentose found in *RNA* and *ATP*.

ribosomal RNA: see *rRNA*.

ribosome: a small *organelle* that is important in synthesising *proteins*. It is made of a mixture of protein and a special sort of *RNA* called ribosomal RNA (*rRNA*). During *protein synthesis*, a molecule of messenger RNA (*mRNA*) carries the *genetic code* for a particular protein from the nucleus into the *cytoplasm* of the cell. Ribosomes now move along this mRNA molecule. This results in *amino acids* being assembled in the correct order to form a molecule of the protein concerned.

There are two basic types of ribosome. Those found in the cytoplasm of *eukaryotic* cells are slightly larger than those which are found in *prokaryotic* cells.

ribulose bisphosphate (RuBP): the substance that combines with carbon dioxide in the *light-independent reaction* of *photosynthesis*. Ribulose bisphosphate is a five-carbon sugar that has two phosphate groups in its molecule. It combines with carbon dioxide. This results in the formation of two molecules of the three-carbon compound, *glycerate 3-phosphate (GP)*. Glycerate 3-phosphate is then reduced to triose phosphate. Some of the triose phosphate is used to produce *carbohydrates* such as *glucose* and *starch* while the rest produces more ribulose bisphosphate in a cycle of biochemical reactions called the *Calvin cycle*.

Rio Convention on Biodiversity: an international treaty that aims to develop national strategies for sustaining biological diversity. It has three aims:
* *conservation* of biological diversity
* sustainable use
* fair sharing of the benefits of genetic diversity.

RNA: a type of *nucleic acid*. There are a number of different forms of RNA. Two of these are important in *protein synthesis*. During *transcription*, part of the *DNA* in the nucleus unwinds. It acts as a template for the formation of messenger RNA (*mRNA*). This takes the code for the gene concerned from the nucleus into the *cytoplasm*. A second type of RNA, transfer RNA (*tRNA*) is found in the cytoplasm. This molecule is important in assembling *amino acids* in the correct position on the mRNA during the process of translation. These two RNA molecules are both *polynucleotides* like DNA but they differ from each other and from DNA in a number of ways. These are summarised in the table.

Feature	mRNA	tRNA	DNA
Number of polynucleotide strands in the molecule	1	1	2
Shape of molecule	Single strand	Single strand twisted around itself	Double strand twisted to form a helix
Name of sugar present	Ribose	Ribose	*Deoxyribose*
Name of bases present	*Adenine, cytosine, guanine, uracil*	Adenine, cytosine, guanine, uracil	Adenine, cytosine, guanine, thymine
Bases linked by *hydrogen bonds*	No	Yes	Yes

In some *viruses*, for example HIV, the genetic material in the virus is RNA.

RNA polymerase: an *enzyme* important in the process of *transcription*. In this process, one of the DNA strands acts as a template for the formation of a molecule of messenger RNA (*mRNA*). The bases of free mRNA *nucleotides* in the nucleus line up against the complementary bases on the DNA chain. They now join to form an mRNA molecule. RNA polymerase is the enzyme that helps to join these nucleotides together to produce the mRNA molecule.

rod cell: a type of light-sensitive cell found in the *retina* of the eye. Rod cells occur in enormous numbers. There are about 120 million in a human eye. Each rod cell consists of two parts: an outer segment and an inner segment. The outer segment has large numbers of flattened membranous vesicles that contain a light-sensitive pigment called *rhodopsin*. When a photon of light strikes a molecule of rhodopsin, the pigment is broken down chemically in a process known as bleaching. Bleaching leads to *nerve impulses* which take sensory information along the *optic nerve* to the *brain*. The inner segment contains the nucleus of the rod cell as well as large numbers of *mitochondria*). These provide the *ATP* necessary for regenerating the rhodopsin after it has been bleached.

Rod cells are unable to distinguish colour but they are extremely sensitive to low light intensities. The retina contains large numbers of bipolar neurones as well as receptors like the rod cells. A number of rod cells *synapse* with a single one of these neurones. This arrangement, known as *retinal convergence*, is largely responsible for the great sensitivity of rod cells to low light intensity.

root: the part of a plant that holds it in the soil and is involved in the uptake of water and mineral salts. The main external difference between roots and stems is that roots do not have either leaves or buds. There are differences in the internal structure as well. The *xylem* and *phloem* form a core that runs down the middle of the root rather than a ring towards the outside. This makes it able to withstand the pulling forces that are exerted as it anchors the plant in the soil.

The other important function of the root is that it is concerned with the uptake of mineral ions and water. The concentration of most mineral ions is higher inside the cells of the root than in the surrounding soil. Movement of these ions into the root therefore involves *active transport* and results in the root cells having a lower *water potential* than the surrounding soil. Water flows down this water potential gradient into the root cells by *osmosis*.

It is not possible to identify a structure as a root just because it is found below ground. Potato tubers, for example, are modified stems, even though they grow in the soil. This can be confirmed by the presence of tiny scale leaves, each of which is associated with a cluster of buds. There are some roots which arise above ground level, while some unusual rainforest climbing plants even have upwardly growing roots that absorb the water and mineral salts which drip down from the canopy. Some roots are important storage organs and are commercially important. These include carrots, turnips and sugar beet.

root hair: an outgrowth from a cell in the *epidermis* of a *root*. Root hairs are usually found in young roots, in the region just behind the root tip. They provide a large surface area through which the root can absorb water and mineral ions.

root nodule: a small swelling on the root of a *leguminous plant* that results from infection with nitrogen-fixing bacteria such as *Rhizobium*.

root pressure: a force generated by the roots of some plants that pushes water and mineral ions up the *xylem*. It can be demonstrated by cutting off the top of an actively transpiring plant such as a tomato plant. Water oozes from the cut surface of the stem and may continue to do so for some time. The mechanism responsible involves *active transport* of mineral ions into the xylem of the root. There is good evidence for this since the process is halted by respiratory poisons or a lack of oxygen. Accumulation of mineral ions in the xylem leads to a lower *water potential* inside the vessels. Water then enters the xylem from the surrounding tissue by *osmosis*.

rough endoplasmic reticulum: see *endoplasmic reticulum*.

roundworm: a member of the *phylum* Nematoda. Members of this phylum are unsegmented worms. There are many parasitic species of roundworm. Some of these affect humans. Others, such as the potato eelworm, are important crop pests.

RQ: see *respiratory quotient (RQ)*.

rRNA: ribosomal ribonucleic acid or ribosomal RNA, a type of *RNA* found in *ribosomes*. The sequence of *nucleotides* in a ribosomal RNA molecule is encoded by *DNA*. The genes that code for rRNA molecules in different species of organisms are similar and have changed little over a long period of time. Because of this, these *genes* are often used to investigate the evolutionary relationships between different species.

rubella: a viral disease that mainly affects children. Its symptoms include swollen *lymph glands* in the neck and a pink rash. If a pregnant woman is infected with the virus, it can cross the placenta and damage the developing fetus. There is a high probability that the child will be born with congenital rubella syndrome. Symptoms of this condition often include:
- deafness
- eye abnormailities that may include *cataracts*
- congential heart disorders.

It is important, therefore, that girls are vaccinated against the disease before puberty.

rubisco: an *enzyme* that is involved in the *light-independent reaction* of *photosynthesis*. It catalyses the reaction in which a molecule of carbon dioxide is taken up by a molecule of ribulose bisphosphate (RuBP) to give two molecules of *glycerate 3-phosphate (GP)*.

RuBP: see *ribulose bisphosphate (RuBP)*.

rumen: a chamber found at the front of the gut in some mammals. *Microorganisms* live in this chamber and they are able to digest *cellulose*. The relationship between these microorganisms and their hosts is an example of *mutualism*. The microorganisms gain as they receive a constant supply of nutrients, while the mammal benefits because it receives the products formed from the *digestion* of cellulose as well as *amino acids* from digestion of the microorganisms. The diagram overleaf summarises the digestive processes that take place in the rumen of a cow.

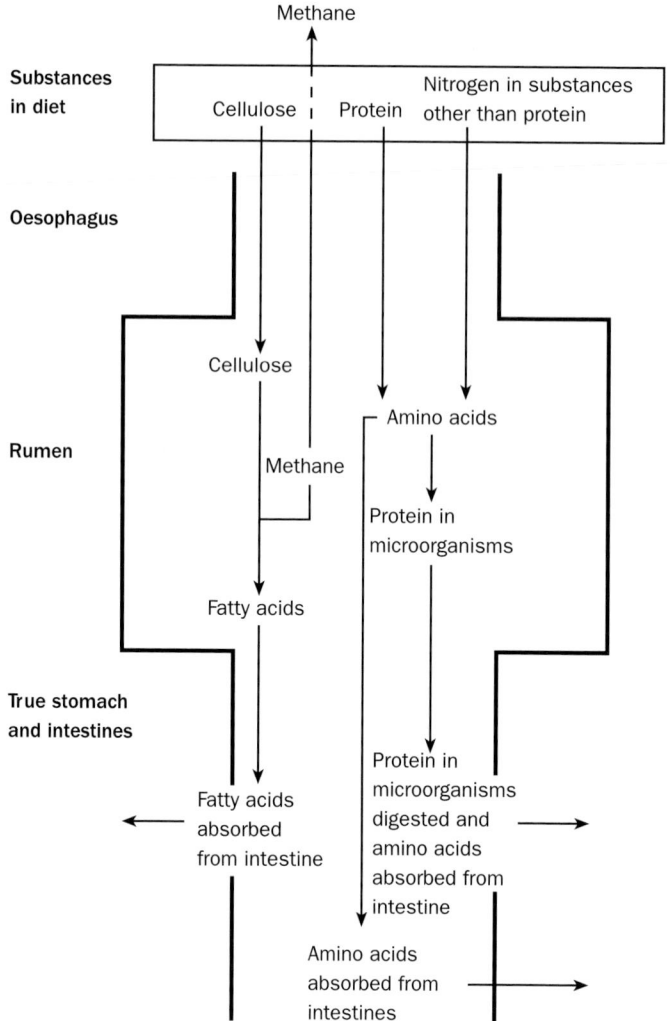

Digestion in the rumen

ruminant: a mammal that possesses a large chamber at the front of the gut called a *rumen*. *Microorganisms* live in the rumen and are able to digest *cellulose*. Ruminants have a complicated way of feeding. Grass and other plant material is swallowed and goes into the rumen. Here, it is mixed with mucus and microorganisms start the process of *digestion* of the cellulose which the food contains. This mixture can be brought back up the *oesophagus* into the mouth where it is chewed as the 'cud'. The cud is then swallowed. It goes down into the *stomach* proper and on through the rest of the gut, completing the normal processes of digestion and *absorption*. Many important domestic animals, including sheep and cattle, are ruminants. As some of the energy in the food they take in is used by the bacteria, they are not such efficient energy converters as non-ruminants like pigs. They can, however, live on foods that few other animals can use. In addition to cattle and sheep, wild mammals such as deer and antelope are ruminants.

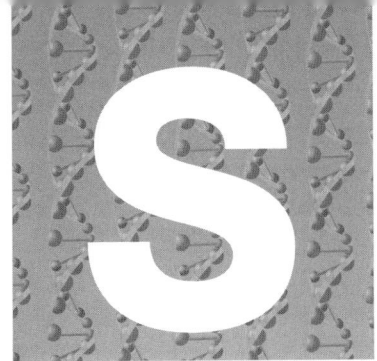

S phase: a stage in the *cell cycle*. It occurs during interphase and is the time when *DNA* is replicated.

saliva: the digestive juice produced by *salivary glands* that open into the mouth. In humans, saliva contains large amounts of *mucus* and an *enzyme*, *salivary amylase*. Salivary amylase starts the process of *starch digestion*. The secretion of saliva is controlled by a *reflex* involving the nervous system. The stimulus is the presence of food in the mouth and it is detected by receptors which are the taste buds. The *salivary glands* respond by increasing the production of saliva. The Russian physiologist Pavlov demonstrated that a *conditioned reflex* was involved as well. The components of saliva differ from animal to animal. *Anticoagulants*, for example, are present in the saliva of blood-sucking animals such as vampire bats and leeches.

salivary amylase: a *starch*-digesting *enzyme* produced in the *saliva*. Salivary amylase is a *hydrolase*. It hydrolyses starch to *maltose*.

salivary gland: one of the *glands* that secretes *saliva*. In humans, there are three pairs of salivary glands. The secretion of saliva is controlled by a *reflex* involving the nervous system. The presence of food in the mouth provides the stimulus. It is detected by the taste buds. A *conditioned reflex* is involved as well.

Salmonella: a *genus* of rod-shaped *bacteria* that cause food poisoning. Food poisoning by *Salmonella* is most likely to arise from food that has been kept in conditions that allow large numbers of bacteria to grow. Once contaminated food has been eaten, the bacteria enter the gut and multiply rapidly. Toxins are released as a result of the breakdown of bacterial cells and these cause gastroenteritis. The symptoms of *Salmonella* food poisoning affect the person concerned 12 to 72 hours after consumption of the contaminated food. The resulting sickness and diarrhoea are often accompanied by fever. In some people, even though the symptoms may have disappeared, bacteria remain in the body and continue to be found in the *faeces*. Such people are known as carriers and can easily contaminate food. Other species of *Salmonella* cause typhoid and paratyphoid fevers.

salmonellosis: an infection of the digestive system by *Salmonella* bacteria.

saltatory conduction: the way in which *nerve impulses* travel along myelinated nerves. The *axons* in many mammalian nerves are surrounded by a fatty material called *myelin*. This does not, however, form a continuous layer. There are small gaps between each length of myelin called *nodes of Ranvier*. Because of the insulating properties of the myelin, it is only at these nodes that the ion movements that result in the next *action potential* can occur. The nerve impulse therefore travels along in a series of jumps from one node to the next. This is saltatory conduction and leads to impulses moving much more rapidly than in non-myelinated nerves.

S

salting: a method of preventing food spoilage by adding dry salt. *Microorganisms* cannot survive in an *environment* that has a high concentration of salt. Water will be drawn out of their cells by *osmosis* and they will die.

sampling: examining a small area or a small number of organisms rather than the whole area or the entire population. In many ecological studies it is necessary to take samples as it would be much too time consuming to examine the whole area. When sampling, you should bear the following points in mind.

- Although limited by the time available, the sample should be as large as possible. Small samples are often unrepresentative of the area or population as a whole.
- Samples must be taken at random or the results will not be representative. They are likely to be biased and statistical tests cannot be used.

SAN: see *sinoatrial node (SAN)*.

saprobiont: an organism that obtains its nutrients from dead or decaying organic matter. Many *bacteria* and fungi are saprobionts. They secrete digestive *enzymes* on to the surface of their food. These enzymes break down complex molecules such as those of *starch* and *protein* into smaller, soluble ones which are then absorbed. Saprobionts play an extremely important role in *nutrient cycles*. They break down organic matter and release simple inorganic molecules such as carbon dioxide and ammonia.

saprophyte: see *saprobiont*.

saprotrophic nutrition: the way in which *saprobionts* obtain their nutrients from dead or decaying matter.

sarcolemma: the membrane that surrounds a muscle fibre.

sarcomere: one of the basic structural units of which skeletal muscles are composed. A sarcomere is a section of a *myofibril* between *Z lines*.

sarcoplasm: the *cytoplasm* in a *skeletal muscle* fibre. It is similar to the cytoplasm in other cells except that it contains *myoglobin* and large amounts of *glycogen*.

saturated fatty acid: a *fatty acid* with a long hydrocarbon tail in which there are no double bonds between the carbon *atoms*. Saturated fatty acids have higher melting points than fatty acids in which there are double bonds. *Triglycerides* which contain saturated fatty acids therefore tend to be solid at temperatures of around 20°C. They are called *fats* and are found mainly in animals.

scanning electron microscope: a type of *electron microscope* that works by reflecting a beam of *electrons* off the surface of the object being examined. It produces a three-dimensional image which shows the surface in great detail.

Schistosoma: a fluke that lives inside *blood vessels* in the walls of the intestine and the bladder. *Schistosoma* is a *parasite* and has many adaptations for its mode of life:

- The adult fluke has suckers that attach it to the wall of a blood vessel.
- The adult fluke is coated itself with substances from its human host. This prevents it being attacked by the host's immune system.
- Male flukes are larger than female flukes. The female fluke lives inside a groove on the surface of a male. This gives permanent contact between male and female and increases the probability of *fertilisation*.
- It has a complex life cycle involving water snails as secondary hosts. This is an adaptation that enables infection of another individual.

Schwann cell: a cell that produces the fatty material called *myelin* which surrounds the *axons* of many mammalian nerve cells. Each Schwann cell wraps round and round the axon, covering it in a layer of myelin. Because these cells occur next to each other along the axon, there are small gaps between the myelin produced by one cell and the myelin produced by the next. These gaps are known as the *nodes of Ranvier*. This arrangement is very important in allowing the fast passage of *nerve impulses*, which jump along from node to node, a process known as *saltatory conduction*.

sclerenchyma: a type of supporting tissue found in plants. Sclerenchyma cells have *cell walls* that are thickened with *lignin*. They do not have living contents. There is no *cytoplasm* present, for example. Two types of sclerenchyma are found. One type consists of long, thin cells tapered at both ends. These are known as fibres and, in the case of crops such as hemp and jute, may be of considerable economic importance. A second type has shorter, more rounded cells which are called sclereids. These form hard protective layers around nuts and other seeds.

sebaceous gland: a gland in the skin of a mammal that opens into a hair follicle. It secretes an oily substance called *sebum*. Sebum plays an important role in making hair and skin waterproof and also has an antibacterial effect.

sebum: an oily substance produced by *sebaceous glands* in the skin of a mammal.

second messenger: a substance found inside a cell that responds to the presence of a *hormone* outside the cell by activating a particular enzyme. The first molecule to be identified with this function was *cyclic AMP*. It is produced when a hormone such as *adrenaline* binds with a receptor site on the *cell-surface membrane* of a cell from the target tissue. Cyclic AMP then activates the *enzymes* that control the biochemical pathways associated with the conversion of glycogen to *glucose*. A single hormone molecule leads to the production of many molecules of cyclic AMP inside the cell. In turn, each of these messenger molecules activates many enzyme molecules and each enzyme molecule affects large numbers of *substrate* molecules. This is referred to as the *cascade effect* and illustrates a second function of cyclic AMP, which is to amplify the effect of the hormone. Other second messengers have now been discovered and among the most important are Ca^{2+} ions.

secondary immune response: the response shown by the immune system when a person is exposed to a particular *antigen* on a second or later occasion. Suppose a person is infected with the virus that causes measles. When appropriate *B cells* come into contact with measles viruses, they divide rapidly and produce two sorts of cells: *plasma cells* and *memory cells*. The plasma cells produce and secrete *antibodies*. The memory cells remain in the blood system. If the same person is exposed again to measles viruses, the memory cells respond rapidly. They multiply and produce more plasma cells. This produces the secondary immune response and usually kills the pathogen before the symptoms of the disease are produced.

secondary metabolite: a substance that is produced by *microorganisms* as they age, rather than when they are still actively growing. When microorganisms are grown in a culture, they soon enter a phase of rapid growth. During this stage, enzyme-controlled reactions produce the substances required for growth. These substances are called *primary*

251

metabolites. Secondary metabolites are substances that are usually produced later in the life cycle when the microorganisms age. *Antibiotics* such as penicillin are examples of useful secondary metabolites that are produced by *biotechnological* processes.

secondary oocyte: a stage in the formation of a mature female *gamete* in a mammal. Secondary oocytes are formed from primary oocytes by the first division of *meiosis*. A secondary oocyte forms shortly after *ovulation*. The second division of meiosis now starts but it stops before it is complete. Only when the secondary oocyte is fertilised, is the second meiotic division completed and a mature female gamete formed.

secondary sexual characteristic: one of the distinguishing features between male and female animals that develops at the time of sexual maturity. In humans, secondary sexual characteristics include the growth of breast tissue, the broadening of the pelvis and distribution of subcutaneous *fat* in females, and the characteristic deepening of the voice and growth of facial and body hair in males. These characteristics develop as a result of the increase in the secretion of the *hormones oestrogen* and *testosterone* at puberty.

secondary structure (protein): the way in which the chain of *amino acids* that make up a polypeptide is coiled or folded. An α helix is formed when the chain is coiled in a spiral shape. Each amino acid in the chain is linked to others by *hydrogen bonds*. Many structural *proteins* are coiled in this way. An example is keratin, a *fibrous protein* found in hair. Another common type of secondary structure is the β-pleated sheet. In this structure, parts of the polypeptide chain run parallel to each other. These are held by hydrogen bonds and form flat sheets. The protein fibroin is the main component of silk. It has molecules which form β-pleated sheets. Because the amino acids in these sheets have been pulled into an extended form, silk cannot be stretched. The secondary structure of different proteins varies but is always determined by the *primary structure*.

secondary succession: *succession* on an area of bare ground left after plant cover has been removed. Secondary succession may take place after a fire or a flood.

secondary thickening: growth of a plant that results in an increase in the thickness of the stem and the roots. There are cells in the root tip and the shoot tip of a young *dicotyledon* that form a *meristem*. These cells divide and elongate and the plant grows taller. Secondary meristems such as the cells that form a layer between the xylem and the phloem now produce new supporting tissue and new conducting cells. This causes the stem to become thicker and is called secondary thickening. Secondary thickening makes up most of the tissue in a mature plant.

seed: the structure that develops from a fertilised ovule in a *flowering* plant. A seed usually contains an embryo and a store of food material. This food store may be in the *cotyledons*, which form part of the actual embryo, or it may be another tissue, the *endosperm*. The embryo and its food store are surrounded by a protective outer layer called the *testa*. There are a number of advantages to plants that are associated with the production of seeds. These include:
- protection of and a supply of nutrients for the developing embryo
- provision of a supply of nutrients that allows the seed to remain dormant. This may enable it to survive harsh conditions between growing seasons, such as winter weather or periods of low rainfall
- an effective means of dispersal. Seeds, or the *fruits* that contain them, often have adaptations associated with dispersal. The seeds of soft fruits, such as blackberries and tomatoes, have outer layers that are resistant to the *enzymes* in the gut of the

animal which eats them. They can therefore be transported long distances, from the plant where they were eaten to the place where they are eventually deposited in the *faeces* of the animal concerned.

seed bank: a collection of seeds used as a way of banking or preserving plant genes. A suitable collection is made containing seeds from as many different varieties and areas as possible. These are kept in cool, dry conditions and will remain alive for many years. Seed banks are therefore very useful in the *genetic conservation* of plants, as each seed contains all the genes that are present in the adult plant. Not only do seeds take up much less room than if the whole plant had to be grown, but they also allow stocks of plants to be held in disease-free conditions. Unfortunately, the seeds of many commercially important tropical species such as rubber and cocoa cannot be stored under these conditions. The only possible way of conserving their genes is either to grow the actual plants or to use *tissue culture*.

selection: the process which results in the best-adapted organisms in a *population* surviving, reproducing and passing their genes on to their offspring. As a population increases in size, many factors in its *environment* become limiting. There may be *competition* for food in animals, or for light, water and mineral salts in plants. Some of the organisms in the population will have *alleles* which mean that they are better adapted to these conditions. These will survive and breed, passing these alleles on to the next generation. This will produce a change in the proportion of alleles in the *gene pool* and may result in evolution. The flow chart shows how *antibiotic*-resistant bacteria may have evolved as a result of selection.

Patient infected with bacteria. These bacteria vary. A few of them will be resistant to a particular antibiotic

↓

Patient treated with the antibiotic. The bacteria that are not resistant die; those that are resistant survive

↓

The resistant bacteria have less competition. They multiply and pass on the resistance gene

↓

Selection has operated. The population has changed from bacteria that were nearly all killed by the antibiotic to bacteria which are mainly resistant to it

There are two basic ways in which selection can operate. These are *directional selection* and *stabilising selection*.

selective breeding: improving a particular variety of crop plant or domestic animal by breeding from organisms with desired characteristics. Plants and animals have been subject to selective breeding since they were first domesticated around 10 000 years ago. Early farmers, for example, tended to harvest and save the seed from cereal plants in which the grains had not fallen from the stalks. These would have provided the seed grain for the following year's crop. The plants that grew from this grain would have inherited the genes for stronger grain attachment. Selective breeding would have led to cultivated cereal plants that did not scatter their grains before the farmer had time to harvest them.

One of the first things that accompanied domestication in animals was a reduction in body size. Smaller cattle, sheep and goats resulted partly from selective breeding. Small parent animals would have been more likely to have produced small offspring. Selective breeding has been a continuous process carried out over many thousands of years. Modern examples include the selective breeding of maize plants to increase the oil and protein content of the grain and of dairy cattle to increase milk yield.

self-fertilisation: *fertilisation* where both male and female *gametes* come from the same organism. Self-fertilisation is *sexual reproduction* because it involves the fusion of gametes, even if they do come from the same organism.

self-pollination: *pollination* in which the pollen comes from the same plant. This should be compared with *cross pollination* where the pollen comes from a different plant. It is quite common among *flowering* plants and has an advantage over cross pollination because it clearly increases the probability of male and female *gametes* meeting. Because the pollen comes from the same individual, however, self pollination results in less genetic variation among the offspring.

semen: the sperm-containing fluid that is produced by male animals. Sperms are produced by the *testes* and stored in the epididymis, a mass of ducts just outside the testis. When a male mates or copulates with a female, some of these sperms are mixed with secretions from various *glands* in the male reproductive system to form a liquid called semen. As well as the sperms themselves, semen contains various substances that are needed to nourish and stimulate them. The volume of semen and the number of sperms it contains differ from animal to animal. In a bull, between 2 and 10 cm^3 of semen containing up to two thousand million sperms per cubic centimetre are produced. Clearly, there are far more sperms than is necessary to fertilise a single egg cell. This makes it possible to collect bull sperm and dilute it for use in artificial insemination. In this way, the semen collected on a single occasion can be used to fertilise many cows.

semi-conservative replication: the way in which a molecule of *DNA* can produce two exact copies of itself. The original molecule unwinds and each of the resulting strands acts as a template for the formation of a new chain. (See also *DNA replication*.)

semilunar valve: a valve that prevents the back-flow of blood. When the muscles of the *ventricle* wall relax during the *cardiac cycle*, the pressure in the left ventricle falls. As it drops below that in the *aorta*, the semilunar valves between the aorta and the ventricle shut and prevent blood in the *artery* from flowing back into the ventricle. A similar mechanism prevents blood from the *pulmonary artery* from flowing back into the right ventricle. Blood in *veins* is at a low pressure and its return to the heart may be helped by semilunar valves. Veins returning blood from the limbs are situated between large muscles. When these muscles contract they squeeze the blood in the veins. Semilunar valves in the walls allow blood to be squeezed towards the heart but do not allow it to go back in the other direction.

seminiferous tubule: one of the long coiled tubes in the *testis* of a mammal. *Spermatogenesis* takes place in these tubules.

sensory neurone: a cell that carries *nerve impulses* from a sense organ to the *central nervous system*.

sepal: one of the segments that forms the calyx of a flower. The sepals often enclose the flower bud.

seral stage: a recognisable stage in a plant *succession*. The diagram summarises some of the stages in succession on sand dunes. Each of the seral stages shown in the diagram is made up of a distinctive *community* of organisms.

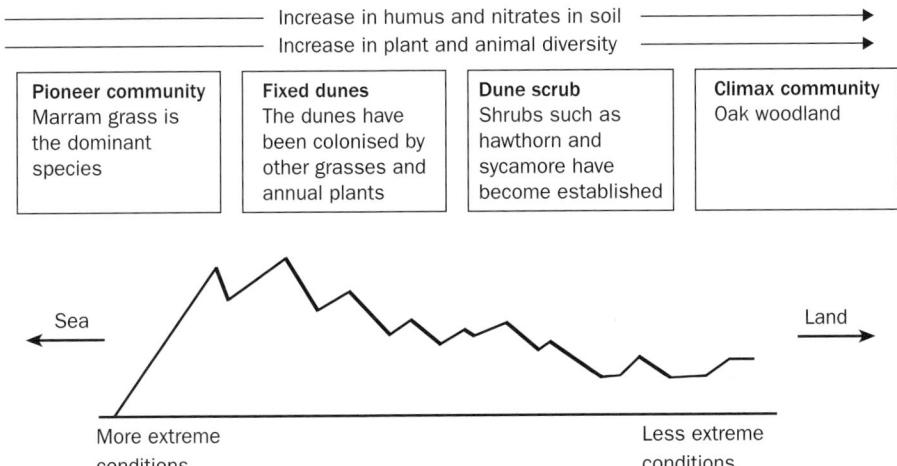

sere: a particular example of a plant *succession*. It is the complete sequence of *communities* from the *pioneer species* stage to the *climax community*.

serial dilution: a technique in which a stock solution or culture is diluted by a known amount each time. This produces a series of dilutions. Serial dilutions are useful in investigations where concentration is the independent variable. They are also useful in microbiological investigations when a scientist wants to count the number of *microorganisms* present in a culture.

serosa: a protective and supportive layer surrounding the *small intestine*.

serotonin: a *neurotransmitter* found in *synapses* in the *central nervous system*. Low concentrations of serotonin may be associated with depression.

serum: the liquid left after blood is allowed to clot and the clot removed. Serum is very similar to *plasma* in its composition. The main difference is that it does not contain *fibrinogen* or some of the other substances involved in *blood clotting*.

sessile: sessile organisms are those that stay in one place. Plants depend on light, water and carbon dioxide in order to photosynthesise. They also require a number of inorganic ions. In suitable areas, these are all available in plentiful supply. A plant can therefore exploit these resources effectively by remaining in the same place.

Animals face slightly different problems. For most of them, food is distributed unevenly. Therefore, it is an advantage for most animals to be motile. The few that are sessile, such as members of the phylum *Cnidaria*, tend to feed on small organisms that float in water currents. They are *filter feeders*. Sessile organisms have adaptations associated with their habit:

- The stimuli to which they respond may come from any direction. Therefore they are usually radially symmetrical (see *radial symmetry*).
- Many sessile organisms have special adaptations involved in the transfer of *gametes* between individuals during *sexual reproduction*.

- There are difficulties involved in colonising new areas. Plants have *fruits* that are often transported considerable distances from the parent. Many sessile animals have young stages that are free-swimming or drift in water currents.

sex chromosome: one of the chromosomes whose presence determines the sex of an organism (see *sex determination*). In humans and other mammals, females have two *X chromosomes* and males have one X chromosome and one *Y chromosome*. Although X and Y chromosomes differ from each other in appearance, they can pair during *meiosis*. This is because part of each chromosome is identical, or *homologous*.

sex determination: in many organisms, sex is determined genetically. In mammals, for example, each body cell contains a pair of *sex chromosomes*. Females have two identical *X chromosomes* while males have one X chromosome and one *Y chromosome*. During *meiosis*, the two X chromosomes of the female can obviously pair with each other. In the male, the X chromosome is able to pair with the Y chromosome. This means that all the *gametes* produced by the female will be the same. They will all contain a single X chromosome. On the other hand, the gametes produced by the male will be different from each other. Half will contain an X chromosome and half will contain a Y chromosome. Since there is an equal probability that the sperm which fuses with the egg cell at *fertilisation* will have an X chromosome or a Y chromosome, it would be expected that equal numbers of males and females will be produced. The same basic mechanism is found in birds but here the male is the homogametic sex. The male bird possesses identical sex chromosomes and the sex chromosomes of the female are different. The female bird is described as the heterogametic sex as she can produce eggs containing either an X chromosome or a Y chromosome.

In addition to the methods of sex determination described above, there are a number of other ways of determining sex. In bees, for example, females are *diploid* and have two sets of chromosomes in each body cell. The males or drones are *haploid*. They only have a single set of chromosomes in each cell. In reptiles such as crocodiles and tortoises, sex is not determined genetically at all. It is the temperature at which the eggs are incubated that determines whether the young will be male or female.

sex linkage: a characteristic is sex linked if the gene that controls it is on one of the sex chromosomes. In most animals, the *Y chromosome* contains few if any genes. Sex-linked genes are, therefore, most likely to be found on the *X chromosome*.

An example of a sex-linked gene is that which controls the production of a pigment in the eye that enables discrimination between red and green. This gene has two *alleles*. The allele for normal vision, R, is *dominant* to that for red-green colour-blindness, r.

The *genotype* of a man with red-green colour-blindness is normally expressed as X^rY. This shows that, because he is male, he has one X and one Y chromosome. The Y chromosome does not contain an allele of this gene. The allele for red-green colour-blindness is r and this is situated on the X chromosome. Only one allele for this gene can be present in a male, as there is only one X chromosome, so he will be red-green colour-blind. The genotype X^RX^r indicates a female since there are two X chromosomes. She is heterozygous and, because the R allele is dominant to the r allele, she will have normal vision.

What are the results of a cross between a red-green colour-blind male and a female with normal vision?

Parental phenotypes:	red-green colour-blind male	female with normal vision
Parental genotypes:	X^{r}Y	X^{R}X^r
Gametes:	X^r Y	X^R X^r

Offspring genotypes:

		Male gametes	
		X^r	Y
Female	X^R	X^{R}X^r	X^{R}Y
gametes	X^r	X^{r}X^r	X^{r}Y

Offspring phenotypes:	X^{R}X^r	Female with normal vision
	X^{r}X^r	Red-green colour-blind female
	X^{R}Y	Male with normal vision
	X^{r}Y	Red-green colour-blind male

sexual reproduction: reproduction that involves the fusion of **gametes**. One of the most important features of sexual reproduction is that the process gives rise to the production of genetically different offspring. The diagram summarises the steps in the process that contribute to genetic variation in the offspring.

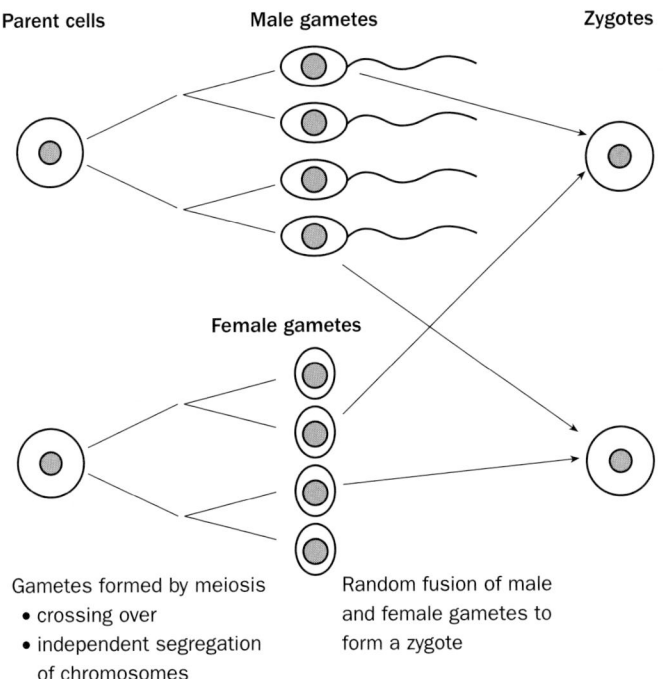

Sexual reproduction and variation

The production of genetically different offspring provides the raw material on which natural *selection* operates.

short-day plant: a plant that flowers when day length is less than a certain amount. Short-day plants include plants such as primroses that flower naturally in the spring as well as those such as chrysanthemums which flower in the autumn. Tropical plants such as poinsettias are also short-day plants.

sickle-cell disease: an inherited condition in which affected people have an abnormal type of *haemoglobin*. There is a slight difference in the amino acid sequence in two of the polypeptide chains that make up this abnormal haemoglobin. As a result, the abnormal haemoglobin crystallises in low concentrations of oxygen, and the red blood cells of an affected person collapse into a distinctive sickle shape. Sickle cells may clump together and block the flow of blood. They are also rapidly broken down, so the affected person suffers from severe *anaemia*. The condition is determined by a single gene with *codominant alleles*. This gives rise to the three *phenotypes* shown in the table.

Genotype	Phenotype	Consequences
Hb^AHb^A	Normal adult haemoglobin	Does not suffer from anaemia but is vulnerable to *malaria*
Hb^AHb^S	Both types of haemoglobin	Shows no symptoms of sickle-cell disease unless in conditions where there is severe oxygen shortage. Has some protection from malaria
Hb^SHb^S	Sickle-cell haemoglobin	Suffers severely from sickle-cell disease. Without medical help likely to die during childhood

The table shows that a person with the genotype Hb^AHb^S has a selective advantage in parts of the world where malaria is common. This is thought to have led to the high frequency of the Hb^S allele among people of African origin.

sieve tube element: a type of cell found in the *phloem* of a plant. Sieve tubes are made up from individual cells arranged end to end. The walls between these cells are perforated with small holes and form the characteristic sieve plates. Sieve tubes do not contain nuclei and have only a little *cytoplasm*.

silage: a grass crop that is fermented and used as food for farm animals, particularly for dairy cattle. The crop is cut and put into a plastic lined pit. It then undergoes anaerobic *fermentation*. Sugars in the grass are converted to acids such as acetic acid and butyric acid.

Simpson's index of diversity (D): a mathematical way of expressing species diversity. It is calculated from the equation:

$$D = 1 - \left(\Sigma \left(\frac{n}{N} \right)^2 \right)$$

where

N = total number of organisms of all species

n = total number of organisms of a particular species

Σ means 'the sum of'.

single circulation: a blood system in which the blood passes through the heart only once in its passage round the body. In fish, for example, blood flows from the single *ventricle* of the heart to the *gills*. After it has become oxygenated it goes to the rest of the body without going back to the heart again. A complete circuit therefore involves only passing once through the heart. A major disadvantage of a circulation like this is that blood is sent to the tissues at a very low pressure, much lower than is the case with a double circulatory system.

single nucleotide polymorphism (SNP): one way in which the sequence of bases in *DNA* varies between individuals that belong to the same species. Single nucleotide polymorphisms may occur in either coding or non-coding DNA. They involve a difference in a single *base pair*.

sink: part of an animal or plant to which transport systems take substances. These organisms are too large to rely on *diffusion* and have mass-flow systems which move substances rapidly from one place to another. Mass-flow systems link exchange surfaces with each other. At some of these exchange surfaces, substances are taken up (sources) while, at others (sinks), they are unloaded. The term sink is usually used when referring to transport by the *phloem* in *flowering* plants. Here, roots and other underground storage organs, buds and developing *fruits* provide examples of sinks.

sinoatrial node (SAN): a small area of muscle in the wall of the right *atrium* of the heart that controls the events of the *cardiac cycle* or heart beat. Heart muscle is not like muscle found in other parts of the body because it contracts and relaxes automatically. It does not need a *nerve impulse* to trigger this action. The sinoatrial node controls and coordinates contraction of the rest of the heart. An impulse spreads from here over the walls of the atria to the *atrioventricular node (AVN)*. This brings about contraction of the atria followed, after a short delay, by contraction of the *ventricles*. The rate at which the sinoatrial node sends impulses to the rest of the heart can be slowed down by nerve impulses coming from the *brain* along the *vagus nerve*. It can be speeded up by impulses reaching the node along the *cardiac accelerator nerve*.

sinusoid: a blood-filled space in the *liver*. The liver is made up of structural units called lobules. Each lobule consists of a number of vertically arranged sheets of cells surrounding a central vein. Blood flows from branches of the *hepatic artery* and *hepatic portal vein* through the spaces, or sinusoids, between the liver cells to the central vein. As it flows past the liver cells, the products of *digestion* are absorbed.

siRNA: see *small interfering RNA (siRNA)*.

skeletal muscle: muscle that, as a result of its contraction, usually brings about the movement of an animal. In mammals, antagonistic pairs of skeletal muscles (see *antagonistic muscles*) are attached to *bones*. They are under conscious control and, because of this, are sometimes called voluntary muscles. A muscle contains many *skeletal muscle fibres*. Muscle fibres are of two different types, *slow muscle fibres* which produce strong, sustained contractions, and *fast muscle fibres* which produce more rapid responses. Each individual muscle fibre is made up, in turn, of a large number of *myofibrils*. A knowledge of the structure of these myofibrils is important in understanding the *sliding-filament model* that explains the way in which a skeletal muscle contracts.

skeletal muscle fibre: one of the individual muscle fibres which make up a *skeletal muscle*. There are two types of muscle fibre, which differ from each other in their properties. These are fast or type 2 fibres and slow or type 1 fibres. The table overleaf summarises some of the important differences between them.

Fast skeletal muscle fibres	Slow skeletal muscle fibres
Produce rapid, powerful contractions, that makes them very important during locomotion.	Allow sustained muscle contractions. Play an important part in maintaining body posture
Muscle contains large amounts of stored glycogen, which acts as the *respiratory substrate*	Muscle does not contain much stored glycogen. *Fat* is often used as the respiratory substrate
Fast skeletal muscle fibres respire anaerobically	*ATP* produced by *aerobic respiration*
Lactate accumulates and an oxygen debt builds up rapidly. Fast skeletal fibres fatigue quickly	Do not fatigue quickly

skeleton: a supporting structure. Skeletons vary from the network of protein filaments that support individual cells, to the very complex systems in mammals. Skeletons have three main functions:

- they provide support. Animals that live on land have much stronger supporting structures than animals that live in the water
- they provide attachments for the muscles and act as a system of levers. This makes locomotion possible
- they have a protective function. In a mammal, for example, the skull protects the *brain* and the ribcage protects the organs in the thorax.

Three types of skeleton are found. An *exoskeleton*, such as that found in the *Arthropoda*, is on the outside of the body and encloses the soft internal structures. An *endoskeleton*, such as that found in mammals and other chordates, is inside the body. The third type of skeleton is the *hydrostatic skeleton*. This is based on fluid contained within the body and is a feature of soft-bodied animals such as worms.

skin: the outer covering of the bodies of mammals and other chordates. In a mammal, the skin consists of an outer *epidermis* made of dead cells. Underneath this is a living layer, the dermis. The dermis contains *blood vessels*, *glands* and nerve tissue. The skin has many important functions:

- it protects the body from injury and prevents the entry of pathogens
- hairs on the skin surface, blood vessels in the skin and sweat glands help to control body temperature
- it is an excretory organ, secreting large amounts of sweat.

sliding-filament model: a model that explains the way in which *skeletal muscle* contracts. The thick *myosin* filaments and the thin *actin* filaments which make up the *myofibrils* slide between each other. As a result, the muscle shortens. The mechanism which allows this to happen involves the formation of cross-bridges between the two types of filament. The head of a myosin molecule attaches to a particular place or attachment site on an actin molecule. The head then rotates and pulls the two filaments into each other. It finally detaches and goes back to its original position. This cycle of attaching, changing position and detaching is repeated many times. **Calcium** ions are important in initiating the process. When the muscle is stimulated to contract, the release of calcium ions makes the attachment sites on the actin molecules available so that the cross-bridges can form. The process of muscle contraction requires a large amount of energy in the form of **ATP**.

slow muscle fibre: see *skeletal muscle fibre*.

slow twitch muscle: see *skeletal muscle fibre*.

small interfering RNA (siRNA): a short double strand of **RNA** that interferes with the expression of a specific **gene** by, for example, preventing **mRNA** from producing a protein. Synthetic siRNA molecules are very important research tools because they can be used to shut down individual genes.

small intestine: part of the gut or **alimentary canal** between the **stomach** and the **large intestine**. It is a long coiled tube that is divided into two parts. The first part, into which **bile** and pancreatic juice are secreted, is called the **duodenum**. The second, much longer, part is the **ileum**. This is the main site of **absorption** of the products of **digestion**.

smallpox: an infectious viral disease that produces a high temperature and causes a rash of large, pus-filled spots. Vaccination has now completely eliminated smallpox.

smooth endoplasmic reticulum: see *endoplasmic reticulum*.

smooth muscle: see *involuntary muscle*.

SNP: see *single nucleotide polymorphism (SNP)*.

solute potential: the pressure produced by solute molecules in a solution. These molecules will be moving around at random and, as a result, some will collide with the membrane that surrounds them. They will exert a pressure on this membrane. This is the solute potential, the symbol for which is the Greek letter psi, ψ_s. Like **water potential**, solute potential has a negative value. The relationship between solute potential, ψ_s, water potential, ψ, and **pressure potential** , ψ_p, is be summarised by the equation:

$$\psi_s = \psi - \psi_p$$

somatic cell: a cell from the body of a multicellular organism, other than a sex cell. In a mammal, cells from the skin, the **lungs** and the intestines are all somatic cells. Each has a diploid number of **chromosomes**. **Gametes** on the other hand are **germ-line cells**. They have **haploid** chromosome numbers.

The technique of somatic cell nuclear transplantation involves removing the nucleus from a somatic cell such as a skin cell. This nucleus is inserted into an egg cell from which the genetic material has been removed. If this egg cell is inserted into the uterus of the same species of mammal, it will develop into a new animal. The new animal will be genetically identical to the animal from which the somatic nucleus was taken.

source: part of a plant or animal from which substances are taken into the transport system. These organisms are too large to rely on **diffusion** and have mass-flow systems which move substances rapidly from one place to another. Mass-flow systems link exchange surfaces with each other. At some of these exchange surfaces, substances are taken up (sources) while, at others, they are unloaded (sinks). A source is the name given to a site where substances are produced and loaded into the transport system. The term source is often used when referring to transport by the **phloem** in **flowering** plants. Here, a leaf is the most obvious example of a source.

spatial summation: see *summation*.

speciation: the formation of two or more new species from existing species. This can happen in one of two different ways. Two different species may cross to form a hybrid that is different from both of its parents. This sometimes happens in plants. Modern varieties of wheat, for example, originated from the accidental crossing of two species of wild grass to form a hybrid, an event that took place many thousands of years ago.

The other method of speciation occurs when a single original species gives rise to two new ones. This occurs when populations of the parent species become separated from each other. If the separation is due to some form of geographical barrier, this is known as *allopatric speciation*. On the other hand, if it occurs in two populations that occupy the same geographical area and some other factor prevents their interbreeding, it is *sympatric speciation*.

species: the starting point in the biological system of *classification*. A species is a group of similar organisms that are able to breed together and produce fertile offspring. At first sight it would appear easy to determine whether two organisms belong to different species. There are obvious differences between horses and donkeys. They can breed with each other but the offspring they produce, a mule, is sterile. Clearly, horses and donkeys belong to different species. However, there are examples where it is much more difficult to decide whether two organisms belong to the same or different species. Individual members of some species can look very different from each other; there may be considerable differences due to sex or geographical distribution. In addition, some species hardly, if ever, reproduce sexually, making it even more difficult to decide. Originally, classification of different organisms into species was based on visible features. More recent approaches draw on a wider range of evidence. This includes similarities and differences in *DNA* and *proteins*.

species diversity: a way of describing the number of species in a *community*. It takes into account how common or rare they are. We will take a sand dune as an example. It has very few species of plants growing on it, although one of them, marram grass, happens to be common. A sand dune has a low species diversity. On the other hand, a tropical rainforest has a huge variety of different species. It has a much higher species diversity. In general terms, the harsher the environment, the fewer the species that are present and the lower the species diversity. As with many ideas in ecology, it is useful to be able to give species diversity a numerical value. There are many different ways of doing this but one relatively simple one is to calculate an *index of diversity*.

sperm: a male *gamete* from a mammal or other animal. In mammals, sperm cells or spermatozoa are formed in the *testes* by the process of *spermatogenesis*. A mature mammalian sperm consists of a head containing a *haploid* nucleus. In front of the nucleus is the *acrosome*. This is a sac-like structure that contains *enzymes*. Just before *fertilisation* the membrane surrounding the acrosome bursts and these enzymes are released. Their digestive action helps to separate the cells which still surround the egg, making it possible for fertilisation to take place. A middle piece behind the head is packed with *mitochondria*. These provide the *ATP* necessary for the sperm to swim with the aid of its *flagellum*.

sperm bank: a collection of frozen sperm cells used as a means of banking or preserving animal genes. Sperms are collected and, after suitable treatment, deep frozen in liquid nitrogen. They can survive in these conditions for long periods of time and may be used to fertilise eggs many years later. As with **seed banks**, sperm banks enable a wide variety of genetic material to be stored in a convenient form. They are valuable in the **genetic conservation** of domestic animals and of species which are in danger of extinction. Considerably greater success has been achieved storing sperms than storing eggs or embryos.

spermatogenesis: the process in which sperms are formed. The **testes** in a mammal contain a large number of coiled tubules. The sperms are formed in these **seminiferous tubules**. Around the outside of each one is a layer of cells called the germinal **epithelium**. The cells of the germinal epithelium divide by **mitosis** and produce a large number of cells, each of which is capable of increasing in size then dividing by **meiosis** to produce four sperms. The basic sequence of mitosis, growth and meiosis is the same as in the formation of **gametes** in a female during **oogenesis**. There are, however, some important differences in the way in which meiosis actually takes place. The process by which sperms are formed is also similar to that leading to the formation of female gametes in that it is controlled by **hormones** released from the **pituitary gland**. The process of spermatogenesis produces large numbers of sperms. This increases the probability of a male gamete meeting and fertilising a female gamete.

spermatophyte: a member of the **Angiospermophyta**.

sphincter muscle: a ring of muscle which goes round the wall of a tubular organ. Its relaxation or contraction allows the organ to be opened or closed. There are a number of sphincter muscles along the gut of a mammal. The pyloric sphincter, for example, regulates the flow of the contents of the **stomach** into the **small intestine**, and the anal sphincter opens when **faeces** are eliminated from the body. Other sphincter muscles lie between many **arterioles** and the **capillaries** which they supply. Their opening and closing regulates the flow of blood to the organ concerned.

sphygmomanometer: an instrument that measures arterial blood pressure. A cuff is put round the upper arm and inflated. The pressure is then slowly released. Using a **stethoscope** to listen to the pulse in the wrist, the **systolic blood pressure** and **diastolic blood pressure** can be measured. The systolic pressure is the pressure when the **ventricles** are contracting. The diastolic pressure is the pressure when the ventricles are relaxing.

spinal cord: part of the **central nervous system** of a vertebrate that is enclosed in and protected by the backbone or vertebral column. A cross-section through the spinal cord shows that it is made up of two distinct regions. There is a central area which is shaped rather like the letter H. This is called the grey matter and consists of a mass of nerve cells. Surrounding this is the white matter, which contains the **axons** that transmit **nerve impulses** up and down the spinal cord. Apart from conducting impulses, the spinal cord plays an important part in coordinating many **reflexes**.

spinal nerve: pairs of nerves that come from the **spinal cord**. They pass between each of the vertebrae and go to different parts of the body. One is on the right, and one is on the left. Each of these spinal nerves has a **dorsal root** that contains **sensory neurones** and a ventral root that contains the **motor neurones**.

spindle: a system of **protein** fibres or **microtubules** that is involved in organising **chromosomes** during cell division. In **mitosis**, the spindle starts to form as the chromosomes become visible during prophase. It is fully formed by the end of metaphase at which stage it fills the space that was originally occupied by the nucleus of the cell. The chromosomes are attached to the spindle by their **centromeres** and are arranged across the equator of the cell. Finally, during **anaphase**, the pair of **chromatids** that make up each chromosome are pulled apart by the spindle fibres to the poles of the cell.

spiracle: one of the openings in the side of an insect's body through which respiratory gases diffuse. The spiracles lead into a series of fine tubes known as **tracheae** which take air directly from the outside of an insect to its cells. The spiracles of most insects have a mechanism that allows them to open and close. They normally remain open for short periods of time. This allows efficient gas exchange but reduces the amount of water that is lost. Both an increase in the concentration of carbon dioxide and a fall in the oxygen concentration act as stimuli for the opening of the spiracles.

spirillum: a bacterium with spiral-shaped cells. One of these bacteria is *Treponema pallidum*, the organism which causes syphilis.

spirometer: an instrument that measures the volumes of gas breathed in and out of the *lungs*. In its simplest form it relies on the subject breathing through a mouthpiece connected by a series of tubes and valves to a set of bellows. As air enters or leaves, the movement of the bellows is recorded by a pen on a moving chart. Records obtained from a spirometer enable the various **lung capacities** to be determined. If the carbon dioxide in the exhaled air is absorbed by a substance such as soda lime, the total volume of gas in the bellows gradually decreases as the oxygen is used in respiration. This allows the rate of oxygen consumption to be measured.

spleen: an organ that forms part of the immune system. It is found in the abdomen. Apart from its role in producing **lymphocytes**, the spleen:

- acts as a reservoir for stored red blood cells. These can be released into the blood system when they are needed
- destroys worn-out red blood cells and platelets.

spongy mesophyll: cells that are found underneath the palisade cells in a leaf. They are irregularly shaped cells and, in between them, there are large air spaces. These spaces play an important role in gas exchange between the leaf and the surrounding atmosphere.

sporangium: an organ produced by fungi and plants that contains **spores**.

spore: a tiny structure, often only single celled, that is produced during reproduction and is dispersed from the parent to produce a new individual. Spores are produced by a variety of organisms and are found in all **kingdoms** except the animal kingdom. They vary in the way that they are formed and in the details of their structure. For example, spores may result either from **sexual** or **asexual reproduction**, and they may or may not be surrounded by a hard protective wall. Many spores are produced in enormous numbers and allow rapid increases in population or colonisation of new areas. Those with hard outer coverings are often resistant to extreme physical conditions.

sporophyte: the diploid, **spore**-producing stage in the life cycle of a plant. Plant life cycles show an **alternation of generations** in which a spore-producing stage, or sporophyte, alternates with a **gamete**-producing stage, or **gametophyte**. The sporophyte produces the spores. These grow and produce a gametophyte. In primitive plants such as mosses and liverworts, the gametophyte is the **dominant** stage. The sporophyte is just a small capsule on the end of a stalk which develops on the moss gametophyte. It is completely dependent on the gametophyte. In *flowering* plants, however, the sporophyte is the dominant stage in the life cycle and the gametophyte is very much reduced. The adult plant, whether it is a cabbage or an oak tree, is a sporophyte.

squamous epithelium: see *epithelium*.

stabilising selection: selection which operates against the extremes in a population. Birth mass in humans provides an example. The graph shows the pattern of variation in birth mass. Although there are many causes of variation in birth mass, some are genetic. Very small babies are less likely to survive. There is also a higher death rate among very large babies. Consequently selection is in favour of babies whose birth mass is around the mean, and genes responsible for very large and very small babies will tend to be eliminated. Stabilising selection, therefore, prevents evolutionary change.

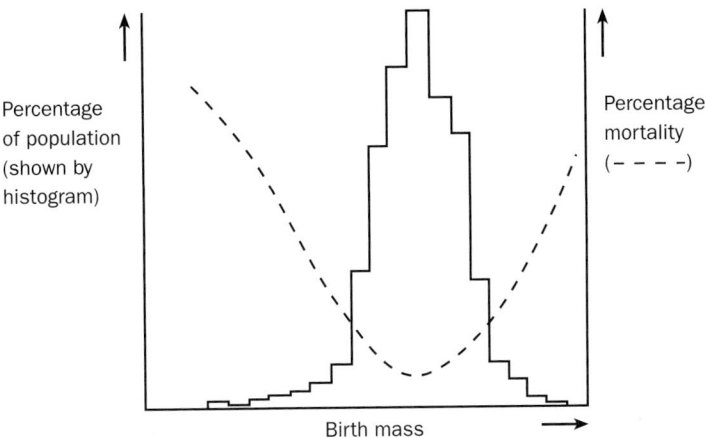

Stabilising selection and human birth mass

stain: a substance used to show up part of a cell or tissue so that it can be seen clearly when examined with a microscope. When a light microscope is used, the stain is a coloured substance. A basic stain has a coloured cation. It stains the more acid parts of cells such as the nuclei which contain large amounts of *nucleic acid*. An example of a basic stain is haematoxylin, which stains nuclei purple. Acid stains have coloured anions and they are useful for staining *cytoplasm*. Eosin is an acid stain which colours cytoplasm pink.

Different sorts of stains are used with *electron microscopes*. These contain heavy metals such as uranium and lead. They bind to chemical groups that are found in particular cell *organelles*. Since few *electrons* are able to pass through them, the structures they stain appear dark in colour.

stamen: the male parts of a *flower*. Each stamen consists of an *anther* in which the pollen develops and a stalk known as a filament. The anther has two or four lobes and, within each lobe, is a layer of cells which nourishes a mass of developing pollen grains. When ripe, the anthers split and release the pollen.

standard deviation: a mathematical term that gives a measure of how spread out a set of results are. In describing any set of results it is useful to know two things: where the middle is and how spread out the individual measurements are. The spread of results could simply be given as the difference between the largest reading and the smallest one. Unfortunately, this can be misleading as it only gives an idea about these two particular values. The standard deviation, however, is much more useful as it gives us a measure of variation in terms of the distances of all the results from the mean.

Staphylococcus: a *genus* of *bacteria* that are common parasites of humans and other mammals. They occasionally cause serious disease. They are able to tolerate low *water potentials* and, as a result, can survive drying and high salt concentrations fairly well. Two species are commonly associated with humans. S. *epidermidis* is a non-pathogenic species found on the skin. S. *aureus* is often associated with boils and pimples but can cause more serious conditions such as food poisoning, pneumonia and meningitis.

starch: a large, insoluble carbohydrate that is an important energy store in plants. Starch is a *polymer* and consists of a large number of α-*glucose* molecules joined together by *condensation*. The exact chemical composition of starch varies from one species of plant to another as it consists of two main components that may be present in different proportions. *Amylose* forms long straight chains while *amylopectin* has branched chains. Starch has a number of features which make it an ideal storage compound:
- its molecules are tightly coiled, allowing a considerable amount of starch to be packed into a relatively small volume
- it is insoluble. It is therefore easy to store because it does not move out of cells readily, nor will it affect the *water potential* of a cell
- it is readily broken down by *enzymes* to glucose.

The test for starch is to add a drop of *iodine* solution. A blue–black colour is produced.

start codon: a sequence of three bases on a molecule of *mRNA* that signals the start of the protein to be translated.

statins: drugs that lower the concentration of *cholesterol* in the blood of people at risk from cardiovascular *disease*. Statins are *enzyme* inhibitors. They inhibit one of the *enzymes* involved in cholesterol synthesis in the *liver*.

stationary phase: part of a *population growth curve* in which the rate of increase in the population is more or less balanced by the number of organisms dying. As a result the population remains constant.

stele: the cylinder of vascular tissue in the stem or root of a plant. It contains the *xylem* and *phloem*.

stem cell: an unspecialised cell that is capable of dividing by *mitosis*, and continually replacing itself. It also has the potential to differentiate (see *differentiation*) and give rise to one of a range of specialised cells. When an *embryo* is only made up of a few cells, each of these cells is a stem cell and is able to develop into the full range of cell types found in the adult organism. There are also stem cells in adults. These play an important part in repairing tissues and replacing cells.

Biologists can now grow stem cells in cultures. In appropriate conditions, these stem cells can differentiate into specialised cells. Adult stem cells, such as those found in *bone* marrow, are sometimes used in treating some conditions such as leukaemia. Embryonic stem cells may be used for treatments in the future.

sterilisation (microbiology): involves killing all the *microorganisms* present in a particular place. There are a number of physical agents that have a harmful effect on bacteria and other microorganisms. Chemical *disinfectants* and low temperatures will damage and kill many cells but will not necessarily achieve complete sterilisation. Some methods that do include:

- heat. Sterilisation by heat is one of the most commonly used and reliable methods. Unlike the use of chemicals, it does not leave any toxic residues. Its only real disadvantage is that it could damage the material being sterilised. Exposure to moist heat is lethal at lower temperatures and in a shorter time than exposure to dry heat. The use of steam at lower than atmospheric pressures is particularly effective
- ultraviolet radiation. Certain wavelengths of ultraviolet radiation are absorbed by and cause damage to *nucleic acids* and *proteins*
- *ionising radiation*. All types of radiation produced by atomic particles have a similar effect on living tissues. *Electrons* are knocked out of some molecules and gained by others. The ions which are produced as a result have extremely damaging effects. Radiation of this type is widely used in producing sterile syringes and Petri dishes.

sterilisation (reproduction): the use of a surgical operation to produce sterility. In women, a small viewing instrument, or *endoscope*, is inserted through the wall of the abdomen. The lower parts of the two oviducts are then permanently closed with the aid of clips or small plastic rings. In males, sterilisation involves vasectomy. The tubes that take the *sperm* from the *testes* are tied and cut. Although both of these procedures provide an effective means of contraception, they must be considered as permanent. They are difficult to reverse.

steroid: one of a group of *lipids* whose molecules do not contain *fatty acids*. One of the most important steroids in the human body is *cholesterol*, which is made in the *liver*. Apart from being found in *plasma membranes*, cholesterol can be converted into a number of other biologically important steroids such as the sex *hormones oestrogen* and *progesterone*.

stethoscope: an instrument used for listening to sounds inside the body. It consists of a funnel-shaped structure that is applied to the body surface and rubber or plastic tubes which connect this structure to ear pieces. It can be used to listen to sounds associated with the heart and breathing. It can also be used with a *sphygmomanometer* to measure blood pressure.

sticky ends: unpaired base sequences at the ends of a piece of *DNA*. Sticky ends may be produced when DNA is cut with a *restriction endonuclease*. These *enzymes* often cut the DNA so that it has a four-base overhang at each end. *Ligase* enzymes join complementary sticky ends to produce *recombinant* DNA.

stigma: the part of the *flower* that receives the pollen grains. In a wind-pollinated flower, the stigma is usually feathery and hangs loosely out of the flower. It is adapted to trap *pollen* that is present in the air around the plant. However, the stigmas of insect-pollinated flowers are usually enclosed within the flower. They are much flatter and often have sticky surfaces, enabling them to trap the pollen carried on the body of a pollinating insect.

stomach: a large, muscular sac in the front part of the gut of a mammal that stores and digests food. Food is chewed in the mouth and mixed with *saliva*. It then travels down

S

the *oesophagus* into the stomach. The wall of the stomach is unlike other parts of the gut because it has three layers of muscle. Contraction of these muscles layers results in the stomach contents being continually mixed and churned. Cells in the wall of the stomach produce **gastric juice**. This is a mixture of mucus, hydrochloric acid and digestive *enzymes*, among which is the **endopeptidase pepsin**.

stomata: the small pores in the surface of a leaf through which gas exchange normally takes place. They may be found on either surface of the leaf but are more commonly located on the underside. However, in aquatic plants like the water lilies that have floating leaves, stomata are only found on the upper surface. They are important in allowing the carbon dioxide required in **photosynthesis** to diffuse from the atmosphere into the intercellular spaces of the leaf. Unfortunately, as gases diffuse into the leaf, so large amounts of water vapour will be lost via **transpiration**. There has to be a compromise between having the stomata permanently open so that the carbon dioxide can diffuse in and closed so that water loss is minimised. Stomata are surrounded by guard cells that are able to regulate the size of the stomatal opening. Various mechanisms have been suggested to explain how they do this. The **potassium-movement hypothesis** is thought to offer the most likely explanation. **Xerophytes** are plants which live in particularly dry places. The stomata on their leaves are often adapted by being sunken into pits or by being surrounded by hairs. This reduces water loss.

stop codon: a sequence of three bases on a molecule of *mRNA* that signals the end of the protein to be translated.

stratigraphy: the study of the sequence of rock layers. Sedimentary rocks are formed as layers of material settle on top of each other. The older the rock, the lower it will be in a sequence; the more recently the rock was formed, the closer it will be to the surface. The stratigraphical sequence of rocks can be used to compare the age of fossils. The older the fossil, the more deeply it will be buried. This situation is complicated by processes such as weathering. The action of wind and water progressively erodes the land, exposing rocks of different age to the surface. Nevertheless, a good knowledge of the geology of an area still allows the various rocks in which fossils have been found to be placed in a time sequence. Unfortunately, stratigraphy can only provide information about the relative ages of particular fossils. To get an absolute date, other techniques such as **potassium–argon dating** must be used.

Streptococcus: a *genus* of bacteria that form chains of round cells. Some species are important in the production of products such as buttermilk and silage. Others are harmful and cause diseases such as tonsilitis and scarlet fever.

streptokinase: a 'clot-busting' drug. Streptokinase is an **enzyme** produced by *Streptococcus* bacteria. It is given to patients to break up blood clots by dissolving *fibrin*. It is often given immediately after a heart attack to break down blood clots in the **coronary arteries** and prevent serious damage to heart muscle. It is also used to prevent clots forming during **kidney dialysis**.

stretch receptor: a receptor that responds to stretch. Stretch receptors are important in the control of breathing. The *lungs* inflate when a person breathes in and this stimulates stretch receptors in the lungs. They send *nerve impulses* to the breathing centre in the *brain* and these impulses inhibit breathing. Stretch receptors are also found in muscles and in the gut wall.

striated muscle: see *skeletal muscle*.

stroke: see *cerebrovascular accident (CVA)*.

stroke volume: the amount of blood pumped out each time the heart beats. The *cardiac output*, which is the total amount of blood pumped out by the heart per minute, is obtained by multiplying the stroke volume by the heart rate. The relationship between stroke volume, cardiac output and heart rate is given by the equation:

$$\text{stroke volume} = \frac{\text{cardiac output}}{\text{heart rate}}$$

stroma: the part of a *chloroplast* that surrounds the internal membranes. It contains the *enzymes* associated with the *light-independent reaction* of *photosynthesis*.

style: the part of the flower that connects the *stigma* and the *ovary*. Once a pollen grain has landed on the stigma, it produces a *pollen tube* which grows down through the tissues of the style.

suberin: a waxy, waterproof substance found in plant *cell walls*. It is found in the outer cell layers of older roots and stems. The *endodermis* in a plant is a ring of cells between the outer part of the root and the vascular tissue in the centre. A band of suberin, called the *Casparian strip*, runs round the walls of each of these endodermal cells. It prevents water and dissolved substances from getting into the vascular tissue by going through the cell walls and intercellular spaces. As a result, these substances have to pass through the *cytoplasm* of the endodermal cells, which can therefore control their movement into the *xylem*.

submucosa: a layer in the wall of the *small intestine,* separated from the *mucosa* by a thin layer of smooth muscle called the *muscularis mucosa*. It contains *connective tissue* as well as blood and *lymph* vessels.

substitution: see *gene mutation*.

substrate: a substance which is acted on or used by something else. The term is used in a variety of contexts in biology:
- in biochemistry, it is the molecule on which an *enzyme* acts. *Enzymes* are extremely specific in their action. Only molecules with a particular shape will fit the *active site* of the enzyme concerned and form an enzyme–substrate complex
- a *respiratory substrate* is an organic substance that is used as a starting point in *respiration*. Note that oxygen is not a respiratory substrate
- in microbiology, a substrate is a medium on which *microorganisms* can grow. It may be either in a liquid or a solid form and will contain all the necessary nutrients
- *sessile* animals are those that stay in one place. They are attached to a particular substrate or surface.

substrate-level phosphorylation: a chemical reaction that results directly in the formation of *ATP* from *ADP* and phosphate. It does not involve either oxidative phosphorylation or photophosphorylation. In *aerobic respiration*, *substrate*-level phosphorylation occus in *glycolysis* and in the *Krebs cycle*.

succession: the way in which the different species of organisms which make up a *community* change over a period of time. Sand dunes are common in many coastal areas. The sand is unstable, constantly being blown by the wind. Not only that, but there are low concentrations of soil nutrients and the sand dries out rapidly. One of the first plants to colonise this harsh environment is marram grass. In time, the roots of the marram grass bind the sand particles together. Plants die and the resulting dead vegetation breaks down. This increases the amount of humus present and the concentration of important soil nutrients such as nitrates. Marram grass no longer thrives and it is replaced by other species that are better adapted to these changed conditions. There is a succession from pioneer species such as marram grass, through a variety of **seral stages** until, ultimately, a **climax community** is established. With the example chosen, this climax community might be oak woodland. A succession that is prevented in some way from reaching this natural climax is called a **deflected succession**.

sucrase: an *enzyme* that breaks down *sucrose*. It is a *hydrolase* and splits each sucrose molecule to produce a molecules of *glucose* and a molecule of *fructose* by adding a molecule of water. In a mammal, the sucrase *enzymes* are found on the **cell-surface membranes** of epithelial cells in the **small intestine**.

sucrose: a *disaccharide*, that is, a sugar which is made up of two sugar units. These units are *glucose* and *fructose* and they are joined together by **condensation**. Sucrose is a **non-reducing sugar** and therefore will not produce a positive result with **Benedict's test** unless it has first been hydrolysed. Sucrose is an important substance in plants. It is the form in which *carbohydrates* are usually transported in *phloem*.

suicide gene: a gene that causes a cell to kill itself during **apoptosis**.

summation: a process that occurs in **synapses** by which a number of weak stimuli can lead to the generation of a **nerve impulse**. A nerve impulse can only be triggered in the postsynaptic neurone when enough of the **neurotransmitter** has reached the receptor molecules on the postsynaptic membrane. This can happen in two different ways. If a large number of synapses are stimulated at the same time, the total amount of transmitter may be enough to trigger a nerve impulse even if the amount released at an individual synapse is not. This is called spatial summation. Temporal summation results from repeated stimulation of the same synapse.

supernatant: the liquid layer left on top when a suspension is centrifuged (see **centrifuge**). *Cell fractionation* is a process in which cells are broken up and the different types of *organelle* separated from each other. The tissue is broken up and suspended in a buffer solution. When it is spun in a centrifuge at a relatively low speed, the larger, denser organelles such as nuclei and *chloroplasts* fall to the bottom of the centrifuge tube to form a pellet. The smaller organelles remain in the supernatant.

surface area–volume ratio: the relationship between the surface area of an organism and its volume. In biological terms, this is important because it means that the larger an organism, the smaller its surface area in relation to its volume. Look at the diagram. If we take a number of cubes, each a little larger than the previous one, we can work out the surface area–volume ratio for each one. If this information is plotted as a graph, the relationship between the surface area–volume ratio and the size of the cube can be clearly shown. Obviously, animals and plants are not cube shaped, but the principle is just the same whether the basic shape is a cube or that of a living organism.

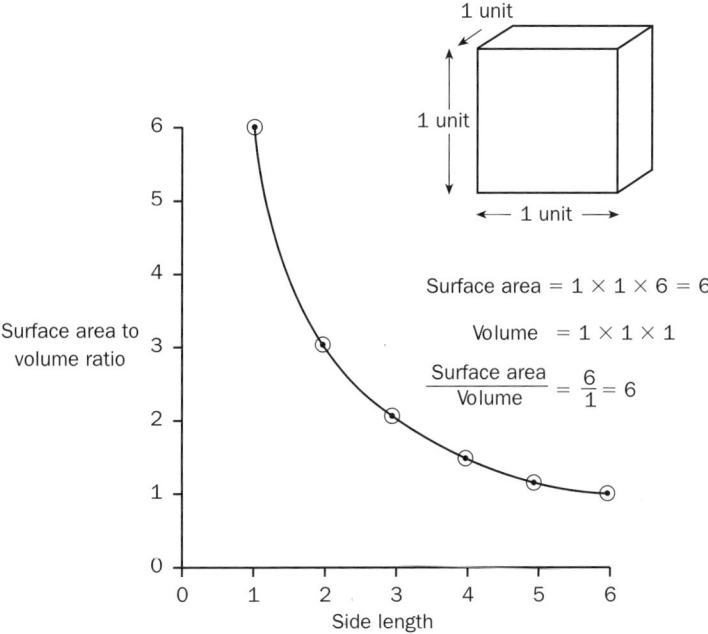

This idea is of enormous physiological importance. A small one-celled organism like amoeba has a very large surface area when compared to its volume. Therefore, it is able to gain all the oxygen it requires by *diffusion* over its general body surface. A larger organism such as a fish, however, has a much smaller surface area to volume ratio. It has a specially adapted gas-exchange surface.

surfactant: a substance that lowers the surface tension of a liquid. If the inside of a polythene bag is wetted slightly, its surfaces stick together. This is due to surface tension, a force which pulls the water molecules together. The *alveoli* in the *lungs* are lined with a layer of liquid. If this liquid had a high surface tension, it would pull the sides of the alveoli together. They would stick to each other and reduce the surface area available for gas exchange. However, the liquid lining the alveoli contains surfactant. This lowers its surface tension and prevents the walls sticking together. There is a lower concentration of surfactant in the lungs of smokers than in those of non-smokers.

Surfactant is important in the lungs of a newborn baby. At birth, a baby takes several strong breaths that inflate the lungs. Without surfactant being present, the lungs would collapse again.

survival curve: in humans, a curve that shows proportion of people that you would expect to be still alive at different times after they were born. This is normally given per 100 000 of the population. You can calculate the average life expectancy from a survival curve. This is the age when 50% of the population is still alive. Two survival curves are shown in the graph overleaf.

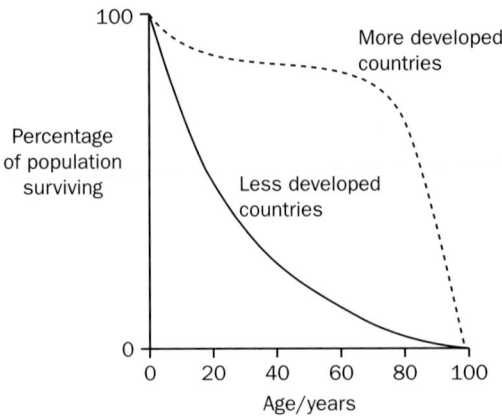

Survival curves for human populations

Curves of this pattern apply to human populations in many less developed countries. They also apply to people who were living in the UK 200 or more years ago. There is a relatively high mortality among children and the curve falls rapidly at first. The curve falls less steeply after this. In more developed countries, a high proportion of children survive to adulthood. The survival curve in this case shows a steep drop towards the end of life.

symbiosis: any nutritional relationship between two organisms of different species. The meaning of this term causes some difficulties since it may also be defined as a relationship between two organisms where both gain a nutritional advantage. It is better to keep the term symbiosis to refer to any nutritional relationship and to use *mutualism* when considering relationships where both of the organisms involved gain an advantage. The diagram summarises some different types of symbiotic relationship.

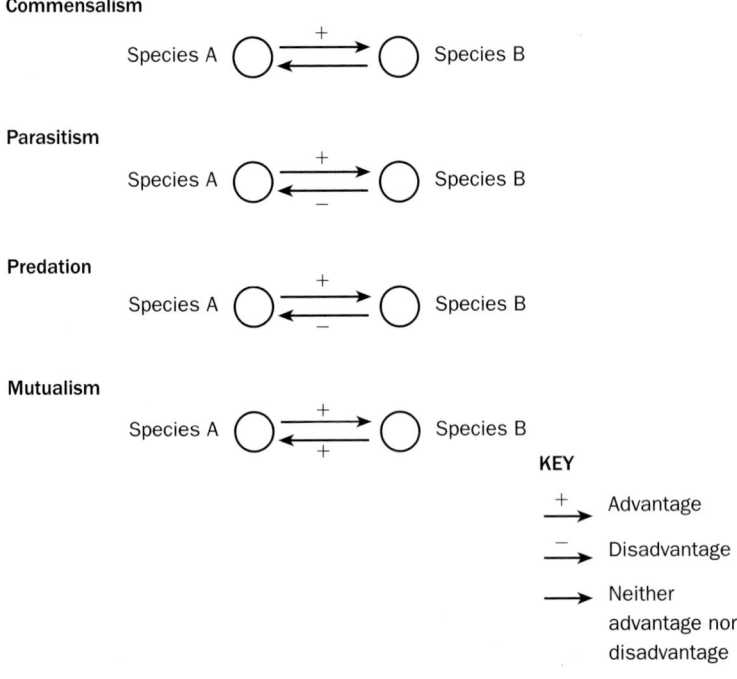

sympathetic nervous system: a division of the *autonomic nervous system* that generally acts in emergencies and controls the functions of the body during times of stress. Some of the important features of the sympathetic system are that:

- some of the *synapses* secrete *noradrenaline* as a neurotransmitter
- effects are usually stimulatory, for example sympathetic stimulation increases the breathing rate and increases the heart rate and *stroke volume*.

sympatric speciation: *speciation* that occurs when some factor prevents two populations of a particular species living in the same area from interbreeding. The mechanism which prevents this is known as an isolating mechanism. In the case of sympatric speciation, it is something other than a geographical barrier. Some examples of isolating mechanisms are given in the table.

Isolating mechanism	Example
Differences in courtship behaviour	The chiffchaff and the willow warbler are small insect eating birds that spend the summer in Britain. They are so similar to each other in appearance that it is very difficult to tell them apart. However, their songs, which form an important part of courtship behaviour, are completely different. Since successful mating will only result from the correct behaviour pattern, chiffchaffs and willow warblers do not interbreed
Failure to transfer *gametes*	Bees can distinguish between flowers of the corn poppy and flowers of other species of poppy. Bees which are visiting corn poppies do not go to the flowers of other species. Pollen is not transferred between different species and interbreeding is prevented
The production of infertile hybrids	The carrion crow is found in the southern part of Britain; the hooded crow in the north. They interbreed and form hybrids but the hybrids are not as fertile as the parents. Selection therefore favours the parents

symplastic pathway: the pathway by which substances go through the *cytoplasm* of plant cells. The contents of neighbouring cells are linked by thin strands of cytoplasm called plasmodesmata that pass through the *cell walls*. Substances are therefore able to pass from cell to cell along this pathway without passing through *cell-surface membranes*. This would require the expenditure of a great deal of energy. Water and mineral ions move through the roots into the *xylem*. They also move out of the xylem and through the leaf tissue. There are three separate pathways along which they can travel: the symplastic pathway described here; the *apoplastic pathway* through the *cell walls* and the intercellular spaces; and the *vacuolar* pathway which involves passing from vacuole to vacuole. The apoplastic pathway is probably the most important of the routes by which water moves through plant tissues, but the symplastic pathway is very important in the transport of mineral ions.

synapse: a junction between nerve cells. Nerve cells do not actually join with each other. There is a very small gap between them, a gap which is only about 20 nm wide. This is called the synaptic cleft. On one side of this is the presynaptic membrane and on the other side, the postsynaptic membrane. A **nerve impulse** arriving at a synapse causes the release of a **neurotransmitter**. This diffuses across the synaptic cleft and produces another nerve impulse on the postsynaptic side. The events which occur in a synapse where the neurotransmitter is **acetylcholine (ACh)** are summarised in the flow chart.

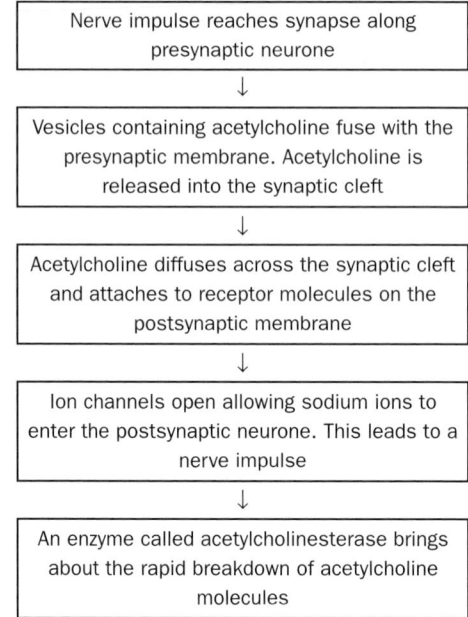

Synapses play an important part in the working of the nervous system. They have some important properties:

- Synapses only transmit information in one direction, from the presynaptic neurone to the postsynaptic neurone.
- The process of **summation** means that a number of weak stimuli can lead to the generation of a nerve impulse.
- The synapse in the flow chart is an excitatory synapse because the effect of the acetylcholine is to stimulate another **action potential**. Inhibitory synapses also exist. In these, the effect of the transmitter is to make it less likely that a new impulse will be produced.

synaptic cleft: the small gap between presynaptic neurone and the postsynaptic neurone at a **synapse**. When a **nerve impulse** arrives at a synapse it causes the release of a **neurotransmitter**. This diffuses across the synaptic cleft and produces another nerve impulse on the postsynaptic side.

synaptic vesicle: a small sac in a presynaptic neurone that is surrounded by a membrane. Synaptic vesicles contain molecules of substances such as **acetylcholine (ACh)** that act as **neurotransmitters**.

syndrome: a combination of different symptoms that is characteristic of a particular disorder. AIDS stands for acquired immune deficiency syndrome. When a person is infected by **HIV**, the virus destroys a type of white blood cell known as a **T helper cell**. As a result, AIDS patients are much more likely to become infected by pathogenic organisms that do not normally affect healthy people. Some conditions, such as athlete's foot, a fungal infection of the skin between the toes, are usually only mild infections in healthy people. They may be severe in AIDS patients. AIDS is referred to as a syndrome because it is associated with a combination of different conditions.

synergistic: where the action of two things together is greater than that of their separate effects added together. Various *enzymes* are added to biological washing powders. On their own, both *lipases* and *proteases* are quite effective at removing stains. Research has shown, however, that addition of both enzymes produces a greater effect than might be expected. Most stains are a mixture of biological compounds and often include both *protein* and *lipid* molecules. The *proteins* that hold the stain together and bind it to the fabric concerned are digested by the proteases. This allows lipases to penetrate more effectively and remove the lipid components of the stain. Drugs can also act synergistically.

synovial joint: a type of joint found between *bones* in the mammalian skeleton. The elbow, knee, shoulder and hip joints are all examples of synovial joints. The bones are held together by *ligaments*. These form a protective capsule round the joint. Lining this capsule is a thin synovial membrane. The cells which make up the synovial membrane secrete a fluid into the capsule of the joint. This fluid acts as a lubricant, helping to ensure smooth movement. The ends of the bones concerned are covered with a smooth layer of *cartilage*. This further reduces the friction between them. *Osteoarthritis* is a disease affecting joints in which this cartilage is lost and the underlying bone is damaged as a result. The condition is painful and affected people experience difficulty in moving the joint concerned.

system: a group of *organs* that work together to carry out one or more functions.

systematics: the placing of organisms into taxonomic groups (see *taxon*) based on their similarities and differences.

systemic herbicide: a substance used to kill weeds. It is transported through the tissues of the plant. Systemic herbicides are synthetic substances that are similar to naturally occurring *plant growth substances*. They are absorbed by leaves and roots and are transported to all parts of the plant where they cause death by interfering with the natural growth pattern. Systemic herbicides have two particular advantages over many other herbicides:

- they have little effect on *monocotyledons* such as cereals and other grasses but they rapidly bring about the death of dicotyledonous weeds (see *dicotyledons*). They are very suitable, therefore, for controlling weeds on lawns or in cereal crops
- contact herbicides only kill those parts of the plant with which they come into contact. Systemic herbicides spread throughout the plant and therefore ensure that the root systems and underground stems of perennial weeds are killed and will be unable to shoot again.

systemic insecticide: a substance used to kill insects. It works by first being absorbed into the tissues of the food plant. Systemic insecticides have two major advantages over other kinds of insecticide that only work when they come into contact with the insect concerned:

- Because the insecticide spreads through the plant's tissues, the whole surface of the plant is poisonous. An insect feeding on any part of the plant will be killed. This means that it is not necessary to spray the entire plant surface, reducing the amount of insecticide that it is necessary to use.

- Only insects which actually feed on the crop that has been sprayed will be killed. There is little risk to predators or insects that are simply resting on the plants.

systole: the stage in the *cardiac cycle* that involves contraction of the heart muscle. There are two steps involved. *Atrial systole* occurs when the muscle of the atrial walls contracts and forces blood into the *ventricles*. *Ventricular systole* is when the walls of the ventricles contract, pumping blood out through the arteries.

systolic blood pressure: the blood pressure in the main arteries measured when the *ventricles* are emptying (*systole*). At this stage in the *cardiac cycle*, blood pressure in the arteries is at its highest.

Are you studying other subjects?

The *A–Z Handbooks (digital editions)* are available in 14 different subjects. Browse the range and order other handbooks at **www.philipallan.co.uk/a-zonline.**

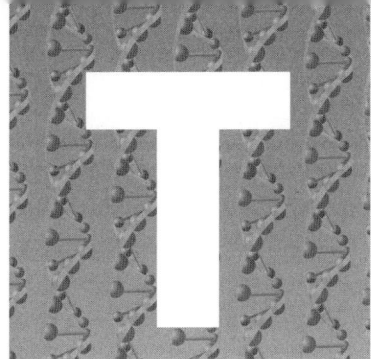

T cell: a type of *lymphocyte* that does not secrete antibodies. There are many different types of T cell and the way that they affect each other and *B cells* is complex. Two of their important functions, however, are that they are responsible for killing cells that have been infected with pathogens and they are involved in controlling the immunological process.

T helper cell: one of many different types of *T cell*. T helper cells synthesise substances that activate or enhance the effects of other cells in the immune system. *HIV* recognises CD4 receptor molecules. These are molecules that are found on the surface of T helper cells. The virus therefore infects and destroys T helper cells. As the number of these cells is reduced, so the ability of the person to fight infection decreases.

T killer cell: a cell from the immune system. T killer cells produce toxic and pore-forming substances. They use these to kill cells that have been infected with pathogens. T killer cells are also called cytotoxic T cells.

T lymphocyte: see *T cell*.

T memory cell: see *memory cell*.

Taenia: a *genus* of *tapeworms*, including species that are human gut *parasites*.

tamoxifen: a drug used to treat breast cancer. Most breast cancers need the female *hormone oestrogen* in order to grow. The cancer cells have oestrogen receptors on their surface. Oestrogen binds to the receptors and causes the cells to divide and the tumour to grow. In the body, tamoxifen is changed into a substance that works by attaching to these hormone receptors and stopping oestrogen from binding.

tapeworm: a worm with a flat, ribbon-like body that lives in the intestines of humans and other animals. Tapeworms are *parasites* and have many adaptations for their mode of life. These include:

- hooks and suckers that attach the head of the adult tapeworm to the gut wall
- a *cuticle* that protects the adult worm from *digestion* by *enzymes* in the host's gut
- well-developed reproductive organs. Tapeworms are hermaphrodites and the eggs in the segments near the end of the body can be fertilised by male *gametes* produced by segments near to the head. Huge numbers of fertilised eggs are produced
- a complex life cycle involving a secondary host. This is an adaptation that enables infection of another individual
- reduction of body systems other than those concerned with reproduction.

target tissue: a tissue on which a *hormone* acts. *Glucagon* is a hormone secreted by the *pancreas*. It is carried by the blood to all organs in the body. However, it only affects cells that have appropriate protein receptors on their *cell-surface membranes*. These cells are found in *liver* and adipose tissue. Liver and adipose tissue are glucagon target tissues.

taxis: a form of behaviour in which an organism moves either directly towards or away from a *stimulus*. A fly maggot, for example, moves away from light. It shows negative phototaxis. On the other hand, adult flies show positive phototaxis. They move towards the light. Other stimuli also produce this form of behaviour. Blood-sucking leeches respond positively towards a source of heat, and *flatworms* respond in the same way to water currents and chemicals. Behaviour patterns like these clearly have survival value. The leech is a *parasite*. By moving towards a source of heat, it is able to find its warm-blooded host.

taxon: a level of *classification*. A taxon (plural taxa) is a group into which an organism is put when it is classified. Taxa include *kingdom*, *phylum*, *genus* and *species*.

taxonomy: the scientific study of *classification*.

TB: see *tuberculosis (TB)*.

temperature coefficient: see Q_{10}.

temperature control: the system by which body temperature is controlled in animals. In order to function normally, animals must keep their body temperatures within certain limits. If an animal's temperature falls too low, biochemical reactions will be too slow for it to remain active. If it goes too high, there is the risk of *enzymes* and other *proteins* being denatured. To a certain extent, all animals exert some control over their internal temperatures. They vary in the way in which they do this. Mammals rely largely on physiological methods. Humans, for example, maintain a temperature within a degree or so of 37°C. If the temperature rises too high, internal mechanisms bring about loss of heat. If it falls too low, other mechanisms increase heat production and reduce heat loss from the body. Humans are, therefore, examples of endothermic animals (see *endotherm*). Crocodiles, on the other hand, rely largely on moving between land and water to remain in the *environment* with the most suitable temperature. They are referred to as ectothermic (see *ectotherm*) as they rely on the external environment to maintain a reasonably constant body temperature.

temporal summation: see *summation*.

tendon: a strip of *connective tissue* that attaches a muscle to a *bone*. Tendons contain a lot of closely packed fibres of a protein called *collagen*. The properties of collagen make it ideally suited to its functions in a tendon. It is flexible but it is resistant to stretching. This is important as it means that when the muscle contracts it will move the bone, not just stretch the tendon.

territory: an area that an animal defends against other animals, usually those of the same species. Many animals, ranging from insects to mammals and birds, have territories but territorial behaviour is very variable, even among individuals of the same species. For example, the spotted hyena lives in clans which fiercely defend feeding territories where their prey are numerous. This is the case in the Ngorongoro crater in northern Tanzania. On the nearby Serengeti plains, however, the antelopes on which the hyenas feed roam extensively, and here the hyenas do not hold territories.

Territories have different functions. In Britain, the pied wagtail defends a winter feeding territory along the side of a river or stream. The bird feeds by moving along the shore and eating small insects that have been washed up. When it has covered the length of the bank that forms its territory, it flies back to the beginning and starts again. The mistle thrush is another bird that defends a feeding territory – often, in this case, a holly bush.

Other animals hold mating territories. One example of an animal that does this is the three-spined stickleback, which defends the area around its nest from other males. Animals mark and defend their territories in different ways. Many mammals use **pheromones** produced from specialised scent **glands**, or mark the boundaries of their territories with **urine** or **faeces**. Birds, on the other hand, frequently use song or visual displays.

tertiary structure (protein): the irregular folding of a polypeptide chain. **Globular proteins** are proteins which consist of polypeptide chains that are folded so that the molecule is roughly spherical and has a compact overall shape. The tertiary structure of the polypeptide chain is determined by its **primary structure**. Its actual shape is held by chemical bonds between different **amino acids**. These bonds include:

- *hydrogen bonds*
- bonds between sulphur-containing amino acids
- ionic bonds.

The tertiary structure of a protein gives the molecule its particular shape. This is very important in the function of the protein in the organism. Some examples of the way in which the shape of a protein molecule is related to its function are given below:

- The **active site** of an **enzyme** molecule has a specific shape. This enables it to bind to one particular **substrate** and, as a result, catalyse a specific reaction.
- Some proteins act as receptor molecules on **cell-surface membranes**. Again, their tertiary structure gives them a specific shape. It is because of this that particular **hormones**, for example, can only affect certain target tissues.
- *Antibodies* are specific. They only respond to particular antigens. This specificity is determined by the precise shape of part of the **antibody** molecule.

Many of the bonds that maintain the tertiary structure of the protein molecule are not very strong and may be broken by heating. If this happens, the protein is denatured and its three-dimensional shape changed. It is no longer able to function.

test cross: a cross between an organism of unknown **genotype** and the relevant **recessive homozygote**. In peas, the **allele** for green pods, G, is **dominant** to that for yellow pods, g. A pea plant with green pods could have one of two possible genotypes, GG or Gg. It is impossible to tell by simply looking at it. The only way of finding out the genotype of a particular plant would be to cross it with a plant with yellow pods. The genetic diagrams show the expected results from this test cross.

If the unknown pea had the genotype GG:

parental phenotypes	green pods	yellow pods
parental genotypes	GG	gg
gametes	G	g
offspring genotypes	Gg	
offspring phenotypes	All with green pods	

If the unknown pea had the genotype Gg:

parental phenotypes	green pods	yellow pods
parental genotypes	Gg	gg
gametes	Gg	g
offspring genotypes	Gg	gg
offspring phenotypes	green pods : yellow pods 1 : 1	

279

By comparing the results of the cross with the predicted ratios, it is possible to determine the genotype of the unknown green-podded plant.

testa: the protective outer layer of a *seed*. When an ovule from a *flowering* plant is fertilised, it develops into a seed. The female *gamete* inside the ovule is fertilised and becomes the embryo. The layers of tissue which surround the ovule form the testa. Because the testa does not actually come from fertilised cells, it has exactly the same genetic make-up as all the other cells in the female parent. It does not contain any of the genes from the male parent.

Some seeds will not germinate even though conditions are otherwise suitable. They are in a state of dormancy. The testa may be associated with dormancy. In some plants, for example, it may contain substances which inhibit growth. In others, it is impermeable to the passage of oxygen and water. Changes have to occur before *germination* can take place.

testes: the organs in a male animal which are responsible for the production of sperms. In mammals, the testis (singular) has two functions. It is packed with a mass of small tubes known as *seminiferous tubules*. Within the seminiferous tubules, the process of sperm formation, or *spermatogenesis*, occurs. When the sperm cells have been formed, they move to the epididymis. This is another mass of small tubes and is found just outside the testis. Sperms are stored in the epididymis. The process of sperm formation in many mammals works most effectively at a temperature a few degrees below body temperature. Because of this the testes are suspended in a sac, the scrotum, outside the body cavity. The second function of the mammalian testis is that it is the site of production of *hormones*. These are secreted by special cells which are found between the seminiferous tubules and known as interstitial or Leydig cells. The most important of the hormones produced in the testes is *testosterone*, the male sex hormone.

testosterone: an important male sex *hormone*. Testosterone is secreted by the cells that are found between the *seminiferous tubules* in the *testis*. These cells are known as interstitial or Leydig cells. Testosterone, like all hormones, is transported by the blood. It has a number of effects on the body. These include:
- stimulating *bone* and muscle growth and initiating the growth spurt at puberty
- promoting the growth of male *secondary sexual characteristics* such as facial and body hair and bringing about the changes in the larynx which cause the voice to deepen
- maintaining the function of the testes throughout the reproductive life of the male.

Testosterone is an anabolic *steroid*; anabolic (see *anabolic reaction*) because it is involved in controlling reactions in the body that are concerned with the building up of tissues, and a steroid because of its chemical nature. Anabolic steroids such as testosterone have been used widely and illegally by sportsmen and women to encourage muscle development.

thalassaemia: a condition in which a person produces abnormal *haemoglobin*. Red blood cells containing this type of haemoglobin do not function properly and the person suffers from *anaemia*. The condition is inherited and controlled by a gene with *codominant alleles*. This gene codes for one of the polypeptide chains that form a molecule of haemoglobin.

THC receptor: a receptor in the *brain* to which THC binds. THC is the main psychoactive drug in marijuana. Scientists have recently discovered that THC is very similar to a group of substances called endocannabinoids. These substances are produced naturally in the body and bind to THC receptors on the presynaptic membranes of some *synapses*. This modifies synaptic transmission.

thermoregulation: see *temperature control*.

thermostable enzyme: an *enzyme* that continues to function at relatively high temperatures without being denatured. Some *microorganisms*, called thermophiles, are able to grow in very hot conditions such as hot springs and around volcanic outlets on the sea floor. Thermophiles are of interest to biotechnologists because of the *enzymes* they contain. Many enzymes are used in industrial processes that are slow, but the reaction rate cannot be increased by heating because the enzymes will denature. Enzymes from thermophiles are, however, thermostable and can be used at higher temperatures. Thermophilic microorganisms are difficult to grow in culture so the commercial production of thermostable enzymes usually involves *genetic engineering*. The relevant gene is transferred to bacteria that can be grown in normal laboratory conditions. Immobilisation, the binding of enzymes to inert particles or surfaces, increases enzyme thermostability.

threshold stimulus: a stimulus that is large enough to trigger an *action potential*.

thrombin: an enzyme that converts *fibrinogen* to fibrin in the process of *blood clotting*.

thrombocyte: see *platelets*.

thromboplastin: a substance involved in *blood clotting*. Thromboplastins are produced by blood *platelets* and damaged tissue. They change the inactive enzyme prothrombin to its active form, thrombin. Thrombin catalyses the reaction in which soluble *fibrinogen* is converted to the threads of *fibrin*. These threads trap the red blood cells and form a blood clot.

thrombosis: a condition in which a blood clot forms inside a *blood vessel*. If this clot or thrombus breaks free, it can travel round the body and block a small *artery*. A heart attack (see *myocardial infarction*) occurs when one of the coronary arteries supplying the heart is blocked. A stroke (see *cerebrovascular accident (CVA)*) may result from the clot interrupting the blood supply to the *brain*.

thylakoid: a system of flattened sacs formed by the *plasma membranes* in a *chloroplast*. At various places, these thylakoids are stacked on top of each other rather like a pile of coins. These are *grana*. The thylakoids contain the *chlorophyll* molecules which trap the light energy in *photosynthesis*. They also contain the various molecules needed for the *light-dependent reaction*.

thymine: a *nucleotide base* found in *nucleic acid* molecules. Thymine is a pyrimidine. This means that it has a single ring of *atoms* in each of its molecules. When two *polynucleotide* chains come together in a *DNA* molecule, thymine always bonds with *adenine*. The atoms of these two bases are arranged in such a way that two *hydrogen bonds* are able to form between them. The base thymine does not occur in *RNA*. *Uracil* is found in its place.

thyroid gland: a gland in the neck of a mammal or other vertebrate that secretes two *hormones*:
- *thyroxine (thyroid hormone)* helps to control the *basal metabolic rate (BMR)* of the body
- calcitonin is a hormone that helps to regulate the blood *calcium* concentration. It stimulates the uptake of calcium by the *bones*.

Thyroxine contains *iodine*. If there is not enough iodine in the diet, the thyroid gland enlarges, producing a condition called goitre.

thyroid stimulating hormone (TSH): a *hormone* produced by the *pituitary gland*. It stimulates secretion of thyroxine by the *thyroid gland*. The concentration of thyroxine in the blood is maintained by a *negative feedback* system involving thyroid stimulating hormone. (See also *thyroxine (thyroid hormone)*.)

thyrotrophic releasing hormone (TRH): a *hormone* involved in the control of thyroxine. It is produced by the *hypothalamus*. TRH is transported in the blood to the *pituitary gland* where it stimulates secretion of another hormone, *thyroid stimulating hormone (TSH)*. The concentration of thyroxine in the blood is controlled by a *negative feedback* system involving both thyrotrophic releasing hormone and thyroid stimulating hormone. (See also *thyroxine (thyroid hormone)*.)

thyroxine (thyroid hormone): a *hormone* produced by the *thyroid gland*. It plays an important part in controlling the *metabolic rate*. It is also involved in the control of body temperature. If the body temperature falls, more thyroxine is produced. The energy released in the *electron transport chain* is normally used to produce *ATP*. One of the effects of thyroxine, however, is to cause some of this energy to be released as heat rather than as ATP. The concentration of thyroxine in the blood is controlled by a *negative feedback* mechanism. This is summarised in the diagram.

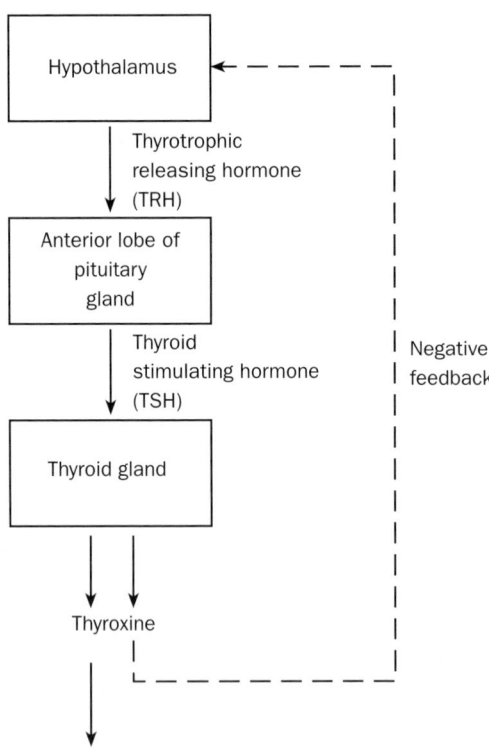

A high concentration of thyroxine in the blood inhibits the production of *thyrotrophic releasing hormone (TRH)* from the *hypothalamus*. As a result, there is a fall in the amount of **thyroid stimulating hormone (TSH)** produced by the **pituitary gland**. This leads to less thyroxine being secreted.

On the other hand, if the concentration of thyroxine falls, there is no negative feedback effect. TRH is secreted by the hypothalamus. This leads to secretion of TSH from the pituitary gland and an increase in the concentration of thyroxine in the blood.

tidal volume: the volume of air taken in and given out at each breath when a person is at rest and breathing quietly. For a normal healthy male, this is about 500 cm^3. (See also **lung capacities**.)

tissue: a group of cells that have a common origin and a similar structure. This enables them to perform a particular function. An example of a simple animal tissue that fits this definition is the layer of cells which lines the **small intestine**. This is an epithelial tissue. Other tissues, however, are often more complex and may consist of several types of cell. Blood is an example of a tissue like this. It consists of **red blood cells**, white blood cells and platelets. The red blood cells, however, all arise from cells in the **bone** marrow, so they have a common origin. They are very similar in structure and this structure is an adaptation to their function in the transport of respiratory gases. In mammals there are four basic tissue types. These are:

- epithelial tissue (see **epithelium**). This covers organs, either on the inside or on the outside. **Glands** are also formed from epithelial tissue.
- **connective tissue**. This tissue has different functions. It can bind other tissues together or it may form **tendons**, **ligaments**, **bone** or blood. Many connective tissues have functions associated with support. In addition, connective tissue plays an important part in the defensive mechanisms in the body
- **muscle** tissue brings about movement and also helps to maintain posture. Muscle tissue is also involved with movement, such as changes in diameter of **blood vessels**, that is associated with organs that are not under conscious control
- nervous tissue. This tissue is made up of specialised cells called **neurones**. These provide links between **receptors** and **effectors**.

Plant cells are also organised into tissues. There are five main types:

- **parenchyma**. This is made up from relatively unspecialised cells that are important in supporting young plants. Parenchyma cells photosynthesise and store substances as well.
- **collenchyma**. A type of supporting tissue whose cells have walls with extra **cellulose** thickening at the corners
- **sclerenchyma**. Supporting tissue in which the **cell walls** are thickened with **lignin**
- **xylem**. A plant tissue responsible for the transport of water and inorganic ions from the roots to the stem and leaves
- **phloem**. A plant tissue that transports the products of **photosynthesis** away from the leaves to other parts of the plant.

In both animals and plants, different tissues are grouped together to form **organs**.

tissue culture: the growth of animal or plant tissues outside a living organism. Artificial **clones** of plants may be produced by tissue culture. Tissue is removed from the parent plant and its surface is sterilised. This tissue, called an explant, is added to a culture medium containing nutrients and **plant growth substances**. The explant develops into a mass of cells called a callus. Pieces of the callus are transferred to a new medium and eventually grow into tiny plants. These plants form a clone because they are all genetically identical.

tissue fluid: the fluid that surrounds the cells in the body. Tissue fluid is formed from *blood plasma*. At the arterial end of a capillary, the pressure of the blood forces water, small organic molecules and mineral ions out through the endothelial cells. This produces tissue fluid. Tissue fluid is very similar to blood plasma in composition but it does not contain blood cells or large protein molecules. At the venous end of the *capillaries*, water re-enters the capillaries by osmosis. The blood at this point has a low *water potential*, due mainly to the presence of protein molecules that do not leave the blood at the arterial end. Water moves back from the higher water potential in the tissue fluid to the lower water potential in the blood plasma. As it moves, it takes with it waste products produced by the cells of the body. Some of the tissue fluid does not go back into the blood. Instead, it enters another system of vessels where it is known as *lymph*.

tonoplast: the *plasma membrane* that surrounds a *vacuole* in a plant cell. It is important in actively transporting ions into the vacuole.

totipotent: a cell that can give rise to all the cell types which make up the organism. Many of the cells found in an adult plant can divide and form new tissues. Cells can be dissected from a growing region from an adult plant and transferred to a tube containing nutrient *agar* to which *plant growth substances* have been added. These cells are totipotent and, in the right conditions, grow into tiny plants containing all the cell types that were in the plant from which they were taken.

toxin: a poison produced by a living organism. Many disease-causing bacteria produce toxins and it is the effect of these substances on the body which produces the symptoms of the disease. *Endotoxins* are inside bacterial cells and are only released into the body of the host when the cell dies and the *cell wall* breaks down. *Exotoxins* are secreted by bacteria as they grow.

Toxocara: a *genus* of *roundworms*. These roundworms are *parasites* of vertebrates. *Toxocara canis* lives in dogs. It does not normally infect humans, but a person may swallow the eggs of *T. canis* either from food or on hands that may have been contaminated with dog *faeces*. These eggs hatch into larvae (see *larva*). The larvae spread to organs in the body where they damage the tissues.

trace element: a chemical element that is needed in very small amounts for the healthy growth of an organism. (See also *micronutrient*.)

trachea (insect): one of the many fine tubes that take air directly from the outside of an insect to its cells. Many insects have openings along the sides of the body. These are known as *spiracles* and they open into the tracheal system. The individual tracheae are lined with rings of *chitin*, a substance that is also found in the *cuticle* of the insect. These rings provide support, preventing collapse of the tubes. At the same time, they allow considerable flexibility. The tracheae form a series of branching tubules, getting smaller and smaller until they finally reach a diameter of about 2 μm. They then branch into a number of even smaller tubes known as tracheoles. The entire system allows oxygen to reach the respiring cells in the insect by the process of *diffusion*. It is well adapted for this function.

- There is a large surface area over which diffusion can take place. This is provided by the very large numbers of small tubes.

- The difference in concentration is kept as high as possible. Not only is oxygen continually being removed by the respiring cells, but ventilation mechanisms replace the air in the tracheae.
- The diffusion pathway is relatively short. Even in the largest insects, the distance from the outside air to the respiring cells is very short. Effectively, air is taken right to the cells.

trachea (mammal): the windpipe, a tube which takes air down to the *lungs*. It is lined with ciliated epithelial cells. Particles in the air that is breathed in are trapped in *mucus* secreted by *goblet cells* in the airways of the lung. The *cilia* beat continuously and the mucus and the particles which have been trapped are carried upwards into the back of the throat. In the wall of the trachea, there are C-shaped rings of *cartilage* which prevent the tube from collapsing as air is breathed in.

Tracheophyta: the plant division that contains ferns and *flowering* plants. Members of the Tracheophyta share the following features:
- they are multicellular plants that have xylem and phloem
- they have a life cycle with an *alternation of generations*. The diploid *sporophyte* is the dominant stage.

transamination: the way in which *amino acids* can be made by transferring an amino group from an amino acid to part of another molecule. In humans, about half the amino acids that we need must be taken in as part of the diet. These are known as *essential amino acids*. The others can be made from essential amino acids by the process of transamination which takes place in the *liver*.

transcription: the first stage of *protein synthesis* in which a messenger RNA (*mRNA*) molecule is produced from the *DNA* which makes up a particular gene. The main steps in the process are summarised in the flow chart.

The hydrogen bonds holding the two DNA strands together break and the strands separate. This process involves enzymes and other proteins

↓

One of the DNA strands acts as a template for the formation of an mRNA molecule. The bases of free mRNA nucleotides in the nucleus line up against the complementary bases on the DNA strand

↓

The nucleotides now join to form an mRNA molecule. This moves out of the nucleus, through the pores in the nuclear envelope, into the cytoplasm of the cell

transcription factor: a protein that is involved in switching genes on and off. Transcription factors bind to *DNA* and control the process of *transcription*. Some transcription factors halt the process by blocking the action of the enzyme *RNA polymerase*. Others activate transcription by promoting the action of the enzyme.

transect: a line along which organisms are sampled in ecological studies. Transects are particularly useful in situations such as sea shores and sand dunes where conditions and species vary across the area being studied.

transfer RNA: see *tRNA*.

transgenic: a term that describes an organism that contains *DNA* that has been transferred from another organism. Although this can happen by natural processes, it usually refers to DNA that has been transferred by gene technology.

transmission electron microscope: a type of *electron microscope* in which a beam of *electrons* is transmitted through the specimen that is being examined. The specimen has to be cut into very thin sections. These are stained with substances such as the salts of heavy metals which scatter the electrons. An image is produced on a fluorescent screen. Transmission electron microscopes have enabled *cell organelles*, that appear as no more than blurred dots when seen with a light microscope, to be examined in great detail.

transmitter substance: see *neurotransmitter*.

translation: the second stage of protein synthesis in which the base sequence on an *mRNA* molecule is used to control the production of a polypeptide chain by a ribosome. The main steps in the process are summarised in the diagram.

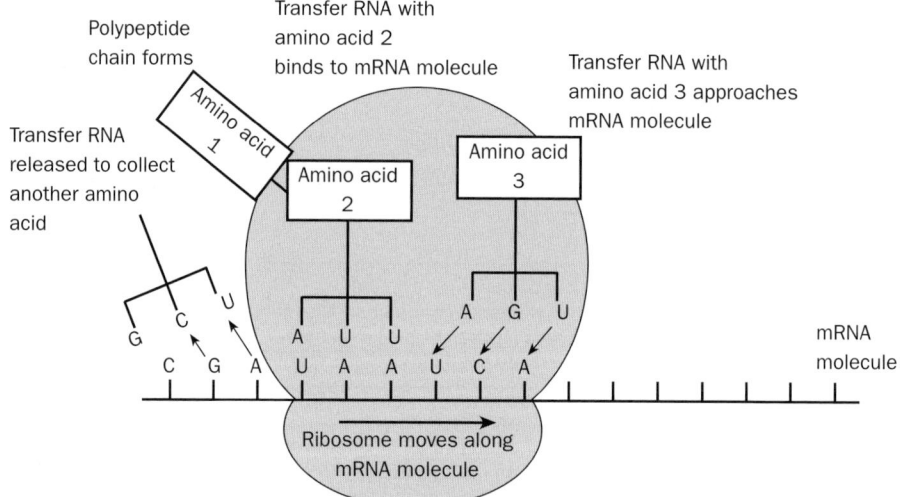

translocation: transport of substances from one part of a plant to another. In *flowering* plants, there are two systems involved:

- Water and mineral ions are transported from the roots to the leaves in the *xylem*. A number of mechanisms contribute to the movement of water inside these vessels. These include *capillary action*, *root pressure* and *cohesion–tension*.
- The products of *photosynthesis* are transported in the phloem. Transport in phloem is in two directions: downwards to the roots, and upwards to the growing buds and *fruits*.

Although these two systems are quite separate from each other, there is evidence from using radioactive tracers that there is some movement between xylem and phloem, particularly in the younger parts of the plant.

transpiration: loss of water from a plant. Although a small amount may be lost from the stem, most of this water escapes from the leaves, through the *stomata* in particular. The air in the spaces in the *mesophyll* of the leaf is saturated with water. There is less water in the air outside the leaf. In other words, the air inside the leaf has a higher concentration of water molecules and therefore a higher *water potential* than the air outside. Water molecules diffuse out of the leaf down this water potential gradient.

Various factors influence the rate of transpiration:

- factors that affect the supply of water to the leaves of the plant. If the soil dries out or roots are damaged by the feeding activity of an insect such as the cabbage root fly, less water will enter the plant. This leads to a reduction in transpiration rate
- factors that influence the opening and closing of the stomata. Approximately 90% of water lost by a plant passes through the stomata. Factors such as light intensity, which have a direct influence on the opening and closing of stomata, will therefore affect transpiration
- factors that affect the water potential gradient between the inside of the leaf and the outside such as wind speed and relative humidity.

The potential loss of water by transpiration is extremely high in dry *habitats*. *Xerophytes* are plants which grow in such places. They usually show a range of adaptations. These include mechanisms that reduce their rates of transpiration.

traumatic brain injury: occurs when the brain is damaged by direct impact such as by a fall or a car accident. When the injury occurs to a specific area of the brain, there may be some loss of function associated with the parts of the body that this area controls.

TRH: see *thyrotrophic releasing hormone (TRH)*.

tricuspid valve: the valve between the right *atrium* and the right *ventricle* in the heart of a mammal. (See also *atrioventricular valve*.)

triglyceride: a *lipid* consisting of three *fatty acid* molecules that are linked by *condensation* to a molecule of glycerol. Triglycerides are the commonest naturally occurring *lipids*. Most of those found in animals are *fats*, that is they are solid at a temperature of about 20°C. This is due to the fact that they contain a high proportion of *saturated fatty acids*. Fats play an important part in energy storage. A given mass of fat will yield a greater amount of energy than the same mass of a carbohydrate such as *starch* or *glycogen*. This is because they have a greater proportion of hydrogen *atoms* in their molecules. In addition to this, fats play an important role in insulating against heat loss.

triose: a sugar that has three carbon *atoms* in each of its molecules. Trioses are important intermediate substances in *respiration* and *photosynthesis*.

triose phosphate: a three-carbon sugar that has a phosphate group attached to it. Triose phosphate is an important intermediate substance in *glycolysis* and *photosynthesis*.

triplet code: a sequence of three bases on a *DNA* molecule that codes for an amino acid. (See also *genetic code*.)

triploblastic: an animal body pattern in which there are three layers of cells. The outer layer of cells is called the *ectoderm*. This gives rise to the animal's *epidermis* and nervous system. The inner layer, or *endoderm*, forms the lining of the gut and digestive *glands*. The remaining layer, between the ectoderm and the endoderm, is the *mesoderm*. From this layer, all the other systems and organs in the animal's body develop. With the exception of the sponges and members of the phylum **Cnidaria**, all animal phyla that you are likely to encounter are triploblastic and have a body pattern which is based on three layers of cells. In some triploblastic animals, a split develops in the mesoderm and forms a body cavity known as a **coelom** (see diagram).

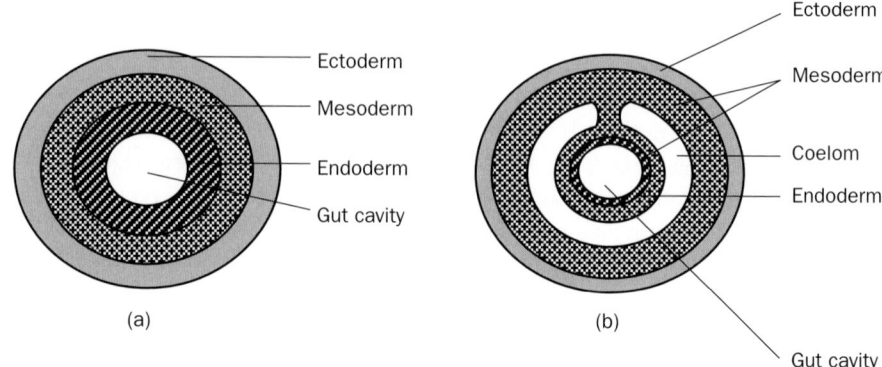

(a) (b)

Sections through triploblastic animals (a) without and (b) with a coelom

triploid: a cell or nucleus in which there are three sets of chromosomes. A number of commercially important plants are triploid. These include many varieties of apples and some bananas. Although they produce larger *fruit*, these varieties are sterile. This is because the *chromosomes* are unable to pair up during the first division of *meiosis*. The *endosperm*, a tissue that is found in seeds where it helps to nourish the developing embryo, is also triploid.

tRNA: a type of *nucleic acid* that is important in assembling *amino acids* in the correct order during *protein synthesis*. A molecule of tRNA consists of a single *polynucleotide* chain. This polynucleotide chain is twisted to form a three-dimensional structure which is often described as being shaped rather like a clover leaf. A loop at the bottom of this molecule has a specific sequence of three bases on it. This is called the *anticodon*. At the other end of the molecule an amino acid may be attached. A tRNA molecule with a particular anticodon will always attach to the same amino acid. The anticodon on the tRNA molecule enables it to line up in the right place on the *mRNA* during *translation*.

trophic level: one of the steps in a *food chain*. In the food chain:

nettle plant → large nettle aphid → two-spot ladybird

there are three trophic levels. The nettle plant is the *producer*. It converts the energy in sunlight into chemical potential energy in the plant. The large nettle *aphid* is the *primary consumer* that feeds on the nettle, while the two-spot ladybird is the secondary consumer. When any of these organisms die, their remains provide energy for *decomposers*. Because energy is lost at each stage by processes such as respiration, it is rare for any food chain to have more than five trophic levels. As with many ecological terms, a trophic level is an over-simplification. Organisms often feed at more than one trophic level. Humans, for example, are both primary and secondary consumers. Insectivorous plants like the Venus fly-trap may even be producers, primary and secondary consumers. In addition, some organisms feed at one trophic level at one stage in their lives and another as they grow. Young partridges are secondary consumers.

They eat insects. The adult birds, on the other hand, are primary consumers and feed mainly on plant material.

tropism: growth made by a plant in response to an external stimulus. If a young seedling, for example, is fixed in a horizontal position, its root will curve and grow downwards. This is caused by unequal growth; the cells on the upper surface grow more than those on the lower surface. Plants respond in a similar way to other stimuli. The response to gravity described in the example above is usually called *geotropism*, while that to light is phototropism.

tropomyosin: a protein found in muscle. When an *action potential* spreads over a muscle fibre, it causes *calcium* ions to be released. These ions bind to the troponin molecules which are attached to *actin* filaments in the muscle. Binding of calcium ions causes a change in the position of molecules of tropomyosin. In a resting muscle, tropomyosin molecules cover the binding sites on the actin filaments to which *myosin* will attach. Myosin can now bind to these sites and form the *actomyosin* cross-bridges that bring about muscle contraction.

troponin: a protein found in muscle. When an *action potential* spreads over a muscle fibre, it causes *calcium* ions to be released. These ions bind to the troponin molecules which are attached to *actin* filaments in the muscle. Binding of calcium ions causes a change in the position of molecules of another protein called tropomyosin. Tropomyosin is also bound to the actin filaments. *Myosin* can now bind to sites on the actin molecules and form the *actomyosin* cross-bridges that bring about muscle contraction.

true-breeding: an organism that has the same *genotype* as its parents. It is therefore homozygous (see *homozygote*) for the *allele* in question.

trypsin: a digestive enzyme produced by the *pancreas*. Trypsin is an *endopeptidase* so it breaks down *proteins* into smaller polypeptides. It does this by *hydrolysis*. Trypsin obtained from animal sources was one of the first *enzymes* to be used in *biotechnology*. Biological washing powder produced in Germany as long ago as 1913 contained small amounts of trypsin. Traditionally, the removal of hair from animal skins in the process of making leather also involved this enzyme. The skins were treated either with chicken or dog *faeces*. Both contain considerable amounts of trypsin and this digested the proteins holding the hair in the skin.

trypsinogen: the inactive form in which the enzyme *trypsin* is secreted. Trypsin is a protein-digesting enzyme secreted by the *pancreas*. If it were released into the pancreas in its active form, it would obviously cause serious damage. It is therefore produced as an inactive substance. Once it enters the *small intestine*, it can be converted into trypsin by *enterokinase*, an enzyme produced by cells in the wall of the intestine. Trypsin itself can also activate trypsinogen. This means that once some trypsin is formed, a chain reaction takes place and all the trypsinogen is rapidly converted to trypsin.

TSH: see *thyroid stimulating hormone (TSH)*.

tubal ligation: an operation in which the oviducts are cut and pinched shut. It is a reliable but permanent method of contraception.

tuberculosis (TB): an *infectious disease* caused by the bacterium *Mycobacterium tuberculosis*. Although it can infect any part of the body, tuberculosis usually infects the *lungs*. The bacterium is most commonly spread by coughing and sneezing. Tiny droplets of mucus carrying the bacteria are sprayed out and may enter the lungs of other people when they breathe in. Inside the lungs, the bacteria multiply producing small lumps called tubercles. In most people, the bacteria are killed by the immune system.

Tuberculosis bacteria are, however, difficult to kill. Some may lie dormant for many years. Eventually, they may start to grow again. This often happens in people with weakened immune systems, such as those suffering from AIDS. The bacteria spread and start to destroy the lung tissue. This produces the common symptoms of tuberculosis: coughing up sputum stained with blood, fever and severe weight loss.

Although tuberculosis may be treated successfully with **antibiotics**, strains of the bacteria have been discovered that are resistant to nearly all the antibiotics commonly used to cure the disease.

tumour: a swelling that may occur in any part of the body. It consists of a mass of abnormal cells which keep on multiplying in an uncontrolled way. There are two main types of tumour. Benign tumours are not cancerous. Although they may grow to a considerable size, they do not actually destroy the tissue in which they are situated or spread to other organs. This is in contrast to a **malignant** tumour. This both destroys the surrounding tissue and can spread through the **blood system** or **lymphatic system** to other sites in the body where it sets up secondary tumours. Malignant tumours are cancerous.

tumour suppressor gene: a gene whose function is to help control cell division. **Mutations** to tumour suppressor genes may prevent them functioning properly. In some cases this leads to uncontrolled cell division and the development of a **tumour**.

turgid: a turgid cell is a cell that is full of water. The **water potential** of a plant cell is influenced by two factors: the concentration of dissolved substances inside the cell and the pressure of the **cell wall** pushing on the contents of the cell. The pressure produced by the concentration of the cell contents is the **solute potential** or ψ_s. The pressure produced by the cell wall is the **pressure potential**, ψ_p. When a cell is fully turgid, solute potential and pressure potential are equal.

At full turgor, the water potential of a cell is zero. Since the water potential cannot be higher than zero, and water always moves from a higher water potential to a lower one, it follows that water cannot move into a turgid cell; it can only move out.

turgor: the pressure that the contents of a cell exert on the **cell wall**. In the parenchyma cells in a young plant stem, water moves into the cells by **osmosis** because there is a greater **water potential** outside than inside. The resulting pressure that the **cytoplasm** exerts on the wall is very important in providing support for the plant.

Turner's syndrome: a genetic condition in which each cell in the person affected has only one **sex chromosome**. This is an **X chromosome**. A person with this condition appears female but has no ovaries and is therefore infertile. **Secondary sexual characteristics** such as facial and body hair are also poorly developed. Turner's syndrome arises from the failure of sex chromosomes to separate properly during **meiosis**.

twin study: a way in which the relative effects of genes and **environment** can be investigated in humans. Monozygotic or identical twins are formed from a single fertilised egg. They are, therefore, genetically identical. Any difference between them must be due to environmental factors. It cannot be due to genetic effects. Dizygotic or non-identical twins develop from two fertilised eggs. They are not genetically identical. Differences between them may be due either to their genes or to their environment.

two-way chromatography: a way of separating substances present in a mixture. In its simplest form, substances are loaded on to a square of filter paper that is suspended with its bottom edge in a solvent and left for a period of time. Individual substances separate out in the usual way. The paper is now turned through 90° and a different solvent used. This separates out substances that are perhaps not separated by the first solvent.

type 1 diabetes: see *diabetes*.

type 2 diabetes: see *diabetes*.

Aiming for a grade A*?

Don't forget to log on to **www.philipallan.co.uk/a-zonline** for advice.

ultracentrifuge: see *centrifuge*.

ultrafiltration: filtration that is helped by the high pressure of the blood. It is a process which is involved in the formation of *tissue fluid*. The pressure of the blood in the *capillaries* helps to force fluid through the capillary walls into the surrounding tissues. Although this process occurs generally in capillaries, the term ultrafiltration is often linked with the first stage of *urine* formation in the *kidney*. The pressure of the blood in the capillaries that make up the *glomerulus* forces fluid through the *basement membrane* into the space in the *renal capsule*. This fluid has the same composition as blood *plasma* except that it does not contain protein. Protein molecules are too large to go through the basement membrane.

ultrasound: sound waves of very high frequency, so high that they cannot be heard by the human ear. Techniques based on ultrasound are frequently used in medicine. An ultrasound beam is aimed at the structure which is being examined and the reflected waves produce an image which shows the various organs. One of its advantages is that it does not subject the patient to the harmful effects of *X-rays*. Partly because of this, it is often used to gain information about a developing *fetus* inside the uterus of its mother.

unsaturated fatty acid: a *fatty acid* with a long hydrocarbon tail in which there are double bonds between some of the carbon *atoms*. Unsaturated fatty acids have lower melting points than fatty acids in which there are no double bonds. *Triglycerides* that contain unsaturated fatty acids therefore tend to be a liquid at temperatures of about 20°C. They are called oils and are found mainly in plants.

uracil: one of the *nucleotide bases* found in *RNA* molecules. Uracil is a pyrimidine. This means that it has a single ring of *atoms* in each of its molecules. In *protein synthesis*, when *mRNA* is formed or when the tRNA *anticodon* and the mRNA codon come together, uracil always pairs with *adenine*. The atoms of these two bases are arranged in such a way that two *hydrogen bonds* are able to form between them. The base uracil does not occur in *DNA*. Thymine is found in its place.

urea: the main form in which nitrogen is excreted in mammals. Breakdown of excess *amino acids* during *deamination* produces ammonia. Unfortunately, ammonia is very toxic and can only be safely excreted if large amounts of water are available to dilute it to safe levels. This is not possible in terrestrial animals such as mammals. They convert ammonia to the much less toxic urea. Urea is a very small molecule that is soluble in water. Because of this solubility, it diffuses readily across *cell-surface membranes* and may be found in small amounts in most body fluids.

ureter: one of a pair of tubes which take *urine* from the *kidney* to the bladder in a mammal. Its walls contain smooth muscle. When this contracts, urine is forced along the ureter.

urethra: the tube that takes *urine* from the bladder of a mammal to the outside. In females it is quite short. In males it is longer and runs through the penis. As well as transporting urine, it also carries semen during copulation.

uric acid: the form in which nitrogen is excreted in some terrestrial animals. It is not as soluble as urea and is excreted more as a sort of paste than a solution. This makes it an ideal excretory product in animals where maximum conservation of water is important. It is the main excretory product in many insects, reptiles and birds.

urine: the fluid produced by the *kidneys*. It contains water in which *urea* and other products of nitrogenous *excretion* are dissolved. Urine also contains significant amounts of sodium chloride as well as traces of many other compounds and ions. The actual composition of urine is variable and can be influenced by a number of features, including climatic conditions and diet. Analysis of urine provides a lot of information about a person's health. People with *diabetes*, for example, may have *glucose* in their urine. Detection of the *hormone human chorionic gonadotrophin (hCG)* in the urine is the basis of some of the tests used to confirm pregnancy.

urinogenital: an adjective meaning 'to do with the *excretion* and reproduction'. The urinogenital system includes the *kidneys*, bladder and reproductive organs.

urticaria: an itchy rash that develops as a response to an *allergen*. Some people, for example, have an *allergy* to strawberries. Allergens present in strawberries cause the release of *histamine* which produces the rash characteristic of an allergy to strawberries.

uterus: the part of the mammalian female reproductive system in which the fetus develops. In many species of mammals the organ has two parts: a right side into which the right *oviduct* opens and a left side into which the left oviduct opens. Humans and a few other species have a single uterus into which both oviducts enter. The uterus has an outer layer composed mainly of muscle and an inner layer called the *endometrium*, which contains many *glands* and distinctive spiral *blood vessels*. The endometrium undergoes a series of changes during the *oestrous cycle*. During pregnancy, the uterus can expand enormously. In humans, the cavity can increase in size by almost 500 times.

A–Z Online

Log on to A–Z Online to search the database of terms, print revision lists and much more. Go to **www.philipallan.co.uk/a-zonline** to get started.

c

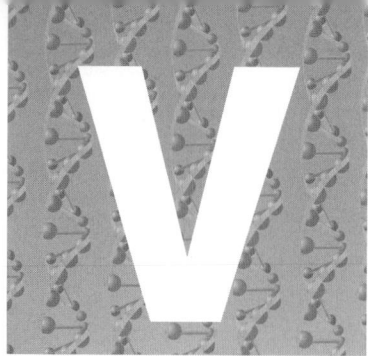

vaccine: a preparation which, when given to a patient, stimulates cells from the immune system to produce antibodies. Vaccines may incorporate whole *microorganisms* that have either been killed or attenuated (see *attenuation*), or they may be based on parts of microorganisms.

vacuolar pathway: the pathway by which substances go through plant cells from *vacuole* to vacuole.

vacuole: an area in the *cytoplasm* of a cell that is surrounded by a *plasma membrane* and contains *cell sap*. Vacuoles are characteristic of many plant cells. In a young cell, there may be several small vacuoles but, as the cell gets older, these join to form a single large vacuole. This may take up as much as 90% of the space in the cell. The cell sap inside the vacuole contains water in which there are dissolved substances such as sugars and mineral ions. They are usually at a much higher concentration than in the surrounding cytoplasm. This means that there is a *water potential* gradient and water moves into the vacuole by *osmosis*. This is important in maintaining the cell in a turgid state and providing support to the plant. (See also *food vacuole*.)

vagus nerve: a nerve that goes from the *medulla* in the *brain* to various internal organs. It is an important part of the *parasympathetic nervous system*. The branch that goes to the heart has an inhibitory effect and slows the heart rate.

vanishing twin syndrome: a fetus in a multiple pregnancy that dies while it is still in the *uterus*. It may die because of a poorly implanted *placenta*, or it may have a genetic or developmental abnormality. It is then partly or completely reabsorbed by the mother. If the vanishing twin dies early in the pregnancy and is completely reabsorbed, there are usually no further complications. If it dies after three months, there is an increased risk of haemorrhage, premature labour or infection.

variation: the differences that exist between living organisms. One of the fundamental characteristics of all living organisms is that they show a considerable range of variation. There is a saying 'as alike as two peas in a pod', but even things that are supposedly this similar differ considerably when examined carefully and in detail. Within a particular species, there are two main causes of variation. Variation can result from the genes which the organism carries, or it may result from the *environment* in which the organism lives. Variation may be described as being either *continuous* or *discontinuous*.

vasa recta: *capillaries* in the *medulla* of the *kidney* that are parallel to the *loop of Henle*. They play an important part in maintaining the concentration gradient in the medulla.

vascular bundle: a strand of *xylem* and *phloem* in a plant.

vasectomy: an operation in which the ducts which take sperm from the **testes** to the **urethra** during sexual intercourse are cut and sealed. It is a reliable but permanent method of contraception.

vascular system: a system that has as its main function the transport of substances from one part of an organism to another. Examples of vascular systems are the **blood system** of a mammal, and **phloem** and **xylem** in a plant.

vasoconstriction: the narrowing of **blood vessels**. The control of body temperature in a mammal involves vasoconstriction. The **arterioles** which supply blood to the network of **capillaries** in the skin have walls containing **involuntary muscle**. In cold conditions, **nerve impulses** from the temperature regulatory centre in the **hypothalamus** cause this muscle to contract. Blood is diverted from the surface capillaries and flows deeper in the skin. As a result, less heat is lost from the body by processes such as radiation.

vasodilation: the dilation of **blood vessels**. The control of body temperature in a mammal involves vasodilation. The **arterioles** that supply blood to the network of **capillaries** in the skin have walls containing **involuntary muscle**. In hot conditions, **nerve impulses** from the temperature regulatory centre in the **hypothalamus** cause this muscle to relax. Blood can then flow to the surface capillaries and heat can be lost from the body by processes such as radiation.

vasomotor centre: a region in the **medulla** of the **brain** that regulates blood pressure. The flow chart summarises the mechanism involved.

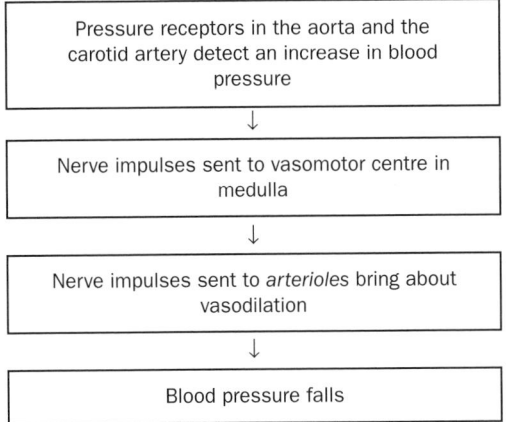

Pressure receptors in the aorta and the carotid artery detect an increase in blood pressure

↓

Nerve impulses sent to vasomotor centre in medulla

↓

Nerve impulses sent to *arterioles* bring about vasodilation

↓

Blood pressure falls

vector: a carrier. In biology, the term may be used in one of two different contexts.

A genetic vector is used in **genetic engineering**. It acts as a carrier. The gene that has been isolated is spliced into the vector **DNA** to produce recombinant DNA. This is then inserted into another organism. There are different genetic vectors and the one that is chosen will depend on the size of the isolated piece of DNA and the type of organism into which it is to be placed. Two of the most frequently used vectors are **viruses** and **plasmids**.

When referring to parasites, a vector still means a carrier. The malarial parasite is a **microorganism** which lives in the blood of its human host. In normal circumstances it would be very difficult, if not impossible, for an organism smaller than a single red blood cell to get from the blood of one person into the blood of another. The transfer is made by a vector, in this case, an *Anopheles* mosquito.

vegetative propagation: a form of *asexual reproduction* found in plants. It often involves the growth of a bud or a stem that eventually becomes separated from the parent to form a new plant. The table gives information about some common examples of vegetative propagation.

Structure	Examples	Notes
Bulb	Onion, daffodil, hyacinth	A bulb is a shoot with a very short stem and fleshy leaves that act as a food store
Rhizome	Some irises, many important weeds such as couch grass and stinging nettles	Rhizomes are underground stems that grow horizontally below the soil surface. Buds on the rhizome give rise to leafy shoots
Sucker	Elm tree	A sucker is a shoot that grows from the existing root system

vein (animals): a *blood vessel* that takes blood from the *capillaries* back to the heart. In mammals, veins usually contain blood with a low concentration of oxygen, but there is an important exception to this. The blood that returns to the heart along the pulmonary vein has come from the *lungs*. Therefore, blood in this vein contains a high concentration of oxygen.

Blood flowing into veins is at a low pressure and its return to the heart may be helped by muscles and valves. The larger veins returning from the limbs are situated between large muscles. When these muscles contract they squeeze the blood in the veins. Valves in the walls allow this blood to be squeezed towards the heart but do not allow it to go back in the other direction. This helps to maintain the flow of blood in the veins. Like all blood vessels, veins have a lining of epithelial cells. Their walls, however, contain much less muscle and elastic tissue than the walls of *arteries*.

vein (plants): a structure found in leaves. Veins help to provide support and contain vascular tissue. This is *xylem*, which supplies the leaf with water and mineral ions, and the *phloem*, which transports the products of *photosynthesis* to the rest of the plant.

vena cava: one of the two main veins that take blood from the other veins to the right *atrium* of the heart.

ventilation: the mechanism by which the air or water in contact with a gas exchange surface is changed. Oxygen is taken up across all gas exchange surfaces by *diffusion*. A ventilation mechanism ensures that the difference in concentration across such a surface is kept as high as possible. If ventilation did not occur, the oxygen concentration in the external medium would gradually fall and diffusion would become less and less effective. Different organisms have different ventilation mechanisms:

- In large, active insects, contraction of muscles flattens the abdomen. When these muscles relax, the abdomen returns to its normal shape. This helps to change the air in the *tracheae* that supply oxygen to the body tissues.
- In fish, water is moved over the *gills* by a pumping mechanism that maintains the flow of water.
- Mammals have a *diaphragm* which separates the thorax from the abdomen. Contraction of the muscles of the diaphragm and of those between the ribs causes the chest cavity to enlarge. This results in a lower pressure in the *lungs* than in the surrounding atmosphere so air is able to move in until the pressures are equal.

ventilation rate: see *pulmonary ventilation*.

ventral root: part of a *spinal nerve* that joins with the *spinal cord*. The spinal nerves contain both sensory and motor *neurones*. Just before each nerve joins with the spinal cord, it splits in two, producing a ventral root and a *dorsal root*. The ventral root contains the motor neurones that carry *nerve impulses* from the spinal cord to *effectors* such as muscles and *glands*.

ventricle: one of the chambers of the heart. The walls of the ventricles are much thicker than those of the *atria*. When the muscle in the ventricle walls contracts, blood is forced out into the arteries. Fish have a *single circulation*. The heart of a fish has only one ventricle. When it contracts, blood is forced out to the *gills* where it picks up oxygen and then continues round the body. Mammals, on the other hand, have a *double circulation*. Their hearts have two ventricles. The right ventricle receives blood from the right *atrium* and pumps this out through the *pulmonary artery* to the *lungs*. Oxygenated blood returns to the left side of the heart from where the left ventricle pumps it to the organs of the body.

ventricular diastole: the stage in the *cardiac cycle* where the muscles of the heart relax. Blood flows into the atria. Some goes into the *ventricles*.

ventricular fibrillation: when the *cardiac muscle* in the walls of the *ventricle* contracts erratically. As a result the ventricles quiver and fail to pump blood into the arteries. Emergency treatment must be given because *brain* damage will result if the blood supply is not restored within 2 or 3 minutes.

ventricular systole: the stage in the *cardiac cycle* where the thick muscular walls of the *ventricles* contract and force blood out of the heart through the *aorta* and pulmonary arteries.

venule: a *blood vessel* that takes blood from *capillaries* to a *vein*.

vertical gene transmission: the passing of *DNA* from the original cell to one of the daughter cells produced when it divides. Bacterial cells copy their DNA then divide into two daughter cells. Each daughter cell gains one copy of the DNA from the parent cell. This is vertical gene transmission.

vesicle: a small sac in the *cytoplasm* of a cell, surrounded by a membrane. The *Golgi apparatus* is an organelle that is responsible for the processing and packaging of substances produced by a cell. These substances are enclosed in vesicles that are continually being pinched off from the flattened sacs which make up the Golgi apparatus. Other vesicles are formed as a result of *phagocytosis* and *pinocytosis*, where particles or small droplets of liquid are taken through the *cell-surface membrane* into the cytoplasm. The membrane surrounds the particles concerned. A vesicle is formed and moves into the cytoplasm.

vessel: one of the *xylem* tubes that transport water and inorganic ions from the roots to the stem and leaves of a plant. Vessels are made of dead cells that have lost their cross-walls and fit together rather like a series of drainpipes. The walls of these cells are thickened and strengthened with a tough, waterproof material called *lignin*. As the vessel gets older, lignin is laid down on the original *cell wall*. At first, this is done in a series of rings or spirals which prevent the vessel from collapsing but still allow the flexibility needed in a young stem. Later this lignin forms an almost complete layer.

viable count: the number of *microorganisms* present in a culture that can multiply and form colonies. A technique used to obtain a viable count is *dilution plating*.

viability: the ability of *seeds* to germinate when exposed to suitable conditions. If seeds are stored under conditions where they are kept cool and dry, they may remain viable for very long periods of time. Some seeds excavated from a grave in South America, for example, were still able to germinate although they were found by carbon dating to be over 600 years old. Seeds from many garden plants, however, lose their viability very rapidly. This is particularly true if the packet of seeds has been opened and left under conditions where the temperature varies and they are exposed to air with a high moisture content. Many important *weeds* have seeds which remain viable for very long periods of time. This enables their seeds to remain buried in the soil for many years and still germinate when they are exposed to suitable growing conditions. Viability is a term which can also be used when referring to *microorganisms*. Microorganisms are viable if they can multiply and form colonies.

vibrio: a bacterial cell that looks like a curved rod. The best-known example is probably *V. cholerae*, the bacterium that causes *cholera*.

villis: small finger-like structures (singular villus) that stick out from the inside of the *small intestine*. Villi are present in enormous numbers and greatly increase the surface area of the intestine. Each villus has a lining of epithelial cells. The *cell-surface membrane* of these epithelial cells is increased by *microvilli*. Together, the villi and microvilli produce an enormous surface area that allows for the efficient *absorption* of the products of *digestion*. Each villus contains a network of *capillaries* into which *amino acids* and simple sugars diffuse, and a *lacteal* which absorbs the products of *fat* digestion.

Villi are also found at the exchange surface between the fetal and maternal parts of the *placenta*. They have a similar function in increasing the surface area and allowing more effective exchange between the blood of the mother and that of the fetus.

virus: an extremely small particle that is only capable of multiplying once it is inside a living cell. A virus consists of a molecule of *nucleic acid*, either *DNA* or RNA, surrounded by a protein coat. Some of the larger viruses have an outer *lipid* layer as well. Outside the cells of their host, they are completely inert. They cannot feed, respire or multiply, so they are best thought of as non-living particles. The life cycle of a virus involves the take-over of the host cell and the use of its biochemical processes to make more virus particles. All viruses are, therefore, to some extent harmful. They cause a wide range of human diseases including the common cold, *influenza*, AIDS and some cancers. They also produce some economically important plant diseases. Viruses may be used as *vectors* in *genetic engineering*.

visceral muscle: see *involuntary muscle*.

visual cortex: the part of the forebrain that processes and interprets visual information.

visual purple: see *rhodopsin*.

vitamin: one of a group of chemically unrelated organic substances that are needed in very small amounts in the diet. Many vitamins are obtained from plant foods but some may be obtained from the tissues of other animals or are produced by *microorganisms* living in the gut. Vitamins can be divided into two groups. *Fat*-soluble vitamins can be stored in the *liver*. Water-soluble vitamins, however, cannot be stored and are removed from the body in the *urine*. It is essential, therefore, to maintain a constant supply of the water-soluble vitamins. Vitamins have different functions, although many of them act as *coenzymes*.

vitamin K: a *fat*-soluble *vitamin* essential for *blood clotting*. It is required by the *liver* for the synthesis of a substance necessary for producing prothrombin.

vital capacity: the total amount of air that can be breathed out following the deepest intake of air possible. It is not as much as the total volume of the *lungs* as there is always a small volume that cannot be expelled. (See also *lung capacities*.)

vitreous humour: a transparent, jelly-like substance between the *lens* and the *retina* of the eye. It is made up mainly of water and a substance called hyaluronic acid. Hyaluronic acid gives it its jelly-like consistency. There are a few *phagocytes* in the vitreous humour. They remove cell debris. The vitreous humour helps to maintain the shape of the eyeball.

voltage-gated ion channel: a *protein* in a *cell-surface membrane* through which ions pass. Channels in these *proteins* open or close as a result of changes in electrical potential difference. When an *action potential* is produced, sodium *ion-channel proteins* in the membrane open. Sodium ions enter and decrease the potential difference across the membrane. This change in potential difference allows even more sodium ions to enter, changing the potential difference still more. The end result is that the potential difference changes from its *resting potential* of around −60 mV to a value of approximately +40 mV. Once it has reached this value, the sodium ion-channel proteins close leaving potassium ion-channel proteins fully open. Positive potassium ions move out of the cell, and the potential difference returns to its resting level.

voluntary muscle: see *skeletal muscle*.

Do you need revision help and advice?

Go to pages 306–314 for a range of revision appendices that include plenty of exam advice and tips.

warfarin: an *anticoagulant*. It is used in the treatment of coronary *thrombosis* and lowers the risk of a blood clot becoming lodged in a branch of the *coronary artery*. Warfarin is also used as a rat poison but many populations of rats are now resistant to it.

waste hierarchy: a set of principles relating to the disposal of waste:
- where possible, waste should be prevented or reduced at its source
- waste material should be reused
- waste materials that cannot be reused in their existing form should be recycled and used as a raw material
- waste that cannot be reused at all should be used as a substitute for non-renewable energy sources
- only waste that cannot be treated in any of the above ways should be removed to landfill sites.

water potential: a measure of the ability of water molecules to move. The diagram shows molecules of water surrounded by a membrane.

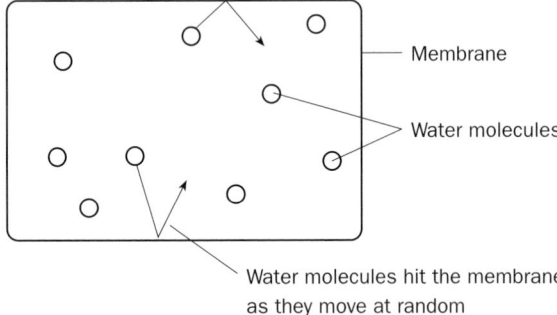

Understanding water potential

These molecules are moving around at random. As a result of this, some collide with the membrane that surrounds them. They exert a pressure on this membrane. This is the water potential, the symbol for which is the Greek letter psi, ψ. Clearly, the higher the concentration of water molecules, the greater the pressure they exert on the membrane, and the higher the water potential. Pure water has the greatest concentration of water molecules and therefore has the highest water potential. The value of the water potential of pure water is, in fact, zero. As all other solutions have a water potential less than this, they will have negative values. The table explains some typical water potential values found in and around a plant.

	Typical value of water potential/MPa	Explanation
Water in the soil around plant roots	−0.03	Water in the soil has small amounts of dissolved substances in it. The concentration of water molecules will be less than in pure water but still very high
A xylem vessel	−0.3	There is a much higher concentration of dissolved substances, therefore a lower concentration of water molecules than in the soil
Humid air in the atmosphere around a leaf	−30	The concentration of water molecules in the atmosphere is very low, even in humid air

Osmosis is the movement of water from a higher (less negative) water potential to a lower (more negative) water potential.

The water potential of a cell is influenced by two other factors, the concentration of dissolved substances inside the cell and the pressure of the **cell-surface membrane** or **cell wall** pushing on the contents of the cell. The pressure produced by the concentration of the cell contents is the **solute potential**, or ψ_s. The pressure produced by the cell-surface membrane or wall is the **pressure potential**, ψ_p. Water potential is the sum of the solute potential and the pressure potential. This can be summarised by a simple equation:

$$\psi = \psi_s + \psi_p$$

The solute potential of a cell from a potato tuber is −0.8 MPa and its pressure potential is 0.5 MPa. Calculate the water potential of this cell.

Using the equation

$$\psi = \psi_s + \psi_p$$
$$\psi = -0.8 + 0.5$$
$$\psi = -0.3 \text{ MPa}$$

weed: a plant growing in a place where it is not required. Weeds are economically important because they severely reduce crop yields by competing for light, water and inorganic ions. The control of weeds is very important in agriculture and there are chemical **herbicides** that may be used for this purpose. Plants which have become important weeds are often those that normally occur early in the process of ecological **succession** and are therefore well adapted to growing in disturbed soil. They share a number of features:
- they have no special **germination** requirements and the seedlings grow rapidly
- seed production begins after a very short growth period and continues for a long time
- weeds produce large numbers of seeds in ideal conditions and at least some in poor conditions
- the seeds of weeds can often live for a long time in the soil and have a variable period of dormancy
- many weeds are **r selected species**.

W

white blood cell: see *leucocyte*.

wilting: a condition in which the leaves and young stems of plants droop. It occurs when the amount of water lost through *transpiration* is greater than that absorbed by the roots. The factors that result in a plant wilting include:

- environmental factors such as high temperature and low relative humidity which promote a rapid loss of water through transpiration
- factors that prevent uptake of sufficient water from the soil. These include a lack of soil moisture and damage to the root system by insect pests or by diseases caused by *microorganisms*.

Are you studying other subjects?

The *A–Z Handbooks (digital editions)* are available in 14 different subjects. Browse the range and order other handbooks at **www.philipallan.co.uk/a-zonline.**

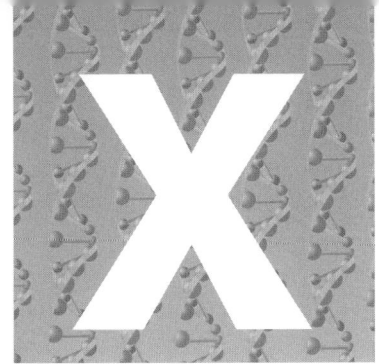

X chromosome: one of the *sex chromosomes*. In humans and other mammals, each body cell in a female contains two identical X chromosomes. The body cells of males contain a single X chromosome and a *Y chromosome*. In most mammals, the X chromosome is larger than the Y chromosome. However, in other animals such as the fruitfly, the Y chromosome is the larger one. Each X chromosome is made up of two distinct regions: a homologous part, which is identical to the same region on the Y chromosome, and a non-homologous part. This is different and found only on the X chromosome. The non-homologous part carries a number of genes. In humans, for example, there are genes associated with *blood clotting* and colour vision. These genes are sex linked. (See also *sex linkage*.)

X-ray: radiation with a very short wavelength that can penetrate tissue through which light cannot pass. X-rays are used in diagnosis. When they are use to identify fractures in *bones*, they are directed at the affected part, behind which is a sheet of photographic film. Bone absorbs X-rays so they do not penetrate to the film. The X-rays that are not absorbed turn the film black. This gives an image in which the bone shows up white on a black background. When X-rays are used with other organs, a contrast medium is used to show up the different parts. Radiologists take care to avoid unnecessary exposure to X-rays because they damage *DNA* and may increase the risk of cancer. Some forms of cancer are treated with radiotherapy. In this process, X-rays are used to kill cancer cells. (See also *tumour*.)

xenotransplantation: transfer of cells, tissues or organs from one species to another. There is a worldwide shortage of human organs for transplantation. It has been suggested that organs from animals such as pigs might be used instead. There are problems with xenotransplantation and one of the biggest of these is rejection by the immune system. Some companies are developing animals that produce human *antigens* on the surface of their cells. It is thought that organs from animals with these antigens may be less likely to be rejected.

xerophyte: a plant that is adapted to dry conditions. Although xerophytes are often found in desert regions, plants showing some similar adaptations may be found in other places. In the Arctic, for example, there is a water shortage in winter as the ground is frozen. Some arctic plants show xerophytic adaptations. General features of xerophytic plants include:

- structural adaptations of the leaves which reduce *transpiration*. Sunken *stomata*, reduced leaf surface area and thickened *cuticles* are often found
- adaptations that enable them to collect and store water. Root systems may cover a very wide area or penetrate deep into the soil. Stems and leaves may be fleshy, storing large amounts of water

- protective mechanisms such as spines. These prevent damage and loss of water-storing tissue by feeding animals
- modified life cycles. Some seeds germinate and the plants grow and reach maturity before all the water from a short wet season evaporates.

xylem: a plant tissue that transports water and inorganic ions from the roots to the stems and leaves. The xylem cells which carry out this function are called *vessels*. They are dead cells that have lost their cross-walls and fit together rather like a series of drainpipes. A number of mechanisms contribute to the movement of water inside these vessels. These are *capillary action*, *root pressure* and *cohesion–tension*. Xylem also plays an important part in supporting plant stems. The vessel walls are thickened and strengthened with a tough, waterproof material called *lignin*. As the vessel gets older, lignin is laid down on the original *cell wall*. At first, this is done in a series of rings or spirals which prevent the vessel from collapsing but still allow the flexibility needed in a young stem. Later the lignin forms an almost complete layer.

Aiming for a grade A*?

Don't forget to log on to **www.philipallan.co.uk/a-zonline** for advice.

Y chromosome: one of the *sex chromosomes*. In humans and other mammals, each body cell in a female contains two identical *X chromosomes*. The body cells of males contain a single X chromosome and a Y chromosome. In most mammals, the Y chromosome is smaller than the X chromosome. However, in other animals such as the fruitfly, the Y chromosome is the larger one. There are very few genes carried on the Y chromosome, but this does not mean that it has no function. In humans, the Y chromosome is important in the *embryo* while it is in the uterus. Its presence brings about the development of the *testes* in the male.

yeasts: single-celled fungi, many of which reproduce asexually by budding (see *asexual*). Some species of yeast can respire anaerobically, converting sugars to ethanol and carbon dioxide. These yeasts are used in baking and in the production of ethanol, either in alcoholic drinks or for use as a *biofuel*. Other yeasts are *pathogens*. Candida albicans, for example, lives *on mucous membranes* and causes thrush.

Z line: a thin line in the centre of the *I band* on a *myofibril*. It holds the thin *actin* filaments in position.

Z scheme: a diagram representing the main steps in the *light-dependent reaction* of *photosynthesis*.

zygote: the cell formed in the process of *sexual reproduction* by the fusion of two *gametes*. The zygote has the *diploid* number of chromosomes. It contains all the genetic information that will be present in the mature organism into which it will eventually grow.

Biology revision lists

Your specification is an important document. It is a contract. It sets out what you have to know and be able to do. It also limits your examiners by making clear what questions can be asked. You will probably be familiar with the main part of the specification in which the content of the different units is set out. There is another section, however, that is even more important. This is the section labelled 'assessment objectives'. It sets out the skills you have to master to achieve success. We can summarise these as:

- learning and understanding biological facts
- applying biological knowledge to new situations
- analysing data.

If you talk to candidates after an examination, you sometimes hear comments like 'I could have done that if they had asked me straight!' The reason that not all questions are 'asked straight' is that examiners have to test other skills as well as knowledge. These skills are set out in the assessment objectives and this is why we have different types of question in a unit test.

The following revision lists have been designed specifically to help with revision of basic facts, but this should really come with a health warning. Learning the facts is essential, but it is only the first step in an effective scheme of revision. You need to practise the skills of applying your knowledge to new situations and analysing data as well.

To revise a particular topic, use the lists of key terms as a starting point. Look up the relevant terms and make sure that you understand the basic principles. These will act as your starting points. You can use the cross-references to allow you to build on the basic ideas explained under the relevant entries.

Using A–Z Online

In addition to the revision lists given below, you can use the A–Z Online website to access revision lists specific to your exam board and the particular exam you are taking. Log on to **www.philipallan.co.uk/a-zonline** and create an account using the unique code provided on the inside front cover of this book. Once you have logged on, you can print out lists of terms together with their definitions, which will help you to focus your revision.

AS topics

1 Cell structure and function
2 Biological molecules
3 Enzymes and enzyme action
4 Gas exchange
5 Transport systems
6 Nutrition and digestion
7 DNA and genes
8 Biodiversity

A2 topics

9 Ecology
10 Respiration and photosynthesis
11 Genes and genetics
12 Homeostasis
13 Excretion
14 Responding to the environment
15 Reproduction

1 Cell structure and function

Active transport

Cell

Cell cycle

Diffusion

Electron microscope

Organelle

Plasma membrane

2 Biological molecules

Biochemical tests

Carbohydrate

Lipid

Polymer

Polysaccharide

Protein

3 Enzymes and enzyme action

Activation energy

Active site

Enzyme

Lock and key model

4 Gas exchange

Diffusion

Fick's law

Gill

Lung

Placenta

Surface area–volume ratio

Trachea

Ventilation

5 Transport systems

Blood system

Blood vessel

Cardiac cycle

Cohesion–tension

Mass transport

Oxygen dissociation curve

Phloem

Tissue fluid

Transpiration

Xylem

6 Nutrition and digestion

Absorption

Alimentary canal

Autotrophic nutrition

Balanced diet

Dietary reference values

Digestion

Heterotrophic nutrition

Ruminant

7 DNA and genes

Cell cycle

DNA

DNA replication

Gene

Gene mutation

Genetic code

Genetic engineering

Protein synthesis

8 Biodiversity

Classification

Genetic conservation

Selection

Species

Species diversity

9 Ecology

Abiotic

Biotic

Ecosystem

Nutrient cycle

Pyramids of number, biomass and energy

Sampling

Succession

Symbiosis

10 Respiration and photosynthesis

Aerobic respiration

Anaerobic respiration

ATP

Chloroplast

Light-dependent reaction

Light-independent reaction

Mitochondrion

Photosynthesis

Respiration

11 Genes and genetics

Dihybrid cross

DNA

Gene

Gene pool

Genetic code

Genetic engineering

Meiosis

Protein synthesis

RNA

12 Homeostasis

Blood glucose pool

Homeostasis

Negative feedback

Osmoregulation

Temperature control

13 Excretion

Antidiuretic hormone (ADH)

Deamination

Excretion

Kidney

Liver

Loop of Henle

Nephron

Renal capsule

Urea

14 Responding to the environment

Action potential

Autonomic nervous system

Auxin

Hormone

Neurone

Plant growth substance

Skeletal muscle

Sliding-filament model

Synapse

15 Reproduction

Asexual reproduction

Fertilisation

Gametogenesis

Oestrous cycle

Sexual reproduction

Revising biology

What is the best way to revise? This is an impossible question to answer. Some people work best late at night; others like to get up early in the morning. Some like silence, others need the radio on. The best advice that anyone can give you is to revise in the way that suits you best. However, to get the most out of the time you have set aside for revision, here are some general points that are worth considering.

Get to know your limitations and difficulties. One way in which you can do this is to make a profile like the one below. You can complete it by looking at the mistakes you have made in tests and mock exams.

	0	1	2	3
Specification topics				
Biological molecules				
Cell structure				
The heart and circulation				
etc.				
Types of question				
Recalling facts				
Interpreting tables and graphs				
etc.				
Examination techniques				
Timing				
Understanding the question				
Answering in enough detail				
Reading the question accurately				
Using appropriate scientific language				

This table has three main sections: the topics that make up the unit in the specification, types of question and examination techniques. You will need to look at your specification or ask your teacher for help in order to fill in the titles for the individual boxes in the first two sections. Across the top is a four-point scale going from 0, which means it is a bit of a disaster, to 3, which means you are pleased with this and you don't think you have any problems.

Look carefully at a test that you want to analyse and shade in the appropriate boxes in the table. In this case, the student concerned obviously thinks he or she is pretty good when it comes to biological molecules and cell structure but is having a lot of problems with the heart and circulation.

You can use a table like this to produce a revision plan. It not only highlights the topics on which you need to concentrate, it also shows you the types of questions and examination techniques you need to practise.

- **You need a revision plan.** Get the coloured pencils out and make a decent job of it. You are probably taking more than one subject, so get the balance right. Build in things that matter. If you have your eighteenth birthday coming up, allow time for the celebrations. Have you taken Friday evenings off? It is important to be realistic rather than idealistic.

- **It is quality not quantity that counts.** There isn't much point in working 5 or 6 hours a night if all you end up doing is gently turning pages over with little going in. It is much better to work in short, manageable sessions during which you keep your mind firmly on the task in hand. Between an hour and an hour and a half is about as much as most people can manage at a time. Then have a break before continuing.

- **Set yourself sensible targets.** Break the subject into bits. It is far more rewarding if, at the end of a revision session, you have achieved what you set out to do. So don't try to 'do biology'; try instead to master protein synthesis or DNA replication.

- **Give yourself a reward.** When you have achieved your target, that's the time to have your cup of coffee, watch something on television or whatever you want to do – not before you start.

- **Revising for an A-level examination is just like preparing for a driving test.** You have things to learn and you have skills to practise. For your driving test you will need to know what particular road signs mean and all about stopping distances at various speeds. You can only learn these. But you have also got to know how to negotiate a roundabout, make a right-hand turn or start on a hill. These are things that require practice. It is the same with A-levels. There are things to learn and these must be learnt, but there are also skills to practise and these can only be mastered by working through examination questions in past papers. So build into your revision programme time to learn and time to practise.

Examiners' terms

Most A-level candidates spend a long time revising and, as a result, they go into the examination with a sound knowledge of basic biology. Unfortunately, some fail to do themselves credit when it comes to turning this hard-won knowledge into marks. One of the reasons for this is that they don't follow the instructions given in the questions. Every year this point is noted by the examiners who make comments such as 'Many candidates did not appreciate that the words "describe" and "explain" do not have the same meaning' and 'There was a widespread failure to use the material provided' in their reports.

Use this section from the start of your course to make sure that you get into the habit of giving the required answer. Remember, the examiners are on your side. They want to give you marks for what you know and for what you can do. They can only help you if you help yourself by following the instructions you have been given.

Calculate means calculate. Although no explanation should be required here, two important points need to be made:

- Show your working as clearly as possible. In some cases examiners award marks for the right approach even if the answer is wrong. You can only get these marks, though, if the examiners can follow your explanation.
- Units must always be given.

Compare: point out the similarities and the differences. This is not a term that you will see often; it is usually replaced by 'Give the similarities and differences between'. Make sure that your answer does point out the similarities and differences. Don't describe one feature then the other and leave the examiner to do all the work for you.

Define/give the meaning of usually requires a simple explanation of the technical term provided. In biology, formal definitions are rarely, if ever, required and you should concentrate on explaining the meaning in your own words.

Describe: this term often involves giving a written description of the trends and patterns shown by data presented in a table or graph. Where this is the case:

- Describe the pattern shown by the results. If you are describing data plotted on a graph, for example, don't describe the position of each individual point. Describe the overall pattern. Remember also that curves on graphs often show a change in slope. Your description needs to reflect this.
- Use the labels on the axes of a graph or headings in a table to describe this pattern. Don't just write about 'it' increasing or decreasing.

- Where there is a change in the gradient of a curve, relate this to the figures provided. Answers such as 'The rate of reaction increases to a peak value of 4.2 g h^{-1} at 50°C. It then falls to zero at 65°C' should gain full credit. An explanation is not required.

Describe can also be used in a more general sense in questions such as 'Describe how water enters a plant root.' All you basically have to do in such cases is to give a step-by-step account of what happens. Use the mark allocations to guide you in the amount of detail required.

Evaluate means 'judge the worth of'. Evaluating evidence, for example, requires you to judge how well it supports a particular conclusion. When examiners ask you to evaluate something, they want you to look at it critically. If you have been asked to evaluate evidence from a table of data, you should be able to point out how it supports a conclusion. You should also be able to explain why it doesn't support this conclusion. The key to answering questions that require evaluation is to get the word 'but' into your answer. This will make sure that you are looking at both sides and are genuinely judging the worth.

Explain means 'give a reason why'. A description is not required and won't get you any marks. The key to answering questions that want you to explain something is to start your answer with the word 'Because'. In this way you can be sure that you are giving the reason why. Many candidates regularly lose marks through failing to follow this instruction.

Give two reasons/Give three examples: if a number of reasons or examples are required, these must be different from each other. For example, if a question wants you to name three resources for which weeds and crop plants are competing, it is better to suggest 'mineral ions, water and light' than 'nitrates, phosphates and sulphates'.

Care should be taken to give no more than the required number of points. Many examination boards operate a system of cancelling if you give more points than you have been asked for. If you have been asked for two points and you give three, just to be on the safe side, a wrong answer may cancel one of the correct ones.

List requires a number of one-word answers or a number of brief points.

Name usually requires no more than a single-word answer. There is no need to write out the question first or to incorporate the answer into a long sentence.

Outline: give a brief account. There are two useful guides to how much you should write in answer to questions beginning with this word: the number of lines available and the mark allocation. In general, the number of marks available tells you how many A-level points you should try to get in to your answer. The number of lines is an indication of how much room the examiners feel that you need to make these points.

Sketch: used where a curve has to be added to a graph. In some cases, you need to draw and label the axes as well. Where this instruction is used, it is simply the shape of the curve that is required. There is no need to invent figures and attempt to plot these accurately.

State means give. Many examination boards do not use the word 'state' any more; they replace it with the word 'Give'. 'State' requires a short, concise answer. As with outline, use the number of lines that you have been allowed and the mark allocation to judge what to write.

Suggest: this term has two uses. It is used where there is more than one valid answer or it may be used where it is not expected that you will be able to answer from memory. In general, any sensible answer based on sound biological reasoning will gain marks.

Use the graph/Use the table/Use the diagram involves using the material provided to illustrate a particular point. This is a requirement of the question, so you won't get full marks for a general answer that fails to refer to the relevant material.

You may have just taken your driving test or you may be preparing for it. If this is the case you may have been advised that you should turn your head every time you look in the mirror so that the examiner knows what you are doing. You could use the same idea with questions that ask you to 'Use the graph to explain…' You need to make sure that you let the examiner know that you really are using the data. Try to use the phrase 'The graph shows…' in your answer. This will force you to follow the instruction and demonstrate to your examiner that you really are using the graph in your answer.